Complex Plasmas and Colloidal Dispersions: Particle-resolved Studies of Classical Liquids and Solids

SERIES IN SOFT CONDENSED MATTER   ISSN: 1793-737X

Founding Advisor:
Pierre-Gilles de Gennes
(1932–2007)
Nobel Prize in Physics 1991
Collège de France
Paris, France

Series Editors:
David Andelman
Tel-Aviv University
Tel-Aviv, Israel

Günter Reiter
Universität Freiburg
Freiburg, Germany

Series in Soft Condensed Matter **Vol. 5**

Complex Plasmas and Colloidal Dispersions: Particle-resolved Studies of Classical Liquids and Solids

Alexei Ivlev
Max-Planck-Institut für extraterrestrische Physik, Germany

Hartmut Löwen
Heinrich-Heine-Universität Düsseldorf, Germany

Gregor Morfill
Max-Planck-Institut für extraterrestrische Physik, Germany

C. Patrick Royall
University of Bristol, UK

World Scientific

NEW JERSEY · LONDON · SINGAPORE · BEIJING · SHANGHAI · HONG KONG · TAIPEI · CHENNAI

Published by

World Scientific Publishing Co. Pte. Ltd.

5 Toh Tuck Link, Singapore 596224

USA office: 27 Warren Street, Suite 401-402, Hackensack, NJ 07601

UK office: 57 Shelton Street, Covent Garden, London WC2H 9HE

British Library Cataloguing-in-Publication Data
A catalogue record for this book is available from the British Library.

Series in Soft Condensed Matter — Vol. 5
COMPLEX PLASMAS AND COLLOIDAL DISPERSIONS
Particle-Resolved Studies of Classical Liquids and Solids

ISBN-13 978-981-4350-06-8
ISBN-10 981-4350-06-0

Printed in Singapore.

Foreword

The study of *Soft Condensed Matter* has stimulated fruitful interactions between physicists, chemists, and engineers, and is now reaching out to biologists. A broad interdisciplinary community involving all these areas of science has emerged over the last 30 years or so, and with it our knowledge of *Soft Condensed Matter* has grown considerably with the active investigation of polymers, supramolecular assemblies of designed organic molecules, liquid crystals, colloids, complex plasmas, lyotropic systems, emulsions, biopolymers, and biomembranes, among others. Taking into account that research in *Soft Condensed Matter* involves ideas combining physics, chemistry, materials science as well as biology, this series may form a bridge between all these disciplines with the aim to provide a comprehensive and substantial understanding a broad spectrum of phenomena relevant to *Soft Condensed Matter*.

The present Book Series, initiated by the late Pierre-Gilles de Gennes, comprises independent book volumes that touch on a wide and diverse range of topics of current interest and importance, covering a large number of diverse aspects, both theoretical and experimental, in all areas of *Soft Condensed Matter*. These volumes will be edited books on advanced topics with contributions by various authors or monographs in a lighter style, written by experts in the corresponding areas. The Book Series mainly addresses graduate students and junior researchers as an introduction to new fields, but it should also be useful for experienced people who want to obtain a general idea on a certain topic or may consider a change of their field of research. This Book Series aims to provide a comprehensive and instructive overview of all *Soft Condensed Matter* phenomena.

The present volume of this Book Series, written as a textbook by Alexei Ivlev, Hartmut Löwen, Gregor Morfill and C. Patrick Royall, focuses on the methodological and conceptual overlap which exists between two domains of soft matter, namely complex plasmas and colloidal dispersions. Although a direct comparison and search for analogies in these two fields is rare, the present volume clearly demonstrates that we can make a significant step towards a comprehensive understanding of atomistic physics in strongly correlated media by exploiting the mutual complementarity of the two systems in terms of the individual-particle dynamics. The book presents basic concepts and properties of the two systems and reviews each field. A clear emphasis is made on numerous cross-connections between complex plasmas and colloidal dispensions. From these joint efforts of the two communities it becomes evident that we can significantly improve our fundamental understanding of some generic physical processes based on the synergy effects described.

Within the next few years, our Series on Soft Condensed Matter will continuously grow and eventually cover the whole spectrum of phenomena in *Soft Condensed Matter*. We hope that many interested colleagues and scientists will profit from this book series.

David Andelman and Günter Reiter
Series Editors

Preface

In the past two decades there has been a remarkable progress in our understanding of processes occurring in classical many-body systems at the individual-particle level. On the one hand, an invaluable role is played here by large-scale computer simulations where tremendous development is being witnessed. On the other hand, an equally important crucial contribution has been made by studying *experimental model systems*, with the most prominent examples being *complex plasmas* and *colloidal dispersions*.

In fact, complex plasmas and (charged) colloidal dispersions represent two domains of soft matter with huge methodological and conceptual overlap, which allows us to perform joint research of numerous fundamental problems. This wonderful similarity and, at the same time, the mutual complementarity of the two systems in terms of the individual-particle dynamics makes such combined studies enormously important for the comprehensive understanding of "atomistic" physics underlying the processes in strongly correlated media. However, active collaborations between the two fields have emerged only very recently. Accordingly, systematic combined studies are only just beginning.

Therefore, the major goal of this book – written by scientists representing both complex plasmas and colloidal dispersions – is to bring the two fields together. The book can be divided into two parts: In the first five chapters the reader is presented with the basic concepts and properties of the two systems, and the emphasis is made on numerous cross-connections between the fields. Such an extended "introductory part" should serve to help each community in understanding the other field better. Simultaneously, this provides the necessary basis for the second part, where *particle-resolved studies* of diverse generic phenomena in liquids and solids, which have been performed with complex plasmas and/or colloidal dispersions, are discussed. In this second part each field is reviewed, and their essential similarity and complementarity become particularly evident. The aim here is to demonstrate that joint efforts of the two communities should result in significant synergy effects and help us to improve our fundamental understanding of generic physical processes.

We tried to make the book as self-contained as possible, assuming only a basic knowledge of statistical physics, plasma physics, and colloid physics. At the same time, we included a large number of references, especially for the second part, to provide the reader with sufficient input for further in-depth analysis of a particular problem.

This book would not have been written without the support of many friends, colleagues and collaborators. First of all, it is our pleasure to thank J. Bartnick, J. Bialké, T. Glanz, A. Kaiser, M. Kohl, and T. Kruppa for their great support in preparing the book (in particular regarding figures and references).

We would like to thank D. G. A. L. Aarts, E. Allahyarov, L. Assoud, P. Bartlett, L. Berthier, A. van Blaaderen, P. Brandt, P. Charbonneau, P. Cicuta, L. Couëdel, D. Derks, J. S. van Duijneveldt, R. P. A. Dullens, S. U. Egelhaaf, J. Eggers, R. Evans, V. Fortov, M. Fuchs, G. Gompper, A. Härtel, J. Horbach, A. Imhof, R. L. Jack, G. Joyce, P. Keim, S. Khrapak, W. Kob, R. Kompaneets, M. Leocmach, A. Lipaev, G. Maret, R. Messina, V. Molotkov, V. Nosenko, E. C. Oğuz, T. Palberg, R. Piazza, W. Poon, C. Räth, M. Rex, R. Roth, M. Rubin-Zuzic, K. Sandomirski, S. Sastry, H. Schöpe, F. Sciortino, M. Sperl, V. Steinberg, K. R. Sütterlin, G. Szamel, H. Tanaka, H. Thomas, T. Vissers, T. Voightmann, E.R. Weeks, N. B. Wilding, S. R. Williams, R. Winkler, A. Wysocki, J. Yamanaka, E. Zaccarelli, and S. Zhdanov for numerous recent discussions and collaborations on various topics addressed in this book.

We are grateful to M. Fuchs, J. Horbach, R. L. Jack, and V. Nosenko for critical reading of several sections, and to C. Du, M. Fink, P. Huber, V. Nosenko, H. Thomas, I. Williams, A. G. Yodh, and I. Zhang for permission to reproduce some of their artwork. A special thank you to Chiharu Nakamura for designing the book's cover.

Preface   ix

Finally, we greatly appreciate financial support from the European Research Council (Advanced grant INTERCOCOS, project number 267499) and from the Royal Society.

November 2011
Alexei Ivlev
Hartmut Löwen
Gregor Morfill
C. Patrick Royall

Contents

Chapter 1

Introduction

Many fundamental issues in classical condensed matter physics such as crystallization, liquid structure, phase separation, glassy states, etc. can be addressed experimentally using model systems of individually visible mesoscopic particles ("grains") playing the role of "proxy atoms". The interaction between such "atoms" is determined by the properties of the surrounding medium and/or by external "tuning". The best known examples of such model systems are two different domains of soft matter – complex plasmas and colloidal dispersions.

Dusty, or *complex plasmas* are composed of a weakly ionized gas and charged microparticles. Dust and dusty plasmas are ubiquitous in space – they are present in planetary rings, cometary tails, interplanetary and interstellar clouds, the mesosphere, thunderclouds, they are found in the vicinity of artificial satellites and space stations, etc. (Whipple, 1981; Grün et al., 1984; Goertz, 1992; Hartquist et al., 1992). The physics of (such naturally occurring) dusty plasmas has for several decades been a well established research field on its own. Furthermore, the presence of dust particles plays a critical role in many important industrial processes [e.g., plasma vapor deposition, microchip production, etching, where growth of dust occurs as a matter of course during the production process (Selwyn et al., 1989; Bouchoule, 1999)] as well as in plasma fusion [where the possibility of producing radioactive and toxic dust in the plasma-wall interactions is an important design issue (Federici et al., 2001; Smirnov et al., 2007; Castaldo et al., 2007)]. Apart from that, plasmas containing micron-size dust particles individually visible under optical microscopy are actively investigated in many laboratories (Shukla and Mamun, 2001; Vladimirov et al., 2005; Fortov et al., 2005; Morfill and Ivlev, 2009; Bonitz et al., 2010). After almost a century of study – the first observations of dust in discharges have been reported by Langmuir *et al.* (1924) – the current interest in complex plasmas began in the mid 1990's, triggered by the laboratory discovery of plasma crystals by Chu and I (1994), Thomas *et al.* (1994), and Hayashi and Tachibana (1994). Today, the physics of complex plasmas (this term is used to distinguish dusty plasmas specially designed for such investigations, from naturally occurring systems) is a rapidly growing field of research.

Colloidal dispersions consist of mesoscopic solid particles with typical sizes ranging from nanometers to micrometers, which are suspended in a molecular fluid solvent (Pusey and van Megen, 1986; Palberg, 1999; Anderson and Lekkerkerker, 2002; Frenkel, 2006). They occur in many everyday environments ranging from paint, ink, or milk to cosmetic products or rheologically modified fluids. Colloidal dispersions belong to the material class of soft matter and are therefore susceptible to external perturbations. Like complex plasmas, trajectories of individual particles can be followed in space and time. These properties make colloids ideal for investigating collective phenomena. Beyond their fundamental importance and current usage, colloids have an enormous potential in emerging and future applications, for example in designing smart materials with novel optical, rheological, electric, or magnetic properties.

Particle-resolved studies of both complex plasmas and colloidal dispersions use optical microscopy. This imposes a lower size limit of around 1 μm, which means that particles employed in both systems *have no principal difference*. An example of such particles is presented in Fig. 1.1. While this image shows particles used in complex plasma experiments, similar images of colloidal particles are *indistinguishable*.

In complex plasmas one can easily change the strength of the electrostatic coupling between particles, Γ (ratio of mean energy of pair Coulomb interaction to the thermal particle energy, the so-called "coupling parameter"), by changing plasma parameters (Shukla and Eliasson, 2009; Morfill and Ivlev, 2009). The magnitude of Γ, which is proportional to the squared charge of the microparticles, Q^2, can vary over an extremely wide range: The charge increases linearly with the particle size and can be quite large (e.g., $Q \sim 3 \times 10^3$ electron charges for a 1 μm particle). In addition, the shape of the interaction potential can be *tuned* externally (Kompaneets *et al.*, 2009). These unique features distinguish complex plasmas from many other laboratory plasmas, where the ion charges are low, the interaction potentials are fixed, and the coupling strength is relatively weak.

In *charged* colloidal dispersions the temperature T is typically kept constant (at room temperature), while the magnitude and even the sign of

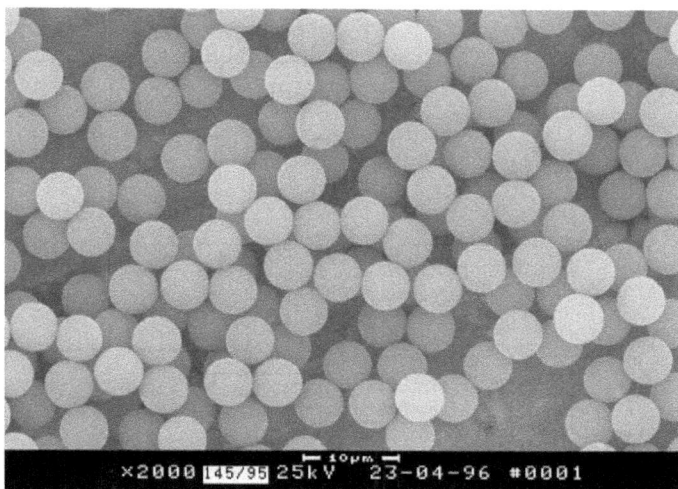

Fig. 1.1   Electron microscopy image of particles typically used for complex plasma experiments. Particles used for colloidal experiments appear identical. Courtesy of *microParticles GmbH*.

the particle charge can be varied up to about $\pm 10^4$ elementary charges
(Yamanaka *et al.*, 1997; Royall *et al.*, 2003). Furthermore, the effective interaction range can be easily tuned via the electrolyte screening, by adding
salt or by de-ionizing the solution (Palberg, 1999; Royall *et al.*, 2003). These
properties allow us to control both the coupling strength Γ and the character of the interparticle interaction, from almost hard-sphere to very soft
plasma-like potentials.

Figure 1.2 demonstrates the broad range of physical parameters (particle
charge Q, number density n and kinetic temperature T) and the resulting
coupling strength Γ accessible in experiments with complex plasmas and
charged colloidal dispersions.[1] This illustration is very helpful in understanding why both systems are particularly well suited for investigations
of states of matter ranging from disordered gases to liquids and to ordered
structures of particles.

---

[1]Here we employ the electronvolt (eV) unit for temperature, which is a natural energy
measure for charged particles: By definition, it is equal to the energy gained by the
electron charge when it passes through an electric potential difference of one volt, i.e.,
1 eV approximately corresponds to 11,600 K.

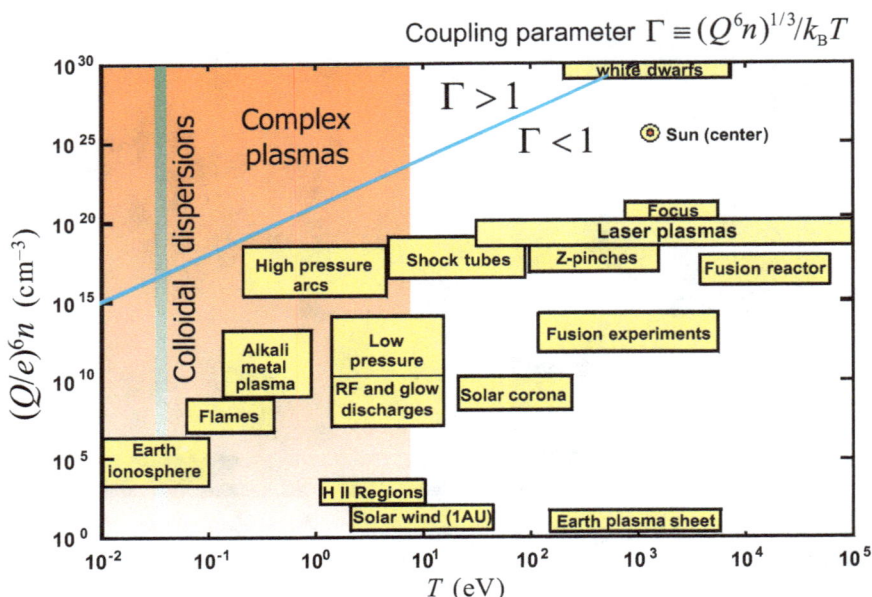

Fig. 1.2   Parameter ranges for different laboratory and naturally occurring charged systems. Complex plasmas and charged colloids allow experimental investigations in the strong coupling regime, including liquid and crystalline states. The blue solid line ($\Gamma = 1$) marks the transition between strongly and weakly coupled regimes.

1.1   Complementarity and Interdisciplinarity

What makes research combining complex plasmas and colloidal dispersions so attractive? The answer is quite simple:

In complex plasmas, the overall dynamic timescales associated with microparticles (e.g., the inverse Einstein frequency) are in the range of tens of milliseconds, yet the microparticles themselves are large enough to be visualized. Thus, the individual trajectories can be obtained by recording with usual CCD cameras and, therefore, fully resolved kinetics can be easily reconstructed. Furthermore, the rate of momentum/energy exchange through interactions between the charged microparticles can massively exceed the damping rate due to friction caused by a dilute ambient gas. Therefore, the motion of individual particles in strongly coupled complex plasmas is *virtually undamped*, which provides a direct analogy to "conventional" liquids and solids in terms of the "atomistic" dynamics. Two other important aspects are that the form of the pair interaction potential can be tuned externally, and that the complex plasma systems are optically thin (scatter little light), so that thousands of particle layers can be visualized, enabling 3D imaging.

In colloidal suspensions, the dynamics of microparticles is fully damped due to the presence of the viscous solvent. Hence the embedding host fluid provides very efficient thermalization of the system, leading to Brownian motion of the individual particles. Therefore, colloidal dispersions can be brought into *equilibrium in a very controlled way*, complementing the complex plasma approach. Otherwise, the colloidal dispersions have the same advantages as complex plasmas: Fully resolved particle trajectories, both in 2D and 3D, can easily be visualized and the pair interactions are tunable. The defining features of complex plasmas and colloidal dispersion are illustrated in Fig. 1.3, which shows the characteristic length scales and composition of both types of system.

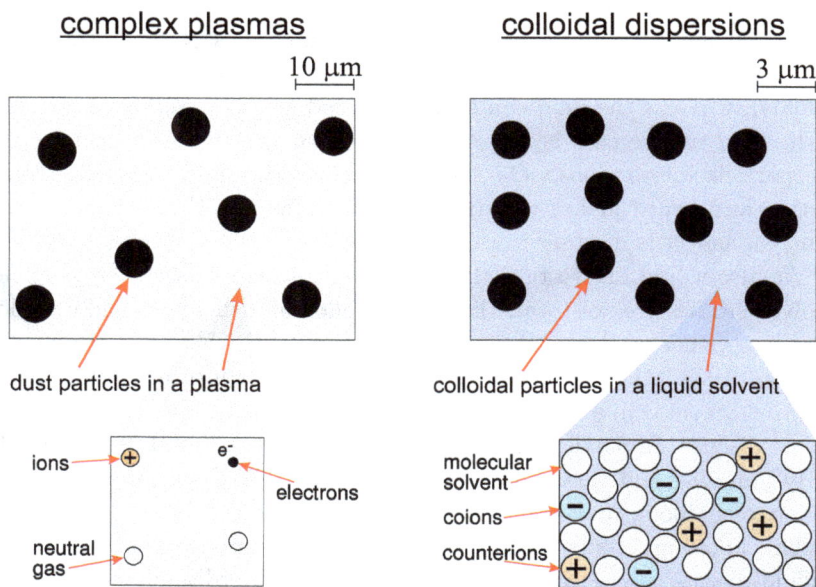

Fig. 1.3   Schematic view of complex plasmas and colloidal dispersions. Dust particles are surrounded by a dilute weakly ionized gas (left panel), while colloidal particles are embedded in a molecular liquid containing also microions (right panel).

Finally, in both complex plasmas and colloidal dispersions individual particles can be easily manipulated in different ways, so that one can perform active controllable experiments to investigate generic processes occurring in liquids or solids at the most fundamental (individual-particle) level.

Because of these properties, combined studies of complex plasmas and colloidal dispersions promise significant synergy and hence bring us more than the sum of the parts: They provide a unique opportunity to go beyond the limits of continuous media down to the fundamental length scale of classical systems – the interparticle distance – and thus to investigate all relevant dynamic and structural processes using the fully resolved motion of individual grains (be they microparticles in complex plasmas or colloids), from the onset of cooperative phenomena to large strongly coupled systems. Hence, the principal aim of such interdisciplinary research is to study generic self-organization processes at the most fundamental individual-particle level, covering the whole range of non-equilibrium and equilibrium phenomena, at a detail not possible until now.

It is important to stress that such particle-resolved studies do not require the identification of individual atoms/molecules of the surrounding fluid. One resolves the individual trajectories of grains (microparticles or colloids), whereas the fluid can still be treated as a continuum. The effect of the fluid in this case is to provide the thermalization and mediate the interparticle interactions. On the other hand, the fluid naturally exerts friction and hence makes the dynamics of grains non-Hamiltonian – this principal aspect is discussed in Chapter 4.

We also would like to take the opportunity to emphasize that particle-resolved studies are not totally limited to colloidal dispersions and complex plasmas. Other systems, notably granular matter, where the strong dissipation of the athermal grains is counteracted by constant driving (such that the system is in a steady state) can provide insight into some phenomena we discuss. In particular, granular matter has been used extensively to study jamming and slow dynamics (see Fig. 9.3). We refer the interested reader to the book by Berthier *et al.* (2011) for further details.

This book represents the first concerted effort to review the current status and discuss the perspectives of the combined interdisciplinary research with complex plasmas and colloidal dispersions, focusing on the complementary approaches developed in these two fields. In Chapters 2 and 3 we summarize basic physical properties which are important for understanding complex plasmas and colloidal dispersions as model systems. Chapter 4 provides a detailed discussion and comparison of the two systems in terms of the individual particle dynamics, illustrating their essential similarities and differences. In Chapter 5 we present a concise summary of experimental methods and hardware currently used for the interdisciplinary research. Chapters 6-11 comprise the core part of the book, describing recent particle-resolved studies of various generic processes in liquid and solid complex plasmas and colloidal dispersions and demonstrating complementarity of this research. In the concluding Outlook we briefly summarize and discuss current "hot topics" and outstanding problems of the interdisciplinary research, where the combined particle-resolved studies are expected to provide crucial new insights.

Chapter 2

Basic Properties of Complex Plasmas

Plasmas have been systematically investigated in laboratories since the beginning of the 20th century. For most of that time, the presence of dust in a plasma was merely seen as a contamination effect, and dusty plasmas were interesting mainly for astrophysicists who developed early theories for the dust charging, interaction, transport, and other processes occurring in cometary and planetary atmospheres, interstellar matter, planet formation, etc. (Grün *et al.*, 1984; Goertz, 1992; Hartquist *et al.*, 1992). In the late 1980s, laboratory research on dusty plasmas became important for industry, e.g., for the fabrication of microelectronics using plasma processes, where dust particles grew during the manufacture process (Selwyn *et al.*, 1989; Bouchoule, 1999).

Plasma crystals were discovered by three groups independently at almost the same time (Chu and I, 1994; Thomas *et al.*, 1994; Hayashi and Tachibana, 1994). This made possible the development of the field where charged grains are used to model classical atomic and molecular systems at the individual-particle level.

It what follows, we briefly review the basic concepts describing the particle charging and interactions in different plasma environments (relevant to typical experimental conditions) and summarize major external forces employed to confine, stabilize, and manipulate microparticles in complex plasmas.

2.1 Charging of Particles

Among the most important parameters of complex plasmas is the particle charge. This determines the interaction of particles with each other, with the surrounding electrons and ions, and with external electric and magnetic fields. Hence all descriptions of complex plasmas necessarily begin with a model of particle charging. We focus mainly on gas-discharge plasmas, where the particles collect electrons and ions from the plasma, so that the charge is determined by the competition between the electron and ion fluxes on the particle surface. Other processes which can also affect charging (e.g., secondary, thermionic and photoelectric emission of electrons from the particle surface) are discussed very briefly. We address topics such as stationary particle charge, charging kinetics, the effects of external electric fields and ion-neutral collisions, and the self-consistent effects caused by the presence of dust.

2.1.1   *Collection of electrons and ions in isotropic plasmas*

In the absence of emission processes, the charge of a dust particle immersed in a plasma of electrons and ions is negative. This is because initially, when the particle is uncharged, the flux of thermal electrons on the particle surface is much larger than the ion flux (the electrons have much higher thermal velocity). The negative charge on the particle leads to repulsion of the electrons and attraction of the ions. The absolute magnitude of the charge grows until the electron and ion fluxes on the particle surface balance. On longer timescales, the charge is practically constant and experiences only small fluctuations around its equilibrium value. The stationary surface potential of the dust particle φ_s is determined by the electron temperature T_e, viz. $-\varphi_s \sim k_B T_e/e$ (where e is the elementary charge). The proportionality coefficient depends on the particular regime which is realized for the electron and ion fluxes to the particle surface.

One frequently used approach to describe the electron and ion fluxes collected by the particle is the orbital motion limited (OML) approximation (Chung *et al.*, 1975; Allen, 1992; Goree, 1994). In the OML approach three major assumptions are employed: (i) The dust grain is isolated in the sense that other grains do not affect the motion of electrons/ions in its vicinity; (ii) Electrons/ions do not experience collisions while approaching the grain; (iii) There are no barriers in the effective potential. Then the cross section for electron/ion collection is determined from the laws of conservation of energy and angular momentum, and is equal to $\sigma(v) = \pi a^2(1 - 2e\varphi_s/mv^2)$ for $e\varphi_s < \frac{1}{2}mv^2$ and zero otherwise. Here m and e are the electron/ion mass and charge (with the appropriate sign) and v denotes the velocity relative to the dust particle of radius a (at infinity). The electron/ion current to the particle surface is determined by the integral over the corresponding velocity distribution function, $I = e \int v \sigma(v) f(v) d^3 v$. For the Maxwellian distribution, $f(v) = n(2\pi v_T^2)^{-3/2} \exp(-v^2/2v_T^2)$, where n and $v_T = \sqrt{k_B T/m}$ are the corresponding number density and thermal velocity, respectively, we get after the integration

$$I_e = \sqrt{8\pi} a^2 e n_e v_{T_e} e^{-z}, \quad I_i = \sqrt{8\pi} a^2 e n_i v_{T_i} (1 + \tau z). \tag{2.1}$$

Here we introduced two dimensionless parameters: $z = |Q|e/ak_B T_e$, which is the normalized magnitude of the particle charge, Q, and $\tau = T_e/T_i$, which is the electron-to-ion temperature ratio. Typically, in low-pressure gas discharge plasmas used in experiments the ionization fraction (i.e., the ratio of the electron/ion to neutral atom densities) is of the order of $10^{-6} - 10^{-7}$; electrons have a temperature about a few eV whereas ions are effectively cooled to room temperature by collisions with neutral atoms, so that $T_i \simeq T_n$ and hence $\tau \sim 10 - 100$; the particle diameter $2a$ usually ranges from ~ 1 μm to ~ 10 μm (colloids used for particle-resolved studies have similar size range, see Sec. 3.1.1). In deriving Eq. (2.1) it was assumed that the particle charge and surface potential are related to each other via $Q = a\varphi_s$. This "vacuum" relation is usually a good approximation for small particles (viz., when a is much smaller than the relevant screening length λ, see Sec. 2.2.1), which is typical for experimental conditions.

In the framework of the OML approximation, the dimensionless surface potential z depends on two parameters – the electron-to-ion temperature ratio, and the gas type (electron-to-ion mass ratio). In Fig. 2.1, values of z are presented for different gases (H, He, Ne, Ar, Kr, Xe) as functions of the temperature ratio τ (Fortov *et al.*, 2005). The particle potential decreases with τ and increases with the gas atomic mass. For typical values of $\tau \sim 10 - 100$, the dimensionless charge is in the range $z \sim 2 - 4$. For a particle with $a \sim 1$ μm and $T_e \sim 1$ eV, the characteristic charge is $|Q| \sim (1-3) \times 10^3$ e.

Note that the above estimates of the charge were made for a "single grain", when the contribution of grain charges into the overall plasma quasineutrality, $en_e + Qn_d = en_i$, is negligible. In reality, a finite number density of microparticles n_d results in $n_i > n_e$ (Havnes *et al.*, 1984), i.e., the ratio of the ion-to-electron flux on a particle increases and hence the absolute magnitude of the charge decreases (in comparison to the single-grain case). The magnitude of this charge depletion effect is characterized by the value of the parameter $P_H = |Z|n_d/n_e$ (often called the "Havnes parameter", see also Chapter 4), where we also introduced the charge number $Z = Q/e$.

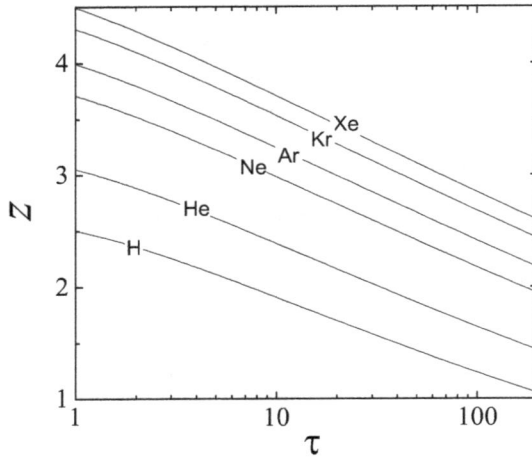

Fig. 2.1   Particle charge in a low-pressure plasma. Dimensionless charge $z = |Q|e/ak_B T_e$ of an isolated spherical particle is shown as a function of electron-to-ion temperature ratio $\tau = T_e/T_i$ for isotropic plasmas of different gases. From Fortov *et al.* (2005).

The evolution of the dust grain charge is governed by the equation $\dot{Q} = I_i - I_e$, so that the stationary charge Q_0 is determined from the current balance, $I_e = I_i$. Let us introduce the relaxation frequency for small deviations of the charge from the stationary value, $\Omega_{ch} = -d(I_i - I_e)/dQ|_{Q_0}$, which fixes the characteristic timescale at which charge varies. Using Eq. (2.1), we obtain (Fortov *et al.*, 2005)

$$\Omega_{ch} = \frac{1+z}{\sqrt{2\pi}} \frac{a}{\lambda_{Di}} \omega_{pi}, \qquad (2.2)$$

where $\lambda_{Di} = \sqrt{k_B T_i/4\pi e^2 n_i}$ is the ion Debye length, and $\omega_{pi} = v_{T_i}/\lambda_{Di}$ is the ion plasma frequency. In deriving Eq. (2.2) the limit $\tau \gg 1$ for the temperature ratio is assumed. The charging frequencies for the thermionic and photoelectric mechanisms were derived by Khrapak *et al.* (1999) and Khrapak and Morfill (2001).

Analysis of charge fluctuations due to the discrete nature of charging was performed by Cui and Goree (1994) and Matsoukas and Russell (1997). The fluctuations can be described as a stationary, Gaussian, and Markovian process [the so-called Ornstein-Uhlenbeck process, which was originally developed to describe the stochastic behavior of the velocity of a Brownian particle (Uhlenbeck and Ornstein, 1930)]. The main properties of charge fluctuations can be easily derived in the framework of the OML approximation (generalization to other charging mechanisms is trivial): The deviation of the particle charge from its equilibrium value, $Q_1(t) = Q(t) - Q_0$, obeys the Langevin equation, $\dot{Q}_1 + \Omega_{ch}Q_1 = q(t)$, where $q(t)$ is the stochastic term describing random acts of electron/ion collections (Matsoukas and Russell, 1997). The function $q(t)$ satisfies the standard normalization: $\langle q(t) \rangle = 0$ and $\langle q(t)q(t') \rangle = 2I_0 \delta(t - t')$, where I_0 is the equilibrium current of electrons and ions [i.e., $I_e(Q_0) = I_i(Q_0) \equiv I_0$, see Eq. (2.1)]. The charge autocorrelation function decays exponentially,

$$\langle Q_1(t)Q_1(t') \rangle = \langle Q_1^2 \rangle \exp\left(-\Omega_{ch}|t - t'|\right), \tag{2.3}$$

and the relative charge dispersion is $\langle Q_1^2 \rangle / Q_0^2 \simeq (1 + z)^{-1}|Z_0|^{-1} \ll 1$ (assuming $\tau \gg 1$).

A remarkable advantage of the OML approximation is that the cross sections are independent of the potential distribution around the grain. This is, however, only true when certain conditions are satisfied (Fortov *et al.*, 2005):

The first condition is associated with finite dust density in experiments. If the mean distance between dust grains is smaller than the characteristic length of interaction between ions/electrons and grains, then the ion/electron trajectories are affected by the presence of neighboring grains, thus influencing the charging (Barkan *et al.*, 1994).

The second condition is more subtle, yet can have profound consequences. In order to understand this, we recall that the problem of an ion/electron orbital motion relative to the grain in the electrostatic potential energy $V(r)$ is equivalent to a one-dimensional problem of the radial motion in the effective potential energy $V_{\text{eff}}(r) = V(r) + (b/r)^2 E_{\text{kin}}$. The second term in $V_{\text{eff}}(r)$ is called "centrifugal", it arises from the conservation of the angular momentum of an ion/electron and is determined by the impact parameter with respect to the grain, b, and the kinetic energy of the impacting ion/electron (at infinity), $E_{\text{kin}} = \frac{1}{2}mv^2$ (Fortov *et al.*, 2005). One can see that if $V(r)$ is negative then $V_{\text{eff}}(r)$ can have a non-monotonic behavior which, in turn, can result in the emergence of a potential energy barrier. However, the OML theory assumes the absence of such a barrier. Hence, for repulsive interactions, i.e., for electrons, the barrier is absent whereas for ions the situation is as follows: The (negative) electrostatic energy scales as $\propto 1/r$ close to the grain, and approaches zero rapidly (exponentially for Yukawa interactions) at large distances (Al'pert *et al.*, 1965; Lampe *et al.*, 2000), so that the (positive) centrifugal part ($\propto 1/r^2$) dominates in these limits; at intermediate distances a competition between the electrostatic and centrifugal parts can create a barrier, which causes reflection of some (low energy) ions approaching a grain and, thus, leads to a decrease in the ion current compared to OML theory (making $|Q|$ smaller). If the fraction of the reflected ions is small then the corrections to OML are also small. For the Yukawa interaction with the effective screening length λ (see Sec. 2.2) one can write the condition of the OML applicability in the form $\sqrt{2z\tau}(a/\lambda) \lesssim \ln[z\tau(a/\lambda)]$ (Khrapak *et al.*, 2004). For typical parameters $z \sim 1$ and $\tau \sim 100$ we obtain that OML is applicable when the ratio of the particle radius to the screening length satisfies the condition $a/\lambda \lesssim 0.2$. This corresponds to grains smaller than ~ 10 μm, which is usual for most laboratory experiments.

The third condition is due to ion-neutral collisions. In the OML approach collisions of electrons and ions with neutrals are neglected, assuming that their mean free paths are large compared to the screening length λ (Goree, 1994). However, theory, numerical simulations, and experiments show that ion-neutral charge-exchange collisions in the vicinity of a dust grain can lead to a substantial increase in the ion current to the surface even when the ion mean free path ℓ_{in} is larger than λ (Zobnin *et al.*, 2000; Lampe *et al.*, 2001a; Khrapak *et al.*, 2005). This increase in the ion current can considerably suppress the grain charge. In the weakly collisional regime for the ions, characterized by the condition $\ell_{in} \gtrsim R_0$ [where the Coulomb radius $R_0 \ll \lambda$ is determined by the condition $V(R_0) \simeq T_i$], the net ion current on the grain is:

$$I_i \simeq \sqrt{8\pi}a^2 e n_i v_{T_i} \left[1 + z\tau + 0.1(\lambda/\ell_{in})z^2\tau^2\right].$$

One can see that for $z \sim 1$ and $\tau \sim 100$ collisions can affect particle charging even when the mean free path is an order of magnitude larger than the screening length. Thus, ion-neutral collisions can play very important role in the particle charging.

2.1.2   *Effect of an external electric field*

A strong electric field is often present in experiments performed with complex plasmas under gravity (see Sec. 2.4), which causes the plasma to flow relative to the dust component. This in turn can affect the particle charge, by changing the collection cross sections and velocity distribution functions of plasma species. Normally, the effect of electron flow is negligible (as compared to their thermal velocity) while for ions the effect is significant. Often, the ion flow occurs in the mobility-limited regime, where the velocity is linearly proportional to the local field, $u_i \propto E$. In this case the electron current to the particle surface is given by Eq. (2.1) while for ions the current can be calculated by integrating the OML collection cross section over the appropriate velocity distribution (Fortov *et al.*, 2005). Model calculations show that the charge is practically constant when u_i is smaller than the ion thermal velocity v_{T_i}, at larger u_i the charge increases by a factor of 2–3 and attains a maximum at $u_i/v_{T_i} \sim 3\sqrt{\tau}$. The range of even larger u_i, where the charge falls off, is normally not relevant to experiments with complex plasmas.

2.1.3   *Other mechanisms of charging*

Note that the collection of ions and electrons from the plasma is not the only charging mechanism. Electrons can also be emitted from the particle surface due to thermionic, photoelectric, and secondary electron emission processes. The secondary emission could be due to energetic electrons with an energy of a few hundred eV or higher interacting with a particle, or because of metastable neutrals. Also, field emission (due to a self-induced field) might be important for sub-micron particles. All these processes are of importance for dust charging in some laboratory experiments, for instance, in thermal plasmas (Fortov *et al.*, 1996a; Nefedov *et al.*, 1997; Khodataev *et al.*, 1998; Nefedov *et al.*, 1999; Samarian *et al.*, 2000) or plasmas induced by UV irradiation (Fortov *et al.*, 1998), in plasmas with photoelectric charging (Sickafoose *et al.*, 2000) or charging by electron beams (Walch *et al.*, 1995), etc. Under certain conditions (e.g., UV radiation) the sign of the charge can invert and become positive, in contrast to the situation discussed previously (see Sec. 5.1.3).

2.2   Interaction Between Microparticles

The interaction between charged microparticles in complex plasmas can be affected, e.g., by plasma absorption on the particle surface, by variability of the particle charges, polarization, etc. Usually, the interparticle interaction (as well as screening and charging) can be calculated in the approximation of "isolated" particles.

2.2.1   *Electrostatic mechanisms in isotropic plasmas*

Assuming for simplicity that two particles have equal charges Q which do not depend on their separation r, the energy of pair interaction is $V(r) =$

$Q\varphi(r)$, where $\varphi(r)$ is the distribution of the electrostatic potential around a particle. Some deviation from this simplified picture will be discussed later. For now, let us focus on the effects which determine the shape of $\varphi(r)$.

The distribution of the electric potential around a small individual spherical particle of radius a and charge Q in isotropic plasmas is often described by the Debye-Hückel (Yukawa) form

$$\varphi(r) = (Q/r) \exp(-r/\lambda), \tag{2.4}$$

where λ is the effective plasma screening length. The exponential screening comes from the redistribution of plasma electrons and ions in the vicinity of the charged particle. This form is *identical* to that of charged colloids (see Sec. 3.2).

In the regime of linear response of electrons and ions, i.e., when $e|\varphi_s|/k_B T \lesssim 1$ is satisfied, their distributions can be linearized. Then Eq. (2.4) follows directly from the solution of the Poisson equation with the boundary conditions $\varphi(\infty) = 0$ and $\varphi(a) = \varphi_s$. The relation between the surface electric field and the particle charge is given by $d\varphi/dr|_{r=a} = -Q/a^2$, the surface potential is $\varphi_s = (Q/a)(1 + a/\lambda_D)^{-1}$. The screening length in this case is equal to the linearized Debye radius $\lambda = \lambda_D$, where $\lambda_D^{-2} = \lambda_{De}^{-2} + \lambda_{Di}^{-2}$.

However, complex plasmas are often characterized by strong ion-dust coupling: The electrostatic energy of the ion interaction with a charged particle can considerably exceed the ion kinetic energy in the vicinity of the particle. Even in this case, close to the particle (up to a distance of a few screening lengths from its surface), the Debye-Hückel form works reasonably well (Kennedy and Allen, 2003; Daugherty *et al.*, 1992). However, λ can deviate from the linearized Debye radius. The exact dependence of λ on plasma and particle parameters is not known for the general case, and so far it was only determined using numerical simulations for a limited number of special cases (Daugherty *et al.*, 1992; Lampe *et al.*, 2003). For a collisionless plasma with $T_e \gg T_i$ the following expression was proposed (Khrapak and Morfill, 2009): $\lambda \simeq \lambda_D(1 + 0.1\beta_{di}^{1/2} + 0.01\beta_{di})$, where β_{di} is the so-called ion scattering parameter (see Sec. 2.3).

Theory predicts a rich variety of screening mechanisms operating in complex plasmas. These mechanisms can delicately affect the charge balance around the particle, and thus have no correspondence with colloidal dispersions where charge is perfectly balanced. An important specific property of complex plasmas is associated with continuous absorption of plasma

species on the particle surface. In particular, in the absence of plasma production (ionization) and loss in the vicinity of a particle, conservation of
plasma flux directed to the particle completely determines the far asymptote of the potential. As a result, at large distances the potential is not
screened exponentially but exhibits a power-law decay. In collisionless plasmas the far asymptote scales as $\varphi(r) \propto r^{-2}$ (Al'pert *et al.*, 1965; Tsytovich,
1997). In the opposite limit of strongly collisional plasma the potential has
a Coulomb-like asymptote $\varphi(r) \propto r^{-1}$ (Su and Lam, 1963; Khrapak *et al.*,
2006).

Recently it was shown that the plasma production and loss processes
can play a crucial role in the long-range behavior of $\varphi(r)$. For an individual
particle this effect was studied by Filippov *et al.* (2007) and Chaudhuri *et al.*
(2008) using the hydrodynamic approach for the case of highly collisional
plasmas. Electron-neutral collisions (impact ionization) were considered as
the main mechanism of plasma production. Plasma loss was either due to
electron-ion volume recombination [which is relevant to high-pressure gas
discharges (Filippov *et al.*, 2007; Chaudhuri *et al.*, 2008)], due to ambipolar
diffusion towards the chamber walls [which occurs in low- and moderate-
pressure gas discharges (Chaudhuri *et al.*, 2008)], their combination (Khrapak *et al.*, 2010a), or collective plasma loss on particles themselves [when
the particle number density is sufficiently high (Else *et al.*, 2009; Chaudhuri
et al., 2010)]. Generally, this results in the emergence of *two* dominating
asymptotes of $\varphi(r)$ – both having the Yukawa form, which we will refer to
as a double-Yukawa repulsive potential (Khrapak *et al.*, 2010a):

$$\varphi(r) = \frac{1}{r}\left(Q_{\mathrm{SR}}e^{-r/\lambda_{\mathrm{SR}}} + Q_{\mathrm{LR}}e^{-r/\lambda_{\mathrm{LR}}}\right). \qquad (2.5)$$

The theory predicts that the length scales λ_{SR} and λ_{LR} can be very different, and therefore we denote them as "short-range" (SR) and "long-range"
(LR). Typically, λ_{SR} is determined by the classical mechanism of Debye-
Hückel screening , so that λ_{SR} is of the order of the plasma Debye screening
length (which is normally smaller than the mean interparticle distance). In
contrast, the magnitude of λ_{LR} is controlled by the balance between the
plasma production and loss. Therefore λ_{LR} can vary over a fairly broad
range and is usually much larger than the interparticle distance. Further-
more, the ratio $Q_{\mathrm{LR}}/Q_{\mathrm{SR}} \equiv \epsilon_Q$ is typically small [both effective charges
are negative (Khrapak *et al.*, 2010a)]. In the special case when plasma
losses are due to ambipolar diffusion only, we have $\lambda_{\mathrm{LR}} \to \infty$, i.e., the
long-range potential is not screened but has a Coulomb-like long-range
asymptote (Chaudhuri *et al.*, 2008). In Sec. 8.3.4 we demonstrate enormous

importance of weak long-range interactions for demixing kinetics, and also give rough estimates for the values of Q_{LR}/Q_{SR} and $\lambda_{LR}/\lambda_{SR}$ expected in experiments.

2.2.2   *Effect of an external electric field*

As we mentioned in Sec. 2.1.2, complex plasmas are often subject to electric fields. This induces an ion flow and, hence, creates a perturbed region of plasma density (and hence of plasma potential) around each particle, caused by downstream focusing of ions – the so-called "plasma wake" which is illustrated in Fig. 2.2. The electrostatic interaction of one particle with the field generated by the wake of another (neighboring) particle *mediates* the interparticle interactions, making them *nonreciprocal* (see Sec. 4.3.1).

One can apply the linear dielectric response formalism [see e.g., Aleksandrov *et al.* (1984)] to calculate the wake potential. This approach is applicable provided the ions are weakly coupled to the particle (i.e., the region of nonlinear electrostatic interaction around the particle is small compared to the plasma screening length). In this approximation, the electrostatic potential created by a point-like charge at rest is defined as

$$\varphi(\mathbf{r}) = \frac{Q}{2\pi^2} \int \frac{e^{i\mathbf{k}\cdot\mathbf{r}} d\mathbf{k}}{k^2 \varepsilon(0,\mathbf{k})}, \qquad (2.6)$$

where $\varepsilon(\omega,\mathbf{k})$ is the plasma permittivity. Using the appropriate model for $\varepsilon(\omega,\mathbf{k})$, one can calculate the anisotropic potential distribution (Nambu *et al.*, 1995; Ishihara and Vladimirov, 1997; Xie *et al.*, 1999; Lemons *et al.*, 2000; Lapenta, 2000; Kompaneets *et al.*, 2007, 2008). This can also be obtained from numerical modeling (Lampe *et al.*, 2000; Melandso and Goree, 1995; Winske, 2001; Lapenta, 2002; Vladimirov *et al.*, 2003; Miloch *et al.*, 2008).

Physically, the generation of plasma wakes in anisotropic plasmas is similar to the generation of electromagnetic waves by a particle which is placed in a moving medium (Ginzburg, 1996), and hence the analogy with the Vavilov-Cherenkov effect can be useful. The potential is no longer monotonic in a certain region downstream from the particle, but has a well pronounced extremum (maximum for a negatively charged particle). Numerical modeling shows that the shape of the wake potential is sensitive to the ion-neutral collisions (Hou *et al.*, 2003) and the electron-to-ion temperature ratio which governs Landau damping (Lampe *et al.*, 2001b). In typical situations, these mechanisms effectively "smear out" the oscillatory wake structure, leaving a single maximum.

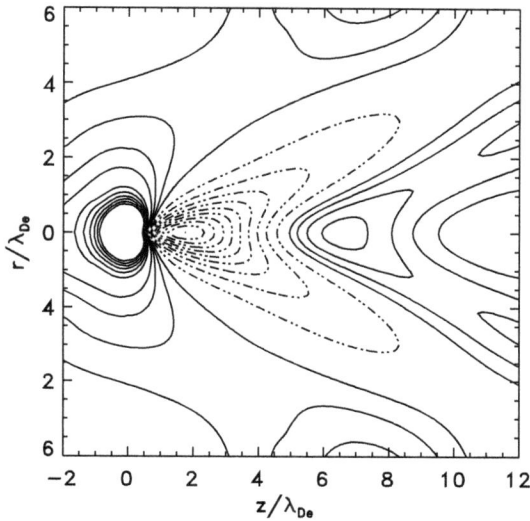

Fig. 2.2 Plasma wake. The figure demonstrates the complex structure of the wake potential $\varphi(\mathbf{r})$ (plasma flows to the right). Calculations are for collisionless ions with a shifted Maxwellian distribution (thermal Mach number $M_T = 7.5$) and for Boltzmann electrons (temperature ratio $T_e/T_i = 25$). The negatively charged grain is at the center of the left-most node, solid and dashed curves indicate contour lines for negative and positive potentials, respectively, distance is in units of λ_{De}. From Lampe *et al.* (2000).

Let us demonstrate how the wake potential depends on the plasma flow. The flow is conveniently characterized by the "thermal" Mach number defined as the ratio of the ion flow velocity to its thermal velocity,

$$M_T = u_i/v_{T_i}.$$

A noticeable anisotropy of the wake structure appears already in a sub-thermal regime, $M_T \lesssim 1$ (which is rather typical for experiments). The potential profile in this case can be calculated from Eq. (2.6) analytically using the Bhatnagar-Gross-Krook (BGK) approach for the ion-neutral collision integral (Schweigert, 2001; Ivlev *et al.*, 2005). The far-field potential has an asymptotic scaling $\propto r^{-3}$ asymptote (Montgomery *et al.*, 1968). By combining this with the near-field Yukawa core, in the case of small collisionality (viz., small ratio of the ion-neutral collision frequency to the ion plasma frequency) we can approximate the potential by the following expression (Kompaneets, 2007):

$$\varphi(r,\theta) = Q\left[\frac{e^{-r/\lambda_D}}{r} - 2\sqrt{\frac{2}{\pi}}\frac{M_T\lambda_D^2}{r^3}\cos\theta\right.$$

$$\left. - \left(2 - \frac{\pi}{2}\right)\frac{M_T^2\lambda_D^2}{r^3}(3\cos^2\theta - 1)\right] + o(M_T^2/r^3), \tag{2.7}$$

where θ is the angle between $\mathbf{r}$ and $\mathbf{u}_i$. Equation (2.7) shows that at large distances microparticles attract each other in a certain range of θ along the flow, and repel in the transverse direction. Such behavior is usually observed in ground-based experiments – particles levitating in, e.g., (pre) sheaths of rf discharges form stable vertical "strings" [see, e.g., Melzer *et al.* (1996) and Fig. 4.5b]. This result also highlights the importance of the self-consistent consideration of the ion kinetics where the ion-neutral collisions are properly taken into account: For instance, the (somewhat arbitrary) use of a shifted Maxwellian distribution (Wang *et al.*, 1981) to model a collisional flowing plasma yields attraction between particles in the transverse direction, which clearly contradicts experimental observations.

In some ground-based experiments particles levitate in the regions where the electric field is so strong that the thermal Mach number can be significantly larger than unity; also, the collisionality can be rather high. Then Eq. (2.7) is no longer applicable. The recent calculations by Kompaneets *et al.* (2007) (based on a more realistic model of a constant cross section for the ion-neutral collision integral) take both effects into account, and the resulting interaction potential is in very good agreement with the direct measurements (performed for $M_T \sim 10$).

A very interesting situation occurs when an (external) ac field is applied. Equation (2.7) shows that the "dipolar" ($\propto M_T$) term in this case is canceled out (assuming that the ac frequency is below the ion-plasma frequency, $\sim 10^7\ \text{s}^{-1}$), i.e., the plasma wakes are "smeared out" along the field axis. This results in the so-called "electrorheological" effect in complex plasmas and allows us to control the interparticle interactions. This effect is also realized in colloidal dispersions (details are discussed in Sec. 11.1).

2.2.3   *Other mechanisms of interaction*

Besides electrostatic effects, there exist other mechanisms that can contribute to interparticle interactions in complex plasmas. These are associated with a specific property of complex plasmas – their thermodynamic openness (non-equilibrium nature) caused by the continuous exchange of matter and energy between the particles and surrounding plasma. For instance, constant plasma absorption on the particle surfaces gives rise to a so-called "ion-shadow" interaction (Lampe *et al.*, 2000; Tsytovich, 1997; Khrapak *et al.*, 2001), which is basically the ion drag force that one particle experiences as a consequence of the ion flux directed to another neighboring particle and vice versa. (Hence, the ion shadowing can in principle be attributed to the general class of wake-mediated interactions, see discussion of the ion drag force in the next section.) The ion-shadow force is always attractive and exhibits $\propto r^{-1}$ scaling with distance. An approximate expression for the ion-shadow potential taking into account the effect of ion-neutral collisions in the weakly collisional plasma regime has been derived by Khrapak and Morfill (2008).

Since at large distances both electrostatic and ion-shadow interactions have the same $\propto r^{-1}$ asymptotes, the total long-range interaction can be either repulsive or attractive, depending on the relative magnitudes of the two contributions. An approximate condition for attraction is $\ell_{in} \gtrsim 4\lambda_D$ (Khrapak and Morfill, 2008), i.e., rather low plasma collisionality is required in order to make the ion-shadow attraction operational in complex plasmas.

A mechanism of interaction similar to ion shadowing can be associated with the neutral component, provided the particle surface temperature is different from the temperature of the surrounding neutral gas (Tsytovich *et al.*, 1998). If the particle surface is hotter, there is a net momentum flux from the particle to neutral gas. This results in the repulsion between two particles. On the other hand, if the particle surface is colder, the net momentum flux generates the attraction. The "neutral shadowing" interaction also decays as $\propto r^{-1}$ and is proportional to the temperature difference between the particle surface and the neutral gas (Tsytovich *et al.*, 1998). Experiments and theoretical predictions demonstrated that in low-pressure gas discharges the particle surface temperature is somewhat hotter that the neutral gas temperature (Swinkels *et al.*, 2000; Khrapak and Morfill, 2006). Recent estimations show that under typical discharge conditions the neutral shadowing is usually weaker than both the ion shadowing and electric interactions (Khrapak and Morfill, 2006).

The shadowing forces are not pairwise, since the interaction between (more than two) particles depends on their mutual arrangement. In this respect, the shadowing interactions appear similar to the hydrodynamic interactions in colloids (see Sec. 3.4), although the physical mechanisms responsible for them are completely different.

2.3   Momentum Exchange Between Species

The momentum exchange between microparticles and other plasma species plays an exceptionally important role in complex plasmas. For example, the momentum transfer in collisions with the neutral gas "cools down" the system of microparticles, introducing some damping. The forces due to the momentum transfer from electrons and ions to the charged grains – the electron and ion drag forces – often determine static and dynamic properties of microparticles, affect wave phenomena, etc. And – most importantly – momentum exchange in grain-grain collisions governs transport, fluid properties, phase transitions, etc. As we pointed out in the Introduction (see also Chapter 4), complex plasmas can be "engineered" as essentially a virtually undamped one-component fluid (when the interactions between the grains dominate), or as a "particle laden gas" (when the interactions with the surrounding gas are of greater importance).

The grain-neutral collisions in complex plasmas typically occur in the free molecular regime, when the Knudsen number $Kn = (\sigma_{nn}n_n a)^{-1}$ (which is the ratio of the neutral mean free path to the particle radius) is quite large. The resulting momentum exchange is characterized by the Epstein damping rate (Epstein, 1924)

$$\nu_{dn} = \delta(8\sqrt{2\pi}/3)(m_n/m_d)a^2 n_n v_{T_n}. \tag{2.8}$$

The value of the numerical factor δ depends on the exact process of neutral scattering from the particle surface. For example, $\delta = 1$ for the cases of complete absorption and specular reflection, whilst $\delta = 1 + \pi/8$ for diffuse scattering with full accommodation. The latter value is more consistent with recent experimental results (Liu *et al.*, 2003).

As regards the grain-grain collisions as well as ion-grain collisions, the regimes of momentum exchange are determined by the "scattering parameter" β, which is the ratio of the bare Coulomb interaction energy at the screening length scale λ to the mean kinetic energy (temperature). Assuming typical experimental conditions, for ion-grain collisions $\beta_{di} = e|Q|/\lambda k_B T_i \sim z\tau(a/\lambda) \sim 0.1 - 30$, and for grain-grain collisions $\beta_{dd} = Q^2/\lambda k_B T_d \sim |Z|(T_i/T_d)\beta_{di} \sim 10^4 - 10^6$ (Khrapak *et al.*, 2002; Hahn *et al.*, 1971). The "weak" scattering regime, $\beta \lesssim 1$, is similar to the Coulomb scattering occurring in conventional plasmas (Barnes *et al.*, 1992) (where the momentum exchange rate does not depend on the sign of the interaction). For "strong" scattering, $\beta \gg 1$, the repulsive interaction is reminiscent to that between hard spheres of radius $\simeq \lambda \ln 2\beta$ (Khrapak *et al.*, 2004; Baroody, 1962), whereas for the attractive interaction the potential barrier appears (see Sec. 2.1.1, discussion of the OML theory) and then the scattering is determined by the position of the barrier.

For grain-grain collisions the regime $\beta_{dd} \gg 1$ is typical, so that the analogy with hard-sphere collisions can be used (Khrapak *et al.*, 2004). The momentum exchange rate is then

$$\nu_{dd} \simeq (4\sqrt{2\pi}/3)n_d v_{T_d}\lambda^2 \ln^2 2\beta_{dd}.$$

The binary collision approach for the interparticle interactions is naturally valid as long as the system is sufficiently dilute. As the number density of microparticles grows the coupling with increasing number of neighbors becomes important, so that eventually the collective modes take over and the momentum exchange rate scales proportional to the relevant eigenfrequency (e.g., Einstein frequency Ω_E, see Sec. 4.1.1). The crossover from the binary to the collective momentum exchange regime is determined by the condition that the effective interaction range, $\simeq \lambda \ln 2\beta_{dd}$, is comparable to the interparticle separation. The leading scaling of this condition naturally coincides with the strong-coupling condition $\Gamma^{(s)} \sim 1$ [see Eq. (7.1)].

The *ion drag force* F_{id} – the momentum transfer from the flowing ions to a charged microparticle[1] – is an inevitable and exceptionally important factor in complex plasmas. The ion flow is usually caused by a "global" large-scale electric field, which can be either due to natural inhomogeneities in a discharge plasma (ambipolar fields) or induced by external sources. For a negatively charged particle the ion drag is pointed in the direction opposite to the electric force (see next section). For subthermal flows ($M_T \lesssim 1$), the expression for F_{id} becomes particularly simple in the regimes of "weak" and "strong" scattering. Calculations for the former regime, which is limited by $\beta_{di} \lesssim 5$, give the following expression (Khrapak *et al.*, 2002, 2003):

$$F_{id} \simeq \sqrt{2/9\pi}(k_B T_i/e)^2 \Lambda \beta_{di}^2 M_T, \tag{2.9}$$

where $\Lambda(\beta_{di}) \simeq -e^{\beta_{di}/2}\text{Ei}(-\beta_{di}/2)$ is the modified Coulomb logarithm integrated over the Maxwellian distribution function (expressed via the exponential integral Ei). Equation (2.9) yields the scaling $F_{id} \propto (Q/\lambda)^2$. In the linear regime $\beta_{di} \ll 1$ the logarithm is reduced to $\Lambda \simeq \ln \beta_{di}^{-1}$, which is identical to the results of the Coulomb scattering theory (Barnes *et al.*, 1992).

---

[1] If the direct momentum transfer (due to the ion absorption on a microparticle) is neglected, then the ion drag force can be equivalently viewed as the electrostatic force of the wake field.

In the opposite regime of strong coupling, $\beta_{di} \gg \beta_{cr} \simeq 13$, one should replace $\Lambda\beta_{di}^2$ with $2\ln^2\beta_{di}$ in Eq. (2.9). In this case the force depends logarithmically on Q and λ. Note that for $M_T \ll 1$ the screening length is determined by ions, $\lambda \simeq \lambda_{Di}$, since the electron temperature is typically two orders of magnitude higher than the ion (neutral) temperature.

2.4   Major External Forces

The most important force on a charged particle in a plasma is, of course, the electrostatic force. This is particularly relevant to ground-based experiments (which are often performed in rf discharges, see Sec. 5.1.1). Figure 2.3 illustrates the typical plasma environment in this case: Microparticles levitate near the *sheath* region – the boundary layer of uncompensated positive space charge (where $n_e \le n_i$) separating a bulk quasineutral plasma and a wall (electrode), generated by the interaction of the plasma with the wall surface (Lieberman and Lichtenberg, 1994; Raizer, 1991). The sheath is characterized by strongly inhomogeneous electric field $E_{sh}(z)$, changing from very large values at the electrode surface to low ambipolar level in the bulk plasma (pre-sheath region), where it is of the order of $k_B T_e/eL \lesssim 1 - 10$ V/cm [$L \sim 1 - 10$ cm is the characteristic size of the discharge chamber, Raizer (1991)]. The ion flow velocity increases towards the electrode while the electron-to-ion density ratio decreases, which makes the charge Q vary with the height as well (see Sec. 2.1.3). The resulting net force acting on a charged particle is then a superposition of the electrostatic force $F_{el} = QE_{sh}$ pointing upwards and the ion drag force F_{id} pushing particles towards the electrode. Thus, the equilibrium levitation height in the (pre) sheath is determined from the force balance $F_{el}(z) = mg + F_{id}(z)$. The equilibrium is stable, since F_{el} increases towards the electrode (unless particles are too heavy) whereas F_{id} decreases (Fortov *et al.*, 2005).

We note that both the ion drag and electrostatic forces are linearly proportional to the electric field E in the limit $E \to 0$ (we assume $M_T \propto E$), so that their ratio in this limit, $(F_{id}/F_{el})_0$, is a constant. Since the forces are opposed, the sign of $(F_{id}/F_{el})_0 - 1$ determines stability of global structures formed by microparticles in the bulk plasma (Morfill *et al.*, 1999; Goree *et al.*, 1999; Samsonov and Goree, 1999). For instance, the void observed in the center of rf discharge plasmas (see Fig. 5.6) is formed when $(F_{id}/F_{el})_0 > 1$, i.e., when the ion drag (pushing particles away from the center) exceeds the confining electric force.

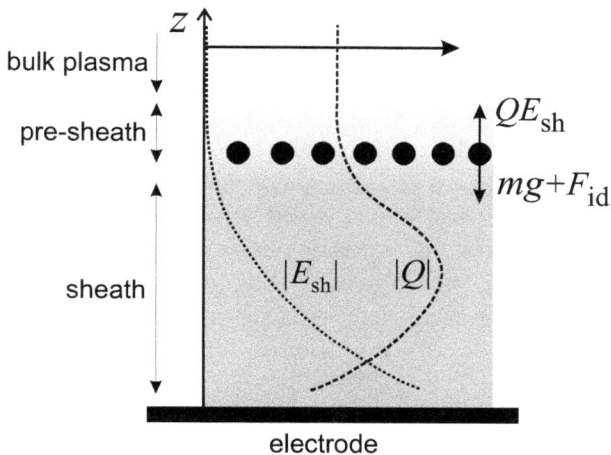

Fig. 2.3   Balance of forces on microparticles in the plasma sheath region. Shown is the qualitative dependence of the (absolute values of) equilibrium particle charge, Q, and the vertical electric field, E_{sh}, on the vertical distance z from the electrode. The gravity force mg is balanced by combination of the electrostatic force QE_{sh} and ion drag force F_{id} (which is determined by E_{sh}).

If a temperature gradient is present in a neutral gas, then the particle experiences a thermophoretic force. This force is due to the asymmetry in the momentum transfer from neutrals and is directed towards lower gas temperatures. In the case of full accommodation of neutrals colliding with the particle surface the thermophoretic force can be approximated by $\mathbf{F}_{th} \simeq -3.3(a^2/\sigma_{nn})k_B \nabla T_n$ (Rothermel *et al.*, 2002). Thus, the thermophoretic force depends on particle radius and gas type (through the cross section of neutral-neutral collisions, σ_{nn}), but does not depend on the gas pressure. For particles of about 1 μm radius and mass density ~ 1 g/cm^3 in argon gas, F_{th} is comparable to the force of gravity at $|\nabla T_n| \sim 10$ K/cm. By contrast, in colloidal dispersions the mechanism of the thermophoretic force is poorly understood (Piazza and Parola, 2008).

A neutral drag force $\mathbf{F}_n$ does not only cause damping, but also plays important role when gas is flowing relative to the particles. In complex plasmas, the gas flow velocity u_n is small compared to the thermal velocity of neutrals v_{T_n} and hence $\mathbf{F}_n = m_d \nu_{dn}\mathbf{u}_n$, where the momentum exchange rate ν_{dn} is given by Eq. (2.8). Neutral drag can be employed to exert controllable stresses and induce shear flows in complex plasmas (see Sec. 10.2).

Finally, the radiation force exerted on a particle by a laser beam provides an exceptionally important mechanism for particle manipulation (see Chapter 10). The force is determined by the radiation intensity I and consists of several components which originate from rather different physical mechanisms (Ashkin and Dziedzic, 1987; Ashkin, 2000). The principal component is the scattering force $F_{\text{scat}} = C_{\text{scat}}(I/c)$, which is due to the refraction of the laser beam in the particle and which is pointed in the direction of the beam propagation. Here, the scattering cross section C_{scat} depends on the scattering regime, viz., on the ratio of the radiation wavelength λ (in this context, such notation cannot cause confusion with the screening length) to the particle radius a. In a general case, $C_{\text{scat}}(\lambda/a)$ is described by Mie theory (Harada and Asakura, 1996): For the Rayleigh regime, when λ/a is large, $C_{\text{scat}} = \frac{128}{3}\pi^5(a^6/\lambda^4)\left(\frac{n_r^2-1}{n_r^2+2}\right)^2$, where $n_r > 1$ is the relative refractive index of the particle. For the opposite geometrical-optics regime C_{scat} naturally tends to the geometrical refractive limit. The second contribution is the gradient force $\mathbf{F}_{\text{grad}} = -2\pi a^3 \left(\frac{n_r^2-1}{n_r^2+2}\right)^2 (\nabla I/c)$, which is due to the particle polarization in an inhomogeneous radiation field (Ashkin, 2000).

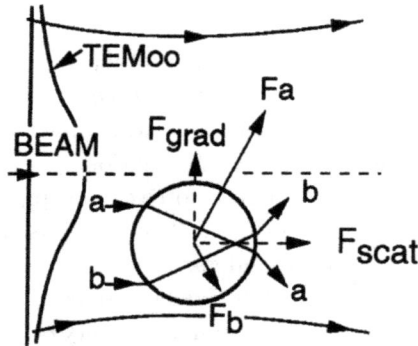

Fig. 2.4   Sketch showing principal components of the radiation force. By considering two rays ("a" and "b") of the laser beam (which is inhomogeneous in the transverse direction), one can decompose the resulting momentum-exchange forces (F_a and F_b) acting on a particle into the scattering component along the beam, F_{scat}, and the gradient component pointed transversely, F_{grad}. From Ashkin (2000).

The effect of the two components of the radiation force is illustrated in Fig. 2.4. In experiments with complex plasmas F_{grad} (which pulls a particle towards the beam focus – the effect utilized in laser tweezers, see Sec. 3.5.5) is usually significantly smaller than F_{scat} and hence plays a minor role. The last (potentially important) contribution is the photophoretic force, which is basically the thermophoretic force driven due to inhomogeneous heating of a particle by laser radiation (Tehranian *et al.*, 2001). For homogeneous radiation this force is parallel to the beam but its direction can reverse, depending on what part of the particle – front or rear – received stronger heating (which depends, in particular, on details of the refraction regime). The relative importance of this force in complex plasmas still needs to be studied.

Chapter 3

Basic Properties of Colloidal Dispersions

Colloidal dispersions are mesoscopic solid particles suspended in a molecular fluid solvent. The almost complete length-scale separation between the size of the colloidal particle and the solvent molecules makes colloidal dispersions analogous to complex plasmas. In both systems the particles can become highly charged, which results in effective screened Coulomb interactions described in the simplest case by a Yukawa pair potential. Therefore charged colloidal dispersions and complex plasmas share very many similarities in their behavior. However, the embedding medium is different – it is a liquid for colloids and a dilute gas for plasmas, which leads to completely different damping mechanisms operating in these two systems (see Chap. 4 for detailed discussion).

The first basic question which needs to be addressed for colloids is their *stability*. For many applications (paints, food, etc.) it is desirable to keep them in solution as long as possible and to avoid their irreversible coagulation. In order to understand colloidal stability quantitatively, the different interaction mechanisms between colloidal particles need to be considered, which leads to the traditional Derjaguin-Landau-Verwey-Overbeek (DLVO) theory – a cornerstone of colloid science. In general, the refractive index/dielectric constant of the colloidal particles differs from that of the solvent, so that there is an effective van der Waals attraction between the colloids. Normally, for particles in the micron-size range these interactions are non-retarded (i.e., they propagate practically instantaneously on relevant time and length scales), so that the potential falls off as the inverse sixth power of the distance between the particles.[1] Integrating the interactions over two spheres of diameter σ separated by the distance r between

---

[1]For retarded van der Waals interactions the potential scales as the inverse seventh power of the distance.

the centers, one derives the effective potential which diverges at contact as $\propto -A_{\mathrm{H}}/(r - \sigma)$, where the amplitude is set by the so-called Hamaker constant A_{H} [see Eq. (3.4)].

Thus, the van der Waals interactions alone would lead to irreversible coagulation where all particles stick together and form a large aggregate which flocculates out of solution – the classic household example is coagulated milk. A stable colloidal suspension can only be formed when the van der Waals minimum at contact is compensated by another repulsive force (or reduced by index matching). There are two practical stabilization techniques which have been exploited, resulting in stable suspensions: *charge stabilization* and *steric stabilization* (see Sec. 3.2 for detailed discussion). If the particles are charged, the Debye-Hückel screening mechanism leads to an effective repulsive screened Coulomb potential added to the van der Waals attraction. This results in an energetic barrier close to contact (which can be as high as $10 - 10^3 \ k_{\mathrm{B}}T$), such that coagulation is prevented on reasonable timescales. The coagulation rate can be increased and controlled by adding salt which causes stronger screening and therefore reduces the height of the energetic barrier (Hansen and Victor, 1985). In the case of steric stabilization polymers are grafted onto the colloidal surfaces. If two spheres approach each other, the polymer brushes overlap leading to an effective repulsion, since the number of configurations of the brushes is restricted close to overlap. In contrast to charge stabilization, steric stabilization causes a short-range repulsion which turns out to be strong enough to overcome the van der Waals attraction.

A crucial caveat in this context is that for 3D particle resolved studies, the colloids and solvent are *refractive-index-matched* (see Sec. 3.2.2). In this case, sterically stabilized colloids can be considered as a reasonable approximation to hard spheres with a purely excluded volume interaction. This makes colloids an ideal model system with simple interactions.

3.1   One-Body Properties

Before we start a detailed discussion of two- and many-body properties of colloidal dispersions, let us consider the most important one-body consequences of immersion of mesoscopic particles in a solvent.

3.1.1   *Dynamics*

Unlike particles in complex plasmas, which exhibit quasi-Newtonian short-time dynamics, colloidal motion is damped far more strongly. The loss of momentum for a particle embedded in a liquid is characterized by the Stokes (viscous) damping rate (Pusey, 1991),

$$\zeta = \frac{3\pi\eta\sigma}{m}, \tag{3.1}$$

where η is the dynamic viscosity and m is the colloid mass. Parameters appropriate to colloids used for confocal microscopy ($\sigma = 2$ μm, $\rho_m = 1$ g/cm^3, and $\eta = 10^{-3}$ Pa·s) lead to $\zeta \sim 10^7$ s^{-1} (in contrast to $\nu_{dn} \sim 1$ s^{-1} in low-pressure complex plasmas). The resulting timescale is clearly too short to be relevant for our purposes and, consequently, colloids are taken to be fully overdamped.

The Brownian motion due to the solvent leads to the following self-diffusion constant for colloids (often referred to as the Stokes-Einstein relation):

$$D_0 = \frac{k_B T}{3\pi\eta\sigma_h},$$

where σ_h is the hydrodynamic diameter [here, we shall neglect any difference between σ_h and the core diameter σ, see Bryant *et al.* (2002) for a more complete discussion]. The mean squared displacement is then $\delta r^2 = 6D_0\delta t$. Setting it equal to σ^2 gives a characteristic diffusive timescale due to Brownian motion,

$$\tau_D = \frac{\sigma^2}{D_0} = \frac{3\pi\eta\sigma^3}{k_B T}, \tag{3.2}$$

which is typically of the order of 1–10 s for the systems we are interested in. (For "overdamped" complex plasmas at high pressures the resulting diffusive timescale is similar; the role of σ in this case is played by mean interparticle distance.) The diffusive timescale τ_D provides us with an important and unambiguous way to define colloidal dispersions: We require that the colloid diffusion should be dominant, i.e., τ_D should be shorter than any timescale associated with other (externally driven) motion. For example, gravity-induced sedimentation is characterized by the velocity

$$v_{\text{sed}} = \frac{\sigma^2 \Delta\rho_m g}{18\eta}, \tag{3.3}$$

where $\Delta\rho_m$ is the difference in mass density between the colloids and solvent and g is the gravitational acceleration. One then defines the sedimentation time $\tau_{\text{sed}} = \sigma/v_{\text{sed}}$. Thus, the "colloidal regime" is where τ_D and τ_{sed} are of the same order, i.e., when the Peclet number $Pe = \tau_D/\tau_{\text{sed}}$ is around unity or less, while systems with very large Pe behave in a similar way to the essentially non-equilibrium granular matter. We note, however, that it is possible for the same particles to exhibit both regimes. For example, colloids of $\sigma \sim 3$ μm may readily be density matched to their solvent, so that $Pe \to 0$ and particles reach thermodynamic equilibrium (e.g., exhibit self-assembly into colloidal crystals). If the density matching is removed, *the very same* particles exhibit hydrodynamic instabilities at $Pe \sim 30$ and behave as granular matter.

3.1.2   *Charging of colloids*

Immersion of colloids in a liquid medium *always* leads to some degree of charging, as illustrated in Fig. 3.1. Entropy results in dissociation of surface groups, or alternatively, absorption of charged species can occur. The degree of charging strongly depends on the medium in which the colloids are dispersed, and often it is so small that the electrostatic interactions can reasonably be neglected. Nevertheless, one should stress that "true" hard spheres (uncharged colloids) do not exist.

The most common medium in which colloids are dispersed is water. In this case one can make a simple estimate of what sort of charging might be expected, by applying Coulomb's law and also noting the typical ion size (a few Angstroms) as well as very high dielectric constant of the solvent ($\simeq 78$). This results in an ion-site binding energy of a few $k_B T$ for monovalent ions, in which case we expect that entropy leads to a large degree of dissociation and strong charging.

While the dielectric constant of water is very high, in the case of many organic solvents it is rather low (around 2). Following our approximate Coulomb analysis, this leads to ionic binding energies of tens or even several hundreds of $k_B T$ and hence makes charging rather weak. On the other hand, the screening is reduced, so that interparticle interactions can remain strong.

At a prescribed added salt concentration which contains both counter- and coions, the colloidal charging process quickly approaches dissociation-association equilibrium. Strictly speaking, the resulting colloidal charge fluctuates (as it does in complex plasmas), but usually these fluctuations are ignored and a fixed charge is assumed (unless it is so small that the discreteness becomes important). It is essential that not only the magnitude, but also the *sign* of charge depends on the thermodynamic parameters such as salt concentration, colloid density, system temperature as well as on the material properties of the colloidal surface and the solvent. The process

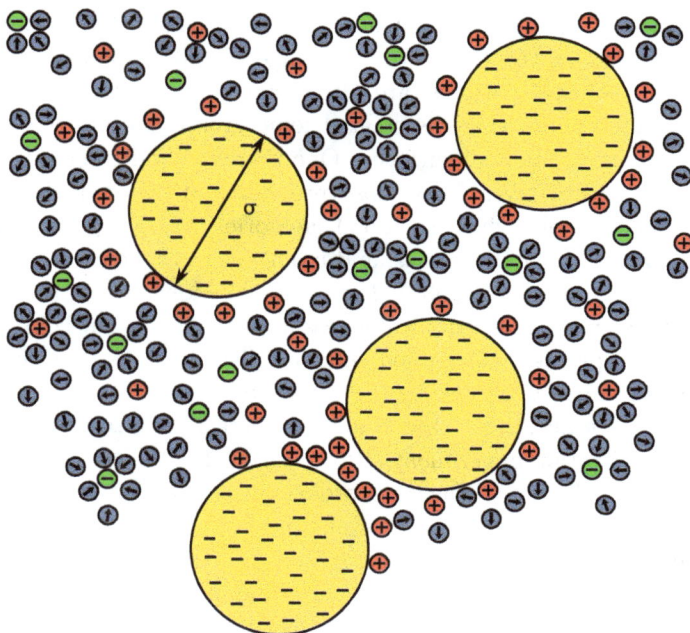

Fig. 3.1   A charged colloidal suspension. The principal components are of macroions (yellow) – charged mesoscopic particles of diameter σ, microscopic counterions (red and green) which are typically monovalent or divalent, and the molecular polar solvent (blue) shown as small particles (the arrows indicate the electric dipole moments). Two macroions repel themselves due to the Coulomb interaction of overlapping electric double layers.

of charge adjustment to the actual environment is called charge regulation (Hansen and Löwen, 2000).

This interaction between a colloid and its counterions has potentially profound consequences for the thermodynamics of the system. In particular, it has been suggested that, under low-salt conditions ion-colloid coupling could lead to phase separation into colloid-rich and colloid-poor phases (van Roij and Hansen, 1997). Such phase separation has indeed been observed experimentally (Dosho *et al.*, 1993), but the whole issue is still debated.

For a given colloid material one expects that the bare charge Q scales approximately with the area, i.e., $Q \propto \sigma^2$, which simply relies on the idea that the surface charge density is fixed. However, to our knowledge, this scaling has never been tested in experiments. A typical charge for a micron-size particle can vary in the (fairly broad) range of $(10 - 10^5)$ e, and therefore charged colloidal particles are often called *macroions* (or polyions). Of course, this implies that dissociated ionic groups are attracted to the oppositely charged colloidal surfaces. One therefore splits charges located in a so-called Stern layer (very close proximity of the colloidal surface, ~ 1 nm) into two (somewhat arbitrary) categories: charged groups strongly attracted by the colloidal surfaces, and mobile charges which are in the solution (Hansen and Löwen, 2000). For the mobile ions, a simple linearized screening theory is usually applicable leading to a Debye-Hückel profile.

A principal problem of charged colloidal dispersions is that the actual (bare) colloid charge Q cannot be measured directly (as is the case in complex plasmas, see Sec. 2.4). One possible approach to resolve this issue is to perform a titration experiment (Russel *et al.*, 1989). Since all ionizable groups on the colloidal surface are probed in this case (which is clearly not the case under actual experimental conditions), the resulting *titration charge* only provides an upper bound for $|Q|$. Another option to access the colloidal charge experimentally is to carry out an electrophoretic experiment, where an external electric field is applied and the resulting drift motion of the colloid is measured. This gives the so-called *electrophoretic charge*, which under proper conditions is very close to the *effective charge* relevant to two-body interactions (Toyotama and Yamanaka, 2011). Nevertheless, since the counterions are also moving in the electric field the measured electrophoretic charge is usually not equal to Q.

Finally, there are experimental possibilities to probe the effective charge which determines interactions between colloids and hence enters the pair interaction potential as a prefactor. This can be done, e.g., by measuring the fluid pair correlation functions or the elastic moduli of crystals. The effective charge is then obtained by performing an optimal fit to a particular interaction, e.g., a Yukawa potential (Royall *et al.*, 2003).

3.2   Two-Body Properties: Pair Interactions and Stability

Colloids are made of a different material to the solvent in which they are immersed. This leads to van der Waals attractions between them, with the non-retarded pair potential (Hamaker, 1937)

$$V_{\text{vdW}}(r) = -\frac{A_{\text{H}}}{12} \left(\frac{\sigma^2}{r^2 - \sigma^2} + \frac{\sigma^2}{r^2} + 2\ln \frac{r^2 - \sigma^2}{r^2} \right), \tag{3.4}$$

where A_H is the Hamaker constant which depends on the relative dielectric constant (as well as on the relative refractive index). As mentioned above, in order to prevent irreversible aggregation some stabilization mechanism is necessary. Since colloids spontaneously acquire an electrostatic charge, *charge stabilization* usually plays a very important role (alternatives include refractive index matching resulting in drastic reduction of A_H, and steric stabilization discussed below). Electrostatic repulsion is crucial in providing some kind of barrier to aggregation. We note that such systems are in fact metastable, since the resulting interactions have a true minimum determined by the van der Waals contact potential, and hence some (finite) aggregation rate is to be expected. However, for practical purposes this can usually be neglected.

If one assumes *a priori* linear Poisson-Boltzmann theory, then the celebrated Derjaguin-Landau-Verwey-Overbeek (DLVO) pair interaction between colloids can be used. This reads (Verwey and Overbeek, 1948)

$$V_Y(r) = \begin{cases} \infty & \text{for } r < \sigma, \\ \epsilon_Y \dfrac{e^{-\kappa(r-\sigma)}}{r/\sigma} & \text{for } r \geq \sigma, \end{cases} \tag{3.5}$$

where $\kappa = \sqrt{4\pi\lambda_B n_{ion}} \equiv \lambda_D^{-1}$ is the inverse Debye screening length expressed via the density of monovalent small ions n_{ion} and the Bjerrum length $\lambda_B = e^2/\varepsilon_r k_B T$ (equal to 0.7 nm for water). The contact potential is given by

$$\epsilon_Y = \frac{Z^2}{(1+\kappa\sigma/2)^2} \frac{\lambda_B}{\sigma} k_B T, \tag{3.6}$$

where $Z \equiv Q/e$ is the colloid charge number. Note that the DLVO interaction has a Debye-Hückel (Yukawa) form. In fact, when the charging (or, more specifically, the electrostatic potential) increases, then linear Poisson-Boltzmann theory breaks down. For instance, rough estimates for water show that this occurs when colloids exceed a few hundred nm in diameter. However, developments in surface chemistry treatment enable a considerable degree of control over the charging (Yamanaka *et al.*, 1997).

Even for elevated degrees of charging, Alexander *et al.* (1984) showed that the interaction still has a Yukawa form with a smaller, *renormalized*, charge (except for small distances; the same is discussed for complex plasmas in Secs. 2.2.1 and 8.3.4). This implies that at high Coulomb coupling the effective charge is renormalized towards much smaller values. Typically, far away from the colloid surfaces linearized screening theory holds, such that simple estimate of the effective saturated (asymptotic) charge number $Z_{\text{eff}}^{\text{sat}}$ is possible (Trizac *et al.*, 2002; Bocquet *et al.*, 2002),

$$Z_{\text{eff}}^{\text{sat}} = (2 + \kappa\sigma)\sigma/\lambda_{\text{B}}. \tag{3.7}$$

This is the effective charge that one would attribute to a single particle in a salty environment if the bare charge is (formally) increased to infinity. According to Eq. (3.7), there are two different regimes: If $\kappa\sigma \ll 1$ (which is typical for complex plasmas but is also realized for colloids in apolar solvents), the leading scaling on the particle diameter σ is linear. In the opposite limit typical for highly salted suspensions, there is a quadratic dependence of $Z_{\text{eff}}^{\text{sat}}$ on σ, as mentioned in the previous section.

A great deal of studies have shown that most of the experimental facts can be conveniently interpreted using the linearized Poisson-Boltzmann theory with a fitted effective charge (Löwen and Kramposthuber, 1993; Bitzer *et al.*, 1994; Baumgartl *et al.*, 2006). This makes the DLVO theory with charge renormalization one of the most successful approaches in describing charged colloidal dispersions (Brunner *et al.*, 2002; Royall *et al.*, 2003, 2006).

3.2.1   *One-component approach and many-body effects*

The DLVO approach brings us to consider an important and fundamental consequence of current methods to tackle colloids. At the very least, colloidal systems feature two components, the solvent and the colloids. There are also other species, such as small ions, or polymers and other macromolecules, or indeed smaller colloids. These are typically much smaller than "regular" colloids and very much larger in number, such that their degrees of freedom are intractable. For example in the DLVO theory, ions are treated as a continuum, and their degrees of freedom are integrated out, leading to an *effective one-component description*. This is crucial when we consider colloidal phase behavior as akin to atomic and molecular behavior. Such a one-component description is a cornerstone of our ideas of using colloids as a model for classical statistical mechanics.

Let us consider the consequences for a moment. Even within the DVLO treatment, we can see that, if the Debye screening length were to change the effective interactions between colloids would change as well. Now the Debye screening length depends on the ionic density, and in deionized systems where concentrations of background salt are low, this is coupled to the colloid concentration. Such effects thus mean that the DVLO interaction, in principle, depends on colloid concentration, which can be measured in experiment (Rojas *et al.*, 2002) (again, very similar to quasineutrality effects in complex plasmas, see Sec. 2.1.1 and also Sec. 7.1.1).

Not only that, but the introduction of a third colloid can affect the interaction between two others. These many-body effects were pioneered in computational work by Löwen and Kramposthuber (1993). Subsequent theoretical work predicted an *attractive* three-body interaction (Löwen and Allahyarov, 1998; Russ *et al.*, 2002). Experimental work also found evidence of such many-body effects (Brunner *et al.*, 2002). In other words, the one-component approach naturally leads to many-body interactions, i.e., a breakdown in pairwise additivity.

3.2.2   *Non-aqueous solvents: weak charging, steric stabilization, and experimental hard spheres*

For non-aqueous solvents in particular, the degree of charging could be reduced to such an extent that aggregation will start (due to the van der Waals minimum). In this case, some other forms of efficient colloidal stabilization are necessary. Usually, *steric stabilization* is arranged for that – particles have a layer of short (typically ~ 10 nm) polymers grafted to the surface, as illustrated in Fig. 3.2. Upon approach of two colloids, these polymers interdigitate, leading to a loss of configurational entropy and hence an effective repulsion between the colloids. This steric stabilization requires that the surrounding fluid is a good solvent for the polymers. Interestingly, it is possible to vary the quality of the solvent (from the point of view of the polymer) by varying the temperature. Thus, one can introduce "sticky sphere" type interactions with varying degree of stickiness (Vrij *et al.*, 1990; Solomon and Varadan, 2001).

However, the canonical hard-sphere model assumes that colloids experience no interactions except an infinite repulsion following overlap. This can be approximated, most famously, in the polymethyl methacrylate (PMMA) system (Cairns *et al.*, 1976; Bryant *et al.*, 2002). The early PMMA colloids were typically half a micron in diameter, too small for optical microscopy, but ideally suited to light scattering. The key hard-sphere experiments were carried out by Pusey and van Megen (1986), which indeed demonstrated that these PMMA particles obey the hard-sphere phase diagram (the most controversial part, namely the entropy-driven freezing, had been predicted by numerical simulations).

It is noteworthy that density matching in non-aqueous solvents can noticeably affect the magnitude of colloid electrostatic charging and hence

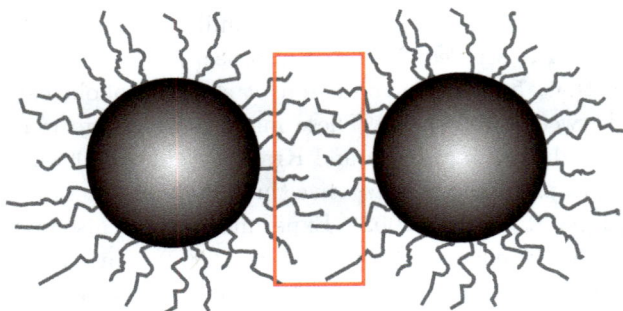

Fig. 3.2   Steric stabilization. Colloids are coated with a thin layer (not drawn to scale) of short polymers. The polymers interdigitate as colloids approach each other, as sketched in the red box. Since chains cannot cross one another and hence fewer configurations can be accessed, this leads to a loss of polymer entropy and therefore causes effective repulsion.

cause soft repulsion. For instance, solvent mixtures including cyclohexyl or cycloheptyl bromide ($C_6H_{11}Br$ or $C_7H_{13}Br$) components, which are added to provide both density and refractive index matching between the PMMA particles and solvent, also induce weak charging. Although this typically leads to charges of only hundreds of e for micron-size colloids (compared to tens of thousands for water), the contact potential ϵ_Y [Eq. (3.6)] can reach 100's of k_BT. Coupled with the ultra-low ion density in these systems ($10^3 - 10^6$ times less than aqueous systems), this results in strong, long-ranged repulsions, often causing crystallization at very low packing fractions (similar to complex plasmas, see Sec. 7.1.1).

One should point out that the PMMA system is not the only candidate for hard-sphere-like behavior. In fact, strongly charged colloids in water can (under the addition of sufficient salt, such that $\kappa\sigma \gg 1$) also form a suitable hard-sphere system. The first measurement of the hard-sphere equation of state was carried out by Piazza *et al.* (1993) with precisely such a strongly screened suspension of polystyrene colloids. However, these systems are not suitable for 3D particle-resolved studies, as the particles and solvent are not index-matched.

3.2.3   *Colloidal microgels*

Colloids need not necessarily have a hard core. Microgel particles shown in Fig. 3.3 are heavily cross-linked polymers. The cross-linking suppresses overlap (normal polymers can fully overlap for a modest interaction energy of $\sim k_\mathrm{B}T$), so that the interaction between such microgels is close to that of hard spheres (Saunders and Vincent, 1999). On the other hand, like polymers, microgels swell in a good solvent. The transition from good to bad solvent can be controlled by temperature (Saunders and Vincent, 1999), and therefore the effective size of such "thermosensitive" particles can be tuned *in situ*. Generally, microgel particles have an intrinsically soft interaction which can often be approximated by an inverse power law $V(r) \propto r^{-n}$, where n varies according to the degree of cross-linking (Senff and Richtering, 1999).

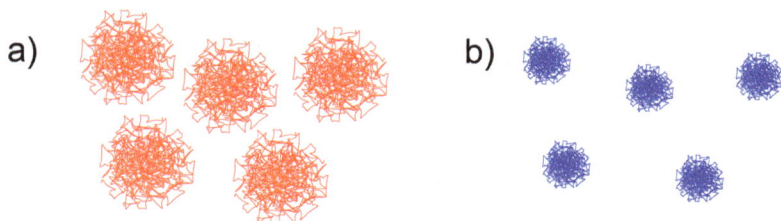

Fig. 3.3   Colloidal microgel particles. These heavily cross-linked polymer coils undergo a swelling transition. Under good solvent conditions microgels are swollen (a), but collapse in a poor solvent (b).

3.3   Interactions Between Different Species

The introduction of a second species has made a remarkably important contribution to our understanding of interactions in colloidal dispersions. In addition to the very short-ranged van der Waals interaction mentioned above, in multispecies colloidal mixtures attraction can be caused by the *depletion interaction*. This was first introduced by Asakura and Oosawa (1954, 1958) and subsequently popularized by Vrij (1976), and therefore is referred to as the AOV interaction.

This depletion interaction is driven by the entropy of the smaller species, as shown in Fig. 3.4. These smaller species can either be colloids or polymers; in the latter case a reasonable approximation is found, so that the polymers are treated as ideal and one can formally integrate them out (Dijkstra *et al.*, 1999). This AOV model leads to a pair interaction between two hard colloidal spheres in a solution of ideal polymers, which reads

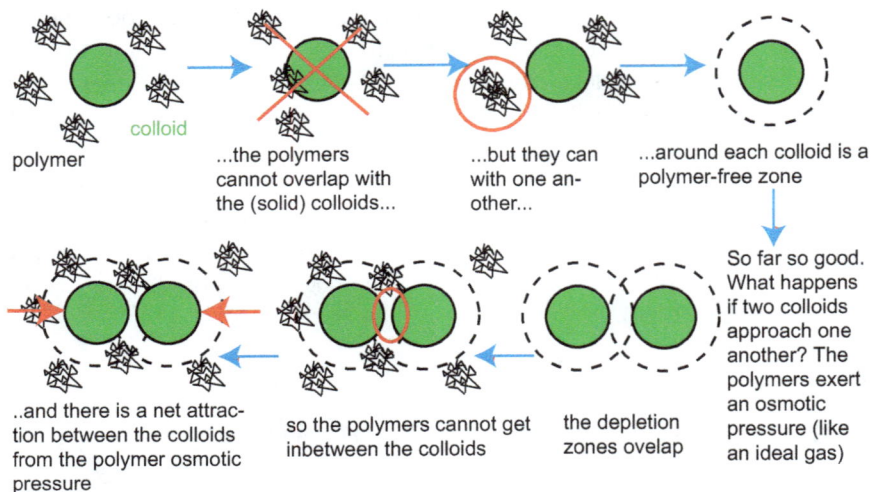

Fig. 3.4   The Asakura-Oosawa-Vrij (AOV) model of colloid-polymer mixtures.   The range of the attraction is given by the polymer size, the strength is determined by the polymer concentration.

$$
V_{\mathrm{AOV}}(r) = \begin{cases} \infty & \text{for } r < \sigma, \\[2ex] \dfrac{2\pi}{3} z_{\mathrm{pr}} R_{\mathrm{g}}^3 \left(\dfrac{1+q}{q}\right)^3 k_{\mathrm{B}} T & \\ \times \left[1 - \dfrac{3r/\sigma}{2(1+q)} + \dfrac{(r/\sigma)^3}{2(1+q)^3}\right] & \text{for } \sigma \le r < \sigma + 2R_{\mathrm{g}}, \\[2ex] 0 & \text{for } r \ge \sigma + 2R_{\mathrm{g}}. \end{cases} \tag{3.8}
$$

Here $q = 2R_{\mathrm{g}}/\sigma$ is the polymer-colloid size ratio, where R_{g} is taken as the polymer radius of gyration and σ is the colloid diameter. The polymer fugacity z_{pr} is equal to the number density of ideal polymers in a reservoir at the same chemical potential as the colloid-polymer mixture. Thus, within the AOV model the effective temperature is inversely proportional to the polymer reservoir concentration.

By replacing polymers with a second colloidal species (binary hard spheres), for low concentration of the smaller species one obtains interactions similar to Eq. (3.8). However, finite packing effects in the smaller species give rise to *oscillatory* effective interactions between the larger species (Roth and Dietrich, 2000; Royall *et al.*, 2007c).

3.4 Hydrodynamic Interactions

Hydrodynamic interactions describe forces which are mediated between colloidal particles via a solvent flow. If a colloidal particle moves relative to the embedding fluid it creates a fluid velocity field around it which then affects the motion of neighboring colloidal particles. Typically, the colloidal velocities are so small that the fluid motion is in the low Reynolds number regime, ensuring a linear relation between the driving force and the resulting drift velocity of a colloid.

It is important to note that there is a fundamental difference between hydrodynamic interactions and effective interactions. The latter affect the static equilibrium properties (such as structural correlations and equilibrium phase transitions) while the former have no impact at all on static equilibrium statistics. In fact, all static properties in equilibrium can be derived from the Boltzmann distribution which is independent of hydrodynamic interactions. However, dynamic correlations – both in equilibrium and non-equilibrium – are strongly influenced by hydrodynamic interactions: These interactions play a major role in processes like the long-time self-diffusion, sedimentation, behavior of colloids under shear and other driving fields, etc. Colloidal hydrodynamic interactions are similar to wake-mediated interactions in complex plasmas – both are nonreciprocal i.e., Newton's third law is violated for the interactions (see Sec. 4.3).

A quantitative description of hydrodynamic interactions starts from a generalized linear relation between particle drift velocities and forces (Doi and Edwards, 1986; Pusey, 1991; Dhont, 2003). Let us consider N *interacting* particles. We use a compact notation for the particle positions,

$$\{x_i\} = \{\mathbf{r}_i\} = \{\underbrace{x_1, x_2, x_3}_{\mathbf{r}_1}, \underbrace{x_4, x_5, x_6}_{\mathbf{r}_2}, \cdots, \underbrace{x_{3N-2}, x_{3N-1}, x_{3N}}_{\mathbf{r}_N}\},$$

and assume a linear relation between the force components F_j acting on the particles and the resulting components of drift velocity v_i (using the same compact notations). The general linear relation is then

$$v_i = \sum_{j=1}^{3N} M_{ij}(\{x_n\})F_j, \tag{3.9}$$

where $F_j = -\partial V_{\text{tot}}/\partial x_j$ and $V_{\text{tot}}(\{x_i\})$ is the total potential energy. The underlying assumption in Eq. (3.9) is that the hydrodynamic interactions act quasi-instantaneously. This is justified by the fact that the timescale at which shear perturbations travel through the suspension within an interparticle distance is much smaller than the diffusive timescale τ_D characterizing Brownian motion. The elements M_{ij} constitute a $3N \times 3N$ mobility matrix and can in principle be obtained by solving the Stokes equations for N spheres with appropriate stick boundary conditions on the particle surfaces.

Now we briefly address the explicit form of the elements $M_{ij}(\{x_n\})$. The linear relationship (3.9) can be rewritten as

$$\mathbf{v}_i = \sum_{j=1}^{N} \mathsf{M}_{ij}\mathbf{F}_j,$$

where M_{ij} is a 3×3 matrix. The direct solution of the linearized Navier-Stokes equations is a difficult problem. Hydrodynamic interactions generally have *many-body character*, and a pair expansion is only possible at low concentrations. Furthermore, hydrodynamic interactions have specific near-field behavior for particles close to contact, due to the presence of diverging lubrication terms.

For systematic calculations in the simplest form, one can consider just a particle pair and perform a multipole-like expansion for large interparticle distance r. This leads to

$$\mathsf{M}_{ii} = \mu_0 \mathsf{I}, \quad \text{and} \quad \mathsf{M}_{ij} = \mathsf{M}_{\mathrm{RP}}(\mathbf{r}_i - \mathbf{r}_j) \quad \text{for } i \neq j, \tag{3.10}$$

with the *Rotne-Prager tensor*,

$$\mathsf{M}_{\mathrm{RP}}(\mathbf{r}) = \mu_0 \left[\frac{3}{4}\frac{R_\mathrm{h}}{r}(\mathsf{I} + \hat{\mathbf{r}}\hat{\mathbf{r}}) + \frac{1}{2}\frac{R_\mathrm{h}^3}{r^3}(\mathsf{I} - 3\hat{\mathbf{r}}\hat{\mathbf{r}}) \right], \quad \hat{\mathbf{r}} = \frac{\mathbf{r}}{r}, \tag{3.11}$$

where R_h is the hydrodynamic radius of the colloid, $\mu_0 = D_0/k_\mathrm{B}T$ is the free mobility, I is the unit matrix, and $\mathbf{rr}$ denotes the dyadic product. The tensor contains the first two leading far-field terms, higher-order expansions beyond r^{-3} are possible (these also include sphere rotation terms). From the Rotne-Prager expression we obtain two special cases: (i) A simpler long-distance approximation is constituted by the Oseen approximation, $\mathsf{M}_{ij} = \mathsf{M}_{\mathrm{O}}(\mathbf{r}_i - \mathbf{r}_j)$, where $\mathsf{M}_{\mathrm{O}}(\mathbf{r}) = \frac{3}{4}\mu_0(R_\mathrm{h}/r)(\mathsf{I} + \hat{\mathbf{r}}\hat{\mathbf{r}})$ is the *Oseen tensor* representing the monopole part of Eq. (3.11). From the Oseen expression it is obvious that the mobility tensor has long-range asymptotics. (ii) If hydrodynamic interactions are ignored completely, one retains only local (diagonal) elements, $\mathsf{M}_{ij} = \mu_0 \delta_{ij}\mathsf{I}$, which corresponds to ordinary Brownian dynamics.

Thus, solvent-mediated hydrodynamic interactions act instantaneously on the colloidal diffusive timescale, have many-body character, and can be decomposed into pairwise tensors for dilute suspensions. Modern approaches have been designed to simulate hydrodynamic interactions efficiently for various geometric constraints. These approaches are based on, e.g., lattice-Boltzmann (Dünweg and Ladd, 2009) or multiparticle collision dynamics (Gompper *et al.*, 2009) techniques.

3.5 Major External Fields

The interactions between colloidal particles are characterized by energies in the range from a few to a few tens of $k_B T$. Since the colloids required for real-space analysis are in the micron size range, yet the interactions between them are comparable to the interactions between *molecules* of a few tenths of a nm, the resulting interaction density is remarkably low: colloidal crystals have shear moduli which are $\sim 10^{12}$ times smaller than those for atomic systems. Such characteristics – which give rise to the term *soft matter* – make colloids extremely susceptible even to weakest external fields.

Below we outline the major types of external fields which have been used to control colloidal dispersions. These fields can generally be divided into two categories: Those which act on individual particles and hence are determined by a position-dependent potential (such as gravity, confinement, dc electric fields, thermal gradients, or shear flows), and those which affect interactions between particles (such as ac electric and magnetic fields). Note that, unlike atomic/molecular materials, all fields which induce particle motion lead to hydrodynamic interactions between the colloids.

3.5.1 *Confinement*

All colloidal systems must be confined. In the most simple representation, the confinement is treated as hard walls with zero field inside the container. Often, when the container dimensions are much larger than the particle size, the boundary effects can be neglected. On the other hand, hydrodynamic correlations can extend over very long distances (Segre *et al.*, 1997), and under gravity such long-range correlations might also appear in colloidal suspensions (Padding and Louis, 2004). Here, the systems whose size in at least one dimension is less than, say, $\sim 100\sigma$ will be considered as confined. Note that the working distance of lenses used for particle-resolved studies typically enables samples to be imaged to a depth of ~ 100 μm, so that all systems studied by 3D microscopy may be regarded, even under our definition, as confined. The confinement effects induced by 3D imaging become particularly important when studying critical phenomena, nucleation and, potentially, a very close approach to the glass transition. For 2D systems, however, the confinement effects are generally less important: The lens is typically mounted perpendicular to the observed monolayer, and then it is often possible to entirely neglect confinement across a sufficiently large system.

3.5.2   *Gravity*

In the absence of density matching, properties of colloidal suspensions can be strongly affected by gravity. The role of gravity as an external field may be traced back to the very dawn of real space analysis, and the Nobel-Prize-winning work of Jean Perrin (1913). The gravitational length ℓ_g defined as

$$\ell_g = \frac{6k_BT}{\pi\Delta\rho_m\sigma^3g},\tag{3.12}$$

is the length scale associated with a potential energy difference of k_BT. By using Eqs (3.2) and (3.3), and also rescaling Eq. (3.12) with the Stokes-Einstein relation we obtain $\ell_g/\sigma = Pe^{-1}$. Hence, if $\ell_g \gg \sigma$ then we have a colloidal system, whereas the opposite limit is analogous to granular matter. In equilibrium and in the dilute limit (where interactions between colloids are neglected), the colloid packing fraction scales as

$$\phi(z) \propto \exp(-z/\ell_g).\tag{3.13}$$

At higher packing, interactions lead to a non-trivial bulk equation of state. Out of equilibrium, hydrodynamic interactions can lead to complex coupled motion [see Russel *et al.* (1989) and also Secs. 3.4 and 4.3.2].

3.5.3   *DC electric fields*

In the simplest treatment, dc electric fields are equivalent to gravity: The role of the mass density difference between colloid and solvent $\Delta\rho_m$ is played by the charge Z, and g is replaced with the electric field E. Using this analogy and measuring the colloidal motion in the electric field (electrophoresis), one can determine the colloid charge [however, there are many other, often uncontrolled effects that complicate the analysis, see, e.g., Russel *et al.* (1989)].

On the other hand, there is a principal difference between gravity and dc electric force. While colloids are strongly affected by gravity, the weight of counterions is practically negligible. Consequently, ions do not sediment but colloids do. This induces an electric field which tends to counteract the gravity, leading to a sedimentation profile which decays much slower than expected, i.e., the colloids become effectively "lighter". This effect was first suggested by Piazza *et al.* (1993), then described theoretically by Löwen (1998); van Roij (2003), and confirmed (Rasa and Philipse, 2004) and directly visualized (Royall *et al.*, 2005) experimentally.

Far away from equilibrium, the distinction between effects of gravity and electric forces becomes particularly evident. The hydrodynamic interactions are screened when the motion of colloids is due to electric field [because the counterions and colloids drift in the opposite directions, see Long and Ajdari (2001)]. Therefore, the resulting solvent flow produced by a superposition of moving particles and counterions is strongly reduced.

3.5.4   *AC electric and magnetic fields*

Alternating electric or magnetic fields have profound consequences for the interactions between the colloidal particles. At low frequencies the induced ion motion leads to the formation of polarization clouds in the vicinity of colloidal particles and results in effective dipolar interactions between colloids (Dhont and Kang, 2010) (the mechanism is similar to that operating in complex plasmas, see Sec. 2.2.2). At higher frequencies, where the ion motion plays a minor role, the induced interactions are determined by the dielectric mismatch between colloids and solvent. (Note that the ac fields used in experiments are typically in the MHz range, while the refractive index matching is for optical frequencies.) For most of the experiments carried out so far the ion-polarization effect is relatively small, so that the dielectric polarization dominates.

For a given (electric) field, the induced dipole moment, $\mathbf{d} = \frac{1}{8}\left(\frac{\varepsilon_p-\varepsilon_s}{\varepsilon_p+2\varepsilon_s}\right)\varepsilon_s\sigma^3\mathbf{E}_{\mathrm{loc}}$, is determined by dielectric constants of the particle, ε_p, and solvent, ε_s, as well as by the local field $\mathbf{E}_{\mathrm{loc}}(\mathbf{r}) = \mathbf{E}(\mathbf{r}) + \mathbf{E}_{\mathrm{ind}}(\mathbf{r})$. Here, we assume that $\mathbf{E}$ is generally a function of position and also take into account the field induced by other dipoles, $\mathbf{E}_{\mathrm{ind}}$, which depends on a particle configuration. The interaction energy of two dipoles with moments $\mathbf{d}_1$ and $\mathbf{d}_2$ located at positions $\mathbf{r}_1$ and $\mathbf{r}_2$ is given by

$$V_{\mathrm{dip}}(\mathbf{r}_1,\mathbf{r}_2) = \varepsilon_s^{-1}\frac{(\mathbf{d}_1\cdot\mathbf{d}_2)r^2 - 3(\mathbf{d}_1\cdot\mathbf{r})(\mathbf{d}_2\cdot\mathbf{r})}{r^5}, \qquad (3.14)$$

Complex Plasmas and Colloidal Dispersions

where $\mathbf{r} = \mathbf{r}_2 - \mathbf{r}_1$ is the relative distance. In the case of magnetic fields, identical interactions are obtained by switching ε_s with μ_s, the magnetic susceptibility of solvent. The role of field-induced interactions is discussed in Sec. 11.1.

3.5.5   *Optical fields*

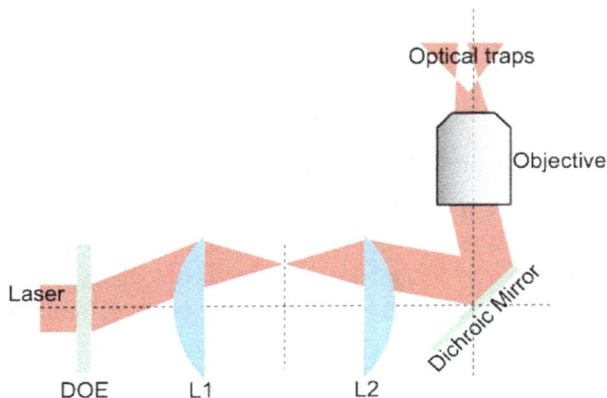

Fig. 3.5   Holographic optical tweezers. Optical tweezers are formed by a focused laser beam. A diffractive optical element (DOE) creates a Fourier transform of the trapping pattern in the back focal plank of the objective. In principle, a trapping pattern with an arbitrary number of traps can be generated, up to ~ 100 traps is readily achieved.

The change in momentum of electromagnetic radiation due to refractive index changes in the medium it passes through imparts a tiny radiation force on that medium (see Sec. 2.4). Ashkin and Dziedzic (1987) used a focused laser beam to optically trap viruses and bacteria, initiating the technique known as optical tweezers. Subsequent developments, notably holographic optical tweezers (see Fig. 3.5) now enable the independent manipulation of hundreds of colloidal particles simultaneously (Grier, 2003). Another approach to generate optical fields is to use crossed laser beams to create a periodic optical potential (Brunner *et al.*, 2002). This can then be tuned to influence the behavior of the colloidal particles in a far more complex way than many other static external fields, since light is easily manipulated.

3.5.6   *Shear*

Colloids can be exposed to a solvent shear flow field. A solvent flow is conventionally produced in various shear cells, e.g., in rotating cone-plate cells or by moving two parallel plates relative to each other. The latter configuration provides the simplest case of a planar Couette flow, where velocity field is a linear function of position between the plates. This situation is termed homogeneous shear flow and is characterized by a time- and position-independent shear rate $\dot{\gamma}$. This is opposite to oscillatory shear, where $\dot{\gamma}$ oscillates in time at a prescribed shear frequency. The viscoelastic response of a material to oscillatory shear reveals its basic rheological properties (Dhont, 2003).

In general, the presence of shear immediately puts the systems into a non-equilibrium situation, where the deviation from equilibrium is conveniently measured in terms of Peclet number $\dot{\gamma}\tau_D$. Real-space experiments with colloidal suspensions, where trajectories of individual particles are resolved, are also possible under shear flow. Such experiments reveal the effect of linear and oscillatory shear on the structural correlations, freezing, fluid demixing, glass transition, etc. (Petekidis *et al.*, 2003; Smith *et al.*, 2007; Besseling *et al.*, 2007; Schall *et al.*, 2007; Koumakis *et al.*, 2008; Wu *et al.*, 2009) (see also Chapter 10).

Chapter 4

Complex Plasmas and Colloidal Dispersions: Similarity and Complementarity

In the two preceding chapters we summarized basic properties of complex plasmas and colloidal dispersions, with the primary focus on a variety of processes which determine the interparticle interactions in these two systems. Some of the interaction mechanisms are common for plasmas and charged colloids (such as the Debye-Hückel screening), some appear to be similar but have very important distinctions in the underlying physics (such as wake-mediated/hydrodynamic interactions operating in plasmas/colloids), and some are unique for each system (e.g., weak long-range forces due to nonlocality in plasma processes, or depletion forces in colloids).

The most important qualitative difference between the interparticle interactions in the two systems is their range: For the micron-size colloids used in particle-resolved studies, the interaction range is usually comparable to or smaller than the particle size and, hence, the hard-core part in the interaction potential is essential. In plasmas the interaction range is much larger than the particle size, so that the potential is very well described by point-like approximation.[1] To a large extent, this is due to the fact that the *surrounding fluid* which mediates interparticle interactions is very different in these two systems (as illustrated in Fig. 1.3): A liquid solvent is many ($\sim 10^6 - 10^8$) orders of magnitude denser than a dilute weakly ionized gas, so that all relevant spatial scales characterizing the interaction range are (usually) significantly smaller for colloidal dispersions.

Generally, the surrounding fluid has a twofold effect: In addition to interparticle interactions (including hydrodynamics), it critically affects the

---

[1]There are, of course, counterexamples – for instance, colloids in a highly deionized solvent (see Sec. 3.2.2) or with induced dipolar interactions (Sec. 3.5.4), or very large dust particles (hollow spheres).

dynamics of particles via the friction force. The friction is linearly proportional to the fluid density, and therefore the characteristic damping timescale for colloids is many orders of magnitude shorter than that for complex plasmas. At the same time, the "atomistic" timescales – which characterize dynamics of individual strongly coupled particles (e.g., the inverse phonon frequency, see below) – can be comparable for both systems. This results in a very interesting complementary hierarchy of dynamic regimes in complex plasmas and colloidal dispersions.

In this chapter we present a detailed analysis of the dynamical hierarchy and demonstrate its impact on generic phenomena occurring in complex plasmas and colloidal dispersions. Furthermore, we compare and discuss mechanisms of nonconservative interparticle interactions operating in these two systems, illustrate their role in the particle dynamics, and analyze their influence on the interdisciplinary research. Before we start the discussion, it is instructive to briefly summarize the most important dimensionless parameters characterizing strongly coupled complex plasmas and colloidal dispersions.

4.1   Important Dimensionless Parameters

4.1.1   *Complex plasmas*

Screened interactions between (effectively) point-like charges in complex plasmas are characterized by two parameters (Fortov *et al.*, 2005; Bonitz *et al.*, 2010; Morfill and Ivlev, 2009): The *coupling parameter* $\Gamma = Q^2/k_{\mathrm{B}}T\Delta$ – a measure of the Coulomb interaction – is naturally normalized by temperature, and the plasma *screening parameter* $\kappa_{\mathrm{p}} = \Delta/\lambda$ relates the two main length scales – the mean interparticle distance $\Delta = n^{-1/3}$ and relevant interaction range (screening length) λ. As we will see in the following chapters, the combination of these parameters provides the most important characterization of equilibrium and non-equilibrium phase states, transport, rheology, etc. The magnitude of the coupling parameter can be as high as $\sim 10^4 - 10^5$ for typical experimental conditions, thus allowing us to span the whole range from gaseous to solid states. The screening parameter in experiments can be made smaller than unity or it can reach values of 5–10. Varying κ_{p} (either by changing the particle density or the screening length) thus enables us to investigate the crossover between the regimes of long-range interactions with many neighbors (smaller κ_{p}) and relatively short-range interactions (larger κ_{p}).

The next important parameter is the damping ratio ν_{dn}/Ω_E which is a measure of importance of neutral gas friction at the "atomistic" timescale (Fortov *et al.*, 2005; Morfill and Ivlev, 2009). The latter is characterized by the relevant eigenfrequency of a strongly coupled system – this can be either the plasma frequency $\omega_p = \sqrt{4\pi Q^2 n/m}$ or, more generally, the Einstein frequency Ω_E which we adopt throughout this book (and which is comparable to ω_p for $\kappa_p \sim 1$). In experiments, ν_{dn}/Ω_E can be made as small as $\sim 10^{-2}$, which indicates that the caged motion of particles is practically undamped and can be treated in the framework of Newtonian dynamics. This allows us to study waves/shocks propagating in liquids and solids, kinetics of heat transport, short-term relaxation processes in supercooled liquids, etc.

The influence of the ambient plasma flow on the electrostatic interaction between particles is determined by the *thermal Mach number* $M_T = u_i/v_{T_i}$, which is the ratio of the ion flow velocity to the ion thermal velocity (see Sec. 2.2.2). For $M_T \lesssim 1$, anisotropic deviations from a spherically symmetric (Debye-Hückel) interaction are relatively weak, and the particle charge is constant to a very good accuracy. When $M_T \gtrsim 1$, the anisotropic wake structure becomes dominant and the resulting (wake-mediated) interactions are highly asymmetric, while the particle charge slowly varies with M_T.

There are several parameters which characterize the charge composition in complex plasmas (see Sec. 2.1.1). The most important among them is the so-called *Havnes parameter* $P_H = |Z|n_d/n_e$, which is the ratio of the volume charge density (number of charges per unit volume) associated with microparticles to the plasma (electron) charge density (Havnes *et al.*, 1984). This parameter can vary from very small ($\ll 1$) to rather large (~ 10) values (Fortov *et al.*, 2005). At sufficiently small P_H charged particles practically do not disturb the ambient plasma, i.e., plasma conditions can be considered as prescribed. At large P_H the major fraction of charges is accumulated at particles and hence the discharge characteristics can be modified significantly: One can observe the effect of charge depletion, when the density of ambient electrons is no longer sufficient to supply constant particle charge, i.e., in this regime charge Q decreases with the particle density. The magnitude of Q, in turn, is characterized by the *normalized charge* $z = |eQ|/k_B T_e a$. For typical experimental conditions this parameter is always of the order of unity (unless P_H is too large), which provides a very convenient scaling of the charge with the electron temperature and particle radius a.

4.1.2   *Colloidal dispersions*

There are several dimensionless parameters which are of particular importance in determining the equilibrium and non-equilibrium behavior of colloids (Brader, 2010; Dhont, 2003; Russel *et al.*, 1989). The choice of these parameters depends on the relative range of the interparticle interactions.

For (almost) hard-sphere interactions of particles with diameter σ, the principal control parameter is the colloidal *packing (volume) fraction* $\phi = \frac{\pi}{6}\sigma^3 n$ (for 2D systems, this is the areal fraction $\phi_{2D} = \frac{\pi}{4}\sigma^2 n_{2D}$). For practical purposes one can conditionally divide the physical range of volume fractions into three ranges: dilute suspensions, $\phi \lesssim 0.1$, representing weakly correlated fluids; intermediate packing approaching the solid state, $0.1 \lesssim \phi \lesssim 0.5$; solid and glassy states limited by $\phi \leq 0.74$. Such (rather arbitrary) division is also useful in discussing various rheological approximations currently available (Brader, 2010).

When the interaction range is comparable to or less than the particle size, the colloidal *screening parameter* $\kappa\sigma$ ($\equiv \sigma/\lambda$) is used to characterize the dispersions. If $\kappa\sigma \ll 1$ (e.g., in highly deionized solvents) or, more generally, the interaction $V(r)$ is long-range (e.g., induced dipolar), the coupling parameter $\Gamma = V(\Delta)/k_B T$ is employed (similar to complex plasmas).

One of the most important dimensionless parameters characterizing the dynamics of individual particles in the presence of inhomogeneous solvent flow is the Peclet number $Pe = \tau_D/\tau_{adv}$ (Dhont, 2003). The relevant diffusive timescale is usually $\tau_D = \sigma^2/D_0$ [see Eq. (3.2)] and the advective timescale is $\tau_{adv} = \dot{\gamma}^{-1}$ (see Sec. 3.5.6). The Peclet number, which is a measure of the importance of advection relative to Brownian motion of a free particle, determines the extent to which the microstructure is distorted away from equilibrium by the flow field. In the limit $Pe \to 0$ Brownian motion dominates and the thermodynamic equilibrium state is recovered. In the strong flow limit, $Pe \gg 1$, solvent-mediated hydrodynamic interactions (see Sec. 3.4) may completely dominate the individual particle dynamics (Brader, 2010).[2]

The increase of the coupling between particles is accompanied by an increase in the structural relaxation timescales of the system, $\tau_{\alpha,\beta}$, which characterize the temporal decay of two-point autocorrelation functions during the alpha- and beta-relaxation (discussed in Sec. 9.1.1). This allows us to define the Weissenberg number as $Wi = \dot{\gamma}\tau_{\alpha,\beta}$. For intermediate and high volume fractions, particularly those close to the glass transition, it is

---

[2]The Peclet number is similarly introduced for the field-induced drift, see discussion of Eqs (3.3) and (10.9).

the Weissenberg number, rather than the Peclet number, which dominates certain aspects of the nonlinear rheological response (Brader, 2010; Dhont, 2003). In the regime of weak correlations (intermediate packing fractions) the structural relaxation timescale is determined by free diffusion and hence $Pe \sim Wi$.

4.2 Dynamic Regimes: Role of the Background

The hierarchy of dynamic timescales in complex plasmas and colloidal dispersions is illustrated in Fig. 4.1. The gradual crossover from Newtonian to Brownian dynamics regimes occurs at timescales of frictional damping. These timescales are determined by the momentum/energy loss rates due to the interaction of individual particles with the surrounding fluid: The damping in complex plasmas occurs in the free molecular regime and is characterized by the Epstein damping rate ν_{dn} [Eq. (2.8)]; in colloidal dispersions the viscous friction is determined by the Stokes damping rate ζ [Eq. (3.1)].

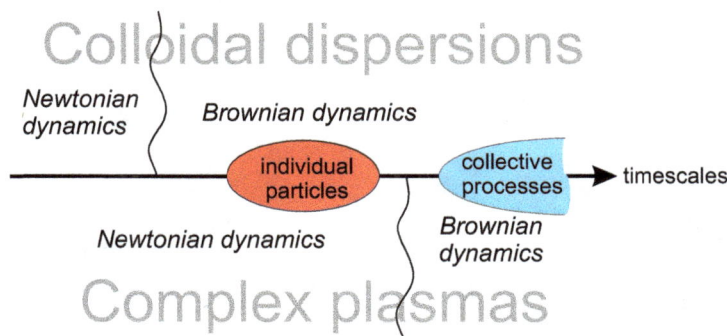

Fig. 4.1   Schematics of dynamic regimes in complex plasmas and colloidal dispersions. The individual ("atomistic") particle dynamics is fully damped in colloidal dispersions and can be virtually undamped in (strongly coupled) complex plasmas. The timescales characterizing collective ("hydrodynamic") processes are in the Brownian dynamics regime for both systems.

As we already pointed out above, the damping timescales in the two media differ by many orders of magnitude, while the timescales of the "atomistic" dynamics (e.g., inverse Einstein frequency Ω_E^{-1} for the caged motion) can be comparable for particles of similar size. For plasma crystal experiments the damping time ν_{dn}^{-1} can be made $\sim 30 - 100$ times smaller than Ω_E^{-1} (for micron-size particles and gas pressure of a few tenths of Pa the damping time is ~ 1 s). In this case, the short-term dynamics is governed by Newtonian equations of motion and hence complex plasmas can be treated as a (practically undamped) one-component system. On the other hand, for colloidal dispersions the typical damping time ζ^{-1} ($\sim 10^{-7}$ s) is much shorter than timescales of all relevant dynamic processes – including "atomistic". Therefore, for our purposes, colloids are in the Brownian regime (see Sec. 3.1.1). The "atomistic" dynamics of the two systems is illustrated in Fig. 4.2, showing the profound difference between characteristic trajectories of the individual particles in the Newtonian (a) and Brownian (b) regimes. As regards collective phenomena (such as hydrodynamic flows, phase transitions and – especially – slow dynamics in liquids), these usually evolve at relatively longer timescales (characterized by, e.g., strain rate, convection time, nucleation rate, etc.) and therefore both complex plasmas and colloidal dispersions are in the Brownian regime.

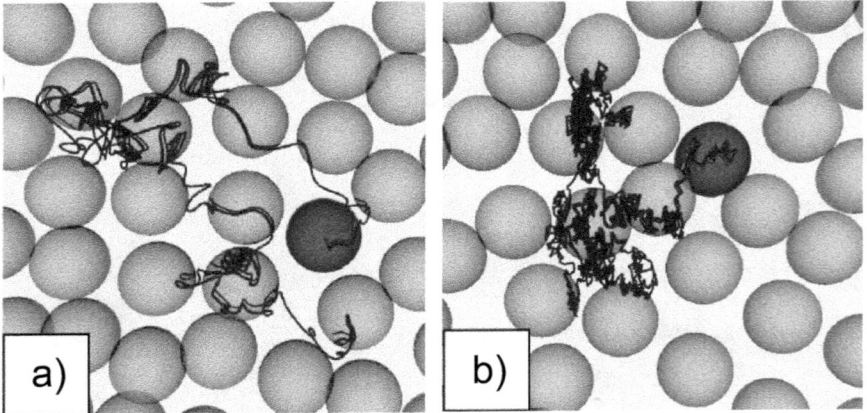

Fig. 4.2   Typical trajectories of a tagged (black) particle for (a) Newtonian dynamics and (b) Brownian dynamics. The interparticle interaction and starting configuration (as represented by the neighboring grey spheres) are the same.

Thus, complex plasmas have very important dynamic complementarity to colloidal dispersions: The "atomistic" motion of microparticles in plasmas is governed by the laws of Newtonian dynamics – as in molecular systems. The accessibility of this virtually undamped short-term regime in complex plasmas allows us to mimic the wealth of generic relaxation/transport processes occurring in conventional liquids and solids outside of equilibrium. As the motion becomes more and more collective – towards hydrodynamics – it gradually crosses over to the Brownian dynamics regime. In this case complex plasmas and colloidal dispersions are essentially equivalent and provide ideal conditions to investigate realm of quiescent states described by the Boltzmann-Gibbs statistics. Hence, complex plasmas significantly extend the variety of dynamic regimes that were accessible so far in experiments with soft matter.

4.2.1 *Kinetics in the Newtonian and Brownian regimes*

The distinction between different dynamic regimes can be most easily understood by employing the Langevin approach to individual particles. By setting the appropriate damping rate ν (which would be ν_{dn} for complex plasmas and ζ for colloidal dispersions), this model approach can be equally applied to both systems. The evolution of momentum $\mathbf{p}_i = m\mathbf{v}_i$ of the ith particle obeys the equation (van Kampen, 1981)

$$\dot{\mathbf{p}}_i + \nu\mathbf{p}_i = \mathbf{F}_i + \mathbf{L}_i, \qquad (4.1)$$

where $\mathbf{F} = \mathbf{F}_{\mathrm{int}} + \mathbf{F}_{\mathrm{ext}}$ is the superposition of forces due to interparticle interactions and external fields. Here, for simplicity we suppose that the interaction force $\mathbf{F}_{\mathrm{int}i} = -\sum_j \partial V_{ij}/\partial \mathbf{r}_i$ is pairwise and reciprocal, $V_{ij} \equiv V(|\mathbf{r}_i - \mathbf{r}_j|)$, the role of nonconservative forces will be discussed in Sec. 4.3. The stochastic Langevin force $\mathbf{L}(t)$ mimics the random kicks of molecules of the surrounding fluid. It is defined as a Gaussian random variable,

$$\langle \mathbf{L}_i(t) \rangle = \mathbf{0}, \qquad \langle \mathbf{L}_i(t+\tau)\mathbf{L}_j(t) \rangle = 2\nu m k_{\mathrm{B}} T \delta_{ij}\delta(\tau),$$

where T is the temperature of the fluid (i.e., of gas or solvent).

The Langevin approach to an individual particle is equivalent to the Fokker-Planck formalism (van Kampen, 1981; Lifshitz and Pitaevskii, 1981). This can be generalized for the whole particle ensemble which is described by the N-particle distribution function (phase-space probability density), $f^{(N)} = f^{(N)}(t, \mathbf{p}_1, \mathbf{r}_1, \dots \mathbf{p}_N, \mathbf{r}_N)$. The corresponding kinetic equation is

$$\frac{\partial f^{(N)}}{\partial t} + \sum_i \left(\mathbf{v}_i \cdot \frac{\partial f^{(N)}}{\partial \mathbf{r}_i} + \mathbf{F}_i \cdot \frac{\partial f^{(N)}}{\partial \mathbf{p}_i} \right)$$
$$= \nu \sum_i \frac{\partial}{\partial \mathbf{p}_i} \cdot \left(\mathbf{p}_i f^{(N)} + mT\frac{\partial f^{(N)}}{\partial \mathbf{p}_i} \right). \qquad (4.2)$$

The r.h.s. is the Fokker-Planck collision operator describing the interaction with the surrounding fluid. When this interaction can be neglected we recover the Liouville equation for the Newtonian dynamics.

Let us first consider the regime of quasi-Newtonian dynamics, when the relevant timescales of the problem are much shorter than ν^{-1} and hence the damping can be neglected to the first approximation. The analysis is usually performed for a reduced number of degrees of freedom, which is obtained by partial integration over momenta and coordinates, $f^{(i)} = \int d\mathbf{p}_{i+1} d\mathbf{r}_{i+1} \dots d\mathbf{p}_N d\mathbf{r}_N f^{(N)}$. For a one-particle distribution function, $f^{(1)} = f^{(1)}(t, \mathbf{p}_1, \mathbf{r}_1)$, we readily derive from Eq. (4.2),

$$\frac{\partial f^{(1)}}{\partial t} + \mathbf{v}_1 \cdot \frac{\partial f^{(1)}}{\partial \mathbf{r}_1} = N \int d\mathbf{p}_2 d\mathbf{r}_2 \frac{\partial V_{12}}{\partial \mathbf{r}_1} \cdot \frac{\partial f^{(2)}}{\partial \mathbf{p}_1}, \tag{4.3}$$

(N is naturally assumed to be large). Thus, the one-particle distribution is affected by pair interactions and hence depends on two-particle correlations via the integral on the r.h.s. of Eq. (4.3). As long as no assumptions are made about the correlations, this equation is exact. It is normally used when the characteristic range of the interaction V_{12} is much smaller than the mean interparticle distance, i.e., when the pair correlations can be neglected and $f^{(2)}$ is simply a product of one-particle distribution functions. By making this essential *superposition* approximation, the r.h.s. of Eq. (4.3) is reduced to the classical *Boltzmann collision integral* describing pair interactions via the collision cross section (the latter is obtained by solving a mechanical problem of two-body interaction for given V_{12}). This approach is commonly used to describe weakly or moderately coupled plasmas (Lifshitz and Pitaevskii, 1981).

As the coupling between particles increases (e.g., by increasing the particle density), the pair correlations become more and more important and eventually the superposition approximation becomes too crude to provide even qualitatively correct description. To take these correlations into account, one has to solve the kinetic equation for a two-particle correlation function which is derived in a similar way as Eq. (4.3). By introducing the relative coordinates $\mathbf{r} = \mathbf{r}_2 - \mathbf{r}_1$ and velocity $\mathbf{v} = \mathbf{v}_2 - \mathbf{v}_1$ and assuming spatial translational invariance we obtain for $f^{(2)}(t, \mathbf{p}, \mathbf{r})$,

$$\frac{\partial f^{(2)}}{\partial t} + \mathbf{v} \cdot \frac{\partial f^{(2)}}{\partial \mathbf{r}} - 2 \frac{\partial V}{\partial \mathbf{r}} \cdot \frac{\partial f^{(2)}}{\partial \mathbf{v}}$$
$$= N \int d\mathbf{p}_3 d\mathbf{r}_3 \left(\frac{\partial V_{13}}{\partial \mathbf{r}_1} \cdot \frac{\partial f^{(3)}}{\partial \mathbf{v}_1} + \frac{\partial V_{23}}{\partial \mathbf{r}_2} \cdot \frac{\partial f^{(3)}}{\partial \mathbf{v}_2} \right),$$

[here $V(r) \equiv V_{12}$]. In this way, one can derive a chain of equations expressing $f^{(i)}$ via $f^{(i+1)}$ – this procedure is called the Bogoliubov-Born-Green-Kirkwood-Yvon (BBGKY) hierarchy. In order to make this approach tractable, the series has to be truncated at a certain step (which depends on the coupling strength/density of a system and which is closely related to the range of the interparticle interactions).

In the Brownian regime the collision operator on the r.h.s. of Eq. (4.2) plays a crucial role. The relevant timescales in this case are assumed to be much longer than ν^{-1}, so that the inertia term in the Langevin equation (4.1) can be neglected. Hence, the particle velocities are assumed to be in close equilibrium with the surrounding fluid (i.e., their local distribution has a Maxwellian form) and therefore it is usually sufficient to consider only the coordinate part of the N-particle distribution function, $\Psi(t, \mathbf{r}_1, \ldots \mathbf{r}_N) = \int d\mathbf{p}_1 \ldots d\mathbf{p}_N f^{(N)}$. The probability density flux $\mathbf{J}_i$ (associated with the ith particle) can be readily obtained by taking the first velocity moment of Eq. (4.2) (and neglecting the time derivative), which gives

$$\mathbf{J}_i = \mu_0 \mathbf{F}_i \Psi - D_0 \frac{\partial \Psi}{\partial \mathbf{r}_i}, \tag{4.4}$$

where the diffusion coefficient $D_0 = v_T^2/\nu$ of a free particle is related to the mobility coefficient via the Einstein relation $D_0 = \mu_0 k_B T$. If the hydrodynamic flow is present in the fluid, the advective term $\mathbf{v}_i \Psi$ must be added to the flux. Then, by integrating Eq. (4.2) over momenta we derive the Fokker-Planck-like equation for Ψ,

$$\frac{\partial \Psi}{\partial t} + \sum_i^N \frac{\partial \mathbf{J}_i}{\partial \mathbf{r}_i} = 0. \tag{4.5}$$

This is the so-called *Smoluchowski equation* which is is widely accepted as an appropriate starting point for the treatment of colloidal dynamics (Dhont, 2003). In particular, a formal solution of this equation for the N-particle distribution Ψ provides powerful method for studying equilibrium and non-equilibrium rheology, which we discuss in detail in Sec. 10.2.

Similar to the Newtonian dynamics regime, one can also employ the BBGKY hierarchy to the Smoluchowski equation. For homogeneous systems, the first relevant measure is the pair correlation function, $g(t, \mathbf{r}) = \int d\mathbf{r}_3 \ldots d\mathbf{r}_N \Psi$. This is governed by the following equation which can be derived from Eqs (4.4) and (4.5) assuming the translational invariance:

$$\frac{\partial g}{\partial t} + 2\mu_0 \frac{\partial}{\partial \mathbf{r}} \cdot \left(k_B T \frac{\partial g}{\partial \mathbf{r}} + \frac{\partial V}{\partial \mathbf{r}} g \right)$$
$$= \mu_0 n \frac{\partial}{\partial \mathbf{r}} \cdot \int d\mathbf{r}_3 g^{(3)} \left(\frac{\partial V_{23}}{\partial \mathbf{r}_2} - \frac{\partial V_{13}}{\partial \mathbf{r}_1} \right), \tag{4.6}$$

where $g^{(3)} = g^{(3)}(t, \mathbf{r}_1, \mathbf{r}_2, \mathbf{r}_3)$ is the triplet distribution function and $n = N/V$ is the mean density. In the equilibrium, the pair correlation function is given by $g_{eq}(r) = e^{-V_{eff}(r)/k_B T}$, where the effective potential is the sum of the pair potential $V(r)$ and the contribution of three-particle correlations given by the r.h.s. of Eq. (4.6).

Finally, the transition from the particle-resolved (kinetic) picture to the coarse-grained (fluid) description is achieved by taking consecutive velocity moments of the kinetic equation. If the higher-order velocity correlations can be neglected (which corresponds to the Maxwellian distribution), the chain can be truncated at the third moment yielding in a set of canonical fluid equations (Morfill and Ivlev, 2009): the continuity equation for mean particle density, momentum equation for mean particle velocity $\mathbf{U}$ (which is transformed to the Navier-Stokes equation for Newtonian fluids), and the heat transport equation for the kinetic temperature of particles T. The presence of the surrounding fluid (of temperature T_b) results in a trivial modification of these equations, as one can readily obtain from Eq. (4.2): The friction yields additional force term $-\nu\mathbf{U}$ in the momentum equation and additional dissipation term $2\nu(T - T_b)$ in the heat transport equation.

4.2.2   *Crossover between the dynamic regimes*

As we pointed out above, in complex plasmas the neutral gas friction plays a minor role in atomistic processes but becomes crucial at much longer timescales related to the overall hydrodynamic flows. This allows us to observe a smooth crossover from the regime of Newtonian dynamics (peculiar to conventional fluids) to Brownian dynamics (operating in colloidal dispersions). Let us consider some characteristic examples illustrating the role of damping and the crossover between the two regimes.

4.2.2.1   *Thermal convection*

Free thermal convection occurs in fluids outside equilibrium and combines typical "fingerprints" of conventional hydrodynamics, like shear flows and the associated dissipation, heat conduction, etc. Let us treat particles as a fluid, so that their convection is described by the Oberbeck-Boussinesq equations (Landau and Lifshitz, 1987; Cross and Hohenberg, 1993) modified due to friction (as discussed in the previous section). As usual, hydrostatic equilibrium assumes constant temperature and zero velocity, with the pressure p changing as $\rho\mathbf{g}\cdot\mathbf{r}$ (where the mass density of particle fluid $\rho = mn$ is constant). The mean particle velocity $\mathbf{U}$ and the deviation of kinetic temperature from the equilibrium, δT, are governed by (Ivlev *et al.*, 2007b):

$$\partial_t \mathbf{U} + (\mathbf{U} \cdot \nabla)\mathbf{U} + \nu\mathbf{U} = -\nabla(p/\rho) + (\eta/\rho)\nabla^2\mathbf{U} - \beta\mathbf{g}\delta T,$$
$$\partial_t \delta T + \mathbf{U} \cdot \nabla\delta T + 2\nu\delta T = \chi\nabla^2\delta T, \tag{4.7}$$

where $\beta = \rho^{-1}(\partial\rho/\partial T)_p$, η, and χ are the thermal expansion, dynamic viscosity, and thermal conductivity of particle fluid, respectively. Equations (4.7) are complemented with the continuity equation $\nabla \cdot \mathbf{U} = 0$. For the Rayleigh problem, when a temperature difference Θ is maintained between two infinite parallel plates separated in z-direction by a distance L, the unperturbed temperature profile (without convection) is determined by $d^2\delta T_0/dz^2 = L_{\mathrm{fr}}^{-2}\delta T_0$, where $L_{\mathrm{fr}} = \sqrt{\chi/2\nu}$ is the length of the temperature decay due to friction. This yields $\delta T_0(z) = \Theta\sinh(\alpha_{\mathrm{fr}}z/L)/\sinh\alpha_{\mathrm{fr}}$, so that the temperature profile is determined by the friction number

$$\alpha_{\mathrm{fr}} = L/L_{\mathrm{fr}}.$$

In addition to this parameter, the problem depends on two "conventional" similarity numbers – the Rayleigh number $Ra = g\rho\beta\Theta L^3/\eta\chi$ and the Prandtl number $Pr = \eta/\rho\chi$.

One can distinguish two limiting cases of convection (Ivlev *et al.*, 2007b): When the particle dynamics is quasi-Newtonian ($\alpha_{\mathrm{fr}} \ll 1$, the temperature profile is practically linear) and when the dynamics is Brownian ($\alpha_{\mathrm{fr}} \gg 1$, the temperature decays exponentially). It turns out that the onset of convective instability remains practically unaffected by friction as long as the friction number is below unity: The critical Rayleigh number for the quasi-Newtonian regime is

$$Ra_{\mathrm{cr}} \simeq 1708 + \alpha_{\mathrm{fr}}^2\left(136 + 22.2Pr^{-1}\right).$$

Remarkably, in the Brownian regime the convection is not eliminated. In this case, all spatial dimensions are simply renormalized by L_{fr} (instead of L in the Newtonian regime). The critical Rayleigh number is then proportional to α_{fr}^3 and can be approximated by

$$Ra_{\mathrm{cr}} \simeq 49\alpha_{\mathrm{fr}}^3 Pr^{-0.35+0.068\ln Pr}.$$

Figure 4.3 shows the asymptotic shape of the convection cell in the limit $\alpha_{\mathrm{fr}} \gg 1$. Thus, we conclude that the classical thermal convection can be also sustained in systems where individual particles experience friction exerted by surrounding fluid. This is valid as long as continuous description of particles at the characteristic *hydrodynamic* scale, $\min\{L, L_{\mathrm{fr}}\}$, can be justified.

This illustrative problem highlights one of the most remarkable properties of complex plasmas: They comprise a unique bridge that links colloidal systems (characterized by fully damped motion of microparticles) with conventional undamped systems. By varying the gas pressure we can control the damping ratio ν_{dn}/Ω_E and thus observe the crossover from the "colloidal" to "molecular" fluid regimes.

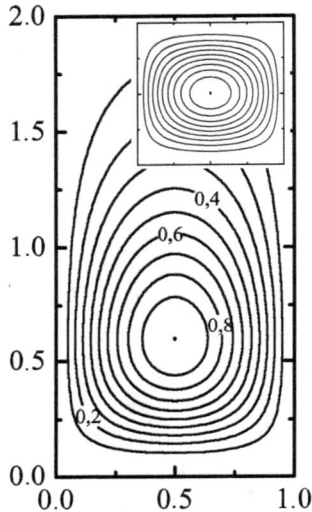

Fig. 4.3   Particle convection cell in the Brownian dynamics regime. Temperature gradient is in the vertical direction, coordinates are normalized by the friction length L_{fr}, the vorticity magnitude is indicated near streamlines. For comparison, the inset shows the classic Rayleigh convection cell (regime of Newtonian dynamics, where the scale is determined by the geometrical length L). From Ivlev *et al.* (2007b).

4.2.2.2   *Phonons*

Another interesting example demonstrating universality of the underlying physics in complex plasmas and colloidal dispersions are phonons. They are sustained in any elastic media[3] – liquid or solid – and damped due to various dissipation processes, which can be either generic (e.g., viscosity operating in any system) or specific (such as a friction due to surrounding fluid, peculiar to complex plasmas and colloids). In complex plasmas, the eigenfrequency characterizing phonons can be much higher than the gas damping rate ν_{dn} (other damping mechanisms are usually negligible)

---

[3]Note that in plasmas elasticity can be associated with a self-consistent electric field created by a wave (e.g., dust-acoustic wave shown in Fig. 5.8).

and therefore the acoustic waves can freely propagate (Fortov *et al.*, 2005; Shukla and Eliasson, 2009; Morfill and Ivlev, 2009). In colloidal dispersions, on the contrary, the viscous damping due to solvent is so strong that no waves can be excited and therefore no phonons in the classical sense can be observed (the typical eigenfrequeny is several orders of magnitude smaller than the damping rate ζ).

Generally, the dispersion relations $\omega(\mathbf{k})$ for phonons in a strongly coupled (liquid or solid) medium are determined by eigenvalues of a dynamic matrix, $\mathsf{D_k}$, which is a measure of the elastic response. This can be derived by considering small perturbations of the form $\propto \exp(-i\omega t + i\mathbf{k}\cdot\mathbf{r})$ for individual particles, so that the elements of $\mathsf{D_k}$ are determined by properties of interparticle interactions. If the interactions are reciprocal then $\mathsf{D_k}$ is Hermitian, and therefore the eigenvalues are always real (i.e., stable). The situation changes if the interactions are nonreciprocal – the eigenvalues in this case can become complex (see Sec. 4.3). Neglecting the influence of the Langevin force in Eq. (4.1) we derive: $\det[\mathsf{D_k} - \omega(\omega + i\nu)\mathsf{I}] = 0$, where I is the unit matrix. Thus, $\omega(\omega + i\nu) \equiv \Omega^2(\mathbf{k})$ are the *eigenmodes* which can be directly obtained from experiments, by measuring thermal particle fluctuations (displacements) and plotting them in the Fourier space.

Figure 4.4 illustrates fluctuation spectra measured in 2D colloidal and plasma crystals [Keim *et al.* (2004); Zhdanov *et al.* (2003), see also papers by Nunomura *et al.* (2002, 2005) for waves in liquid and solid complex plasmas, and by Baumgartl *et al.* (2007, 2008) for eigenmodes in colloidal crystals]. In complex plasmas friction is usually unimportant for the real part of the dispersion relation, Re $\omega(\mathbf{k}) \simeq \Omega(\mathbf{k})$ (while Im $\omega \simeq -\frac{1}{2}\nu_{dn}$ provides weak damping). In contrast, the dispersion relation for colloids is purely imaginary, $\omega(\mathbf{k}) = -i\Omega^2(\mathbf{k})/\zeta$. Thus, phonons in plasma crystals (described by a hyperbolic equation) can freely propagate as weakly damped waves, while in colloidal dispersions a "phonon diffusion" (governed by a parabolic equation) is observed. Note that the magnitudes of Ω^2 in both systems are proportional to $\Gamma T/\Delta^2$, where Γ is the corresponding coupling coefficient.

Thus, irrespective of the magnitude of damping one can directly measure thermal fluctuations in a system of strongly coupled particles. The results of such measurements are the "fluctuation spectra" showing phonon modes – the eigenvalues of the dynamic matrix which are *independent* of the damping rate and are *solely* determined by interparticle interactions (although the notion "dynamic matrix" in a fully damped regime might be somewhat misleading). Such measurements are useful indeed – for instance, from the behavior at small k one can derive the elastic moduli of the crystal (Landau and Lifshitz, 1986): The shear modulus is determined by the transverse mode, $\mu \propto \lim_{k\to 0}(\Omega_t/k)^2 \equiv C_t^2$, whereas the sum of the bulk and shear moduli is determined by the longitudinal mode, $\mu + K \propto \lim_{k\to 0}(\Omega_l/k)^2 \equiv C_l^2$, where $C_{t,l}$ are the corresponding acoustic velocities.

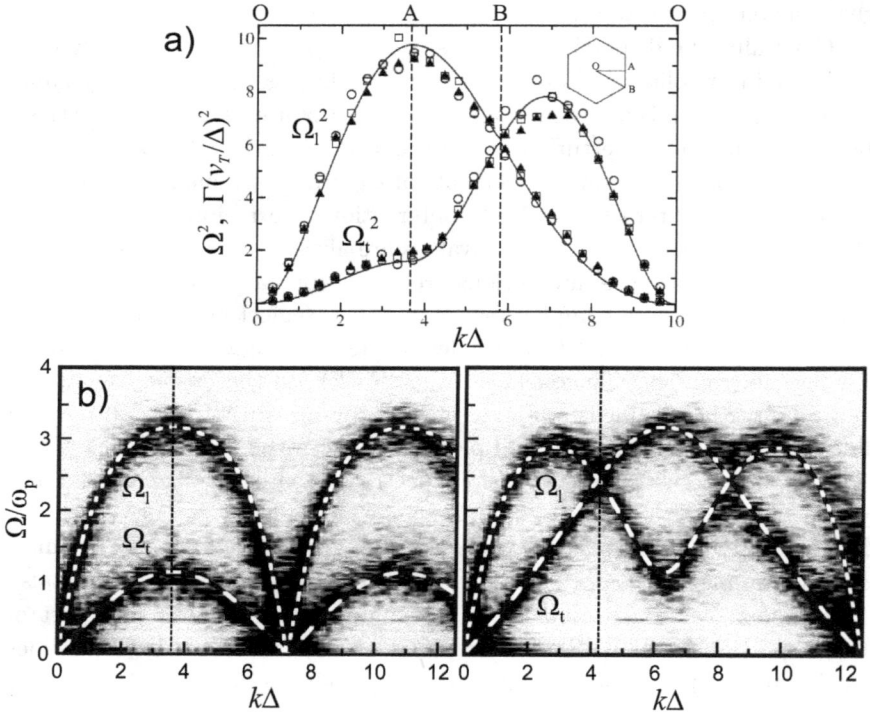

Fig. 4.4   Eigenmodes in 2D hexagonal crystals. (a) Longitudinal (Ω_l^2) and transverse (Ω_t^2) modes in a 2D colloidal crystal (particles with dipolar interactions, induced by external magnetic field perpendicular to the plane). Shown are measurements for wave vector **k** along the lines OA, AB, and OB within the first Brillouin zone (as depicted in the upper right corner). Symbols represent experiments with different coupling parameters Γ, the solid lines are for theory, the modes are in units of $\Gamma(v_T/\Delta)^2$. From Keim *et al.* (2004). (b) The eigenfrequencies $\Omega_{l,t}$ in a 2D plasma crystal (particles with Yukawa interactions), plotted for **k** along the lines OA (left panel) and OB (right panel). Gray scaling represents measured fluctuation spectra (dispersion relations) normalized by the plasma frequency ω_p, white dashed lines show theoretical curves. From Zhdanov *et al.* (2003). In (a) and (b), the vertical (black dashed) lines indicate points A and B at the border of the first Brillouin zone, k is normalized by the lattice constant Δ.

4.2.3   *Influence of particles on surrounding fluid*

The treatment of the particle dynamics is significantly simplified if the effect of particle motion on the *average* fluid flow can be neglected. This is relevant to the situation when there exists a macroscopic flow induced in a system, and characteristics of this flow must be prescribed. The corresponding conditions can be easily obtained for a particular flow configuration.

As a characteristic example, let us consider a shear flow induced in the surrounding fluid and identify conditions when the presence of particles embedded in the flow has no substantial influence on the prescribed shear rate. The flow remains unaffected when the bulk viscous force acting on a unit volume of the fluid, $F_{\text{bulk}} \sim \eta U / L^2$ (here U and L are the characteristic velocity and length scale of the flow, respectively), is much larger than the force F_{p} exerted in a fluid due to the presence of particles.

In complex plasmas, $F_{\text{p}} \sim mn\nu_{dn}U$ (m and n are the particle mass and number density; we consider the extreme case when particles are at rest). Taking into account that the viscosity of neutral gas is $\eta_n \sim m_n n_n v_{T_n} \ell_{nn}$ and using Eq. (2.8) we obtain $na^2 L^2 / \ell_{nn} \ll 1$, where $\ell_{nn} = (\sigma_{nn} n_n)^{-1}$ is the mean free path of neutral-neutral collisions (typically, a few tenths of cm at a pressure ~ 3 Pa). Thus, for systems as large as $L \sim 10$ cm the presence of microparticles has no significant impact on a shear flow induced in a neutral gas, i.e., the prescribed flow is preserved for any typical experimental configuration. This allows for flexible implementation of various types of induced gas flows, in particular – cylindrical Poiseuille flows for rheological studies.

In colloidal dispersions, $F_{\text{p}} \sim 3\pi n \eta \sigma U$, so that the influence of particles can be generally neglected when $n\sigma L^2 \ll 1$. In terms of the colloidal volume fraction, this yields $\phi \ll (\sigma/L)^2$, i.e., the condition can be only satisfied for very low colloid concentrations. Therefore, the only shear flow configuration which can be implemented to investigate rheology of colloidal dispersions is the plane Couette flow which is characterized by a constant velocity shear (affine flow) prescribed by boundary conditions. We note that this assumption is rather restrictive since it excludes from the outset the possibility of inhomogeneous flow, as observed in shear banded and shear localized states. Thus, being physically reasonable for low and intermediate colloid densities this assumption could become questionable when considering the flow response of glassy states (Brader, 2010).

It is worth mentioning that in experiments where strain or stress are applied at the sample boundaries a finite time is required for transverse momentum diffusion to establish the velocity field of solvent. Nevertheless, experiments and simulations of the transition from equilibrium to homogeneous steady-state flow have shown that a linear velocity profile is established long before the steady state regime is approached and hence the transient processes occurring in solvent can be neglected (Brader, 2010; Zausch *et al.*, 2008).

4.3   Role of Nonconservative Interactions

When studying generic classical phenomena occurring in "conventional" (molecular) liquids and solids, the *relevance* of a model system (be it colloidal suspensions, granular media, or complex plasmas) becomes crucial. Of course, careful analysis is required in the context of a given phenomenon (or, class of phenomena), but one can certainly identify essential common principles. In particular, the applicability of the Hamiltonian approach for the analysis of atomistic dynamics is one of these basic principles.

Both in complex plasmas and colloidal dispersions, interactions between particles do not always conserve the symmetry peculiar to conservative systems and hence the particle dynamics cannot always be described by the Hamiltonian equations. The reason for this is the presence of the surrounding fluid – but not only because of "trivial" friction: When the fluid moves relative to particles, this creates asymmetric perturbations around each particle. These perturbations exert additional *nonreciprocal* forces which violate the *actio et reactio* principle. Also, charges on particles can vary in space and time, which might additionally influence their dynamics.

Below we elaborate on nonconservative interaction effects separately for plasmas and colloids, identify the conditions when they are important, and when they can be neglected.

4.3.1   *Wake-mediated interactions*

In complex plasmas the surrounding neutral gas is dilute (free molecular regime), so that perturbations induced by particles in the gas do not affect their interactions. Primarily, plasma species interact via self-consistent electric fields. An external electric field can induce strong ion flow and hence make the screening cloud around each charged particle highly asymmetric – these clouds are usually referred to as "plasma wakes" (see Sec. 2.2.2).

The importance of plasma wakes has been long recognized. The first experiments with complex plasmas were performed in highly inhomogeneous regions of rf plasma sheaths or dc striations (see Sec. 5.1). The magnitude of the vertical electric field in these regions can be so large that ions are accelerated to suprathermal velocities ($M_T \gtrsim 1$). The attractive forces exerted by wakes can cause the particle alignment – the formation of vertical chains along the flow, as illustrated in Fig. 4.5b.

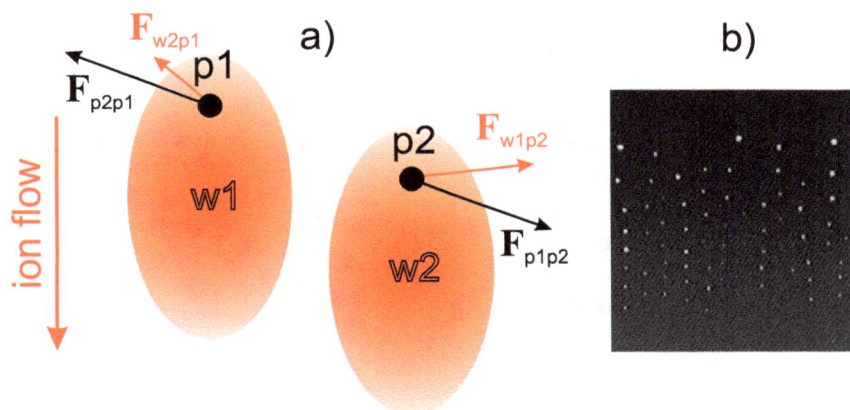

Fig. 4.5   Nonreciprocal wake-mediated interactions. a) The total force exerted on particle "p1" by particle "p2" is the sum of direct particle-particle interaction, $\mathbf{F}_{p2p1}$, and indirect force from wake "w2", $\mathbf{F}_{w2p1}$ (and vice versa for the force on particle "p2"). While the direct forces are reciprocal, $\mathbf{F}_{p2p1} = -\mathbf{F}_{p1p2}$, the wake forces are not, $\mathbf{F}_{w2p1} \neq -\mathbf{F}_{w1p2}$. b) Typical example of aligned chains due to wake-mediated interactions (courtesy of H. Thomas). Shown is the side view with particles of 6.9 μm diameter levitating in a rf sheath.

The wakes play the role of a "third body" in the interparticle interaction and make it nonreciprocal (Melzer *et al.*, 1999), as explained in Fig. 4.5a. This results in a very efficient mechanism of converting the energy of the flowing ions into the kinetic energy of grains (Fortov *et al.*, 2005; Morfill and Ivlev, 2009; Melzer *et al.*, 1996). One can easily demonstrate this by considering a pair of charged grains in the center-of-mass and relative coordinates, $\mathbf{r}_c = \frac{1}{2}(\mathbf{r}_1 + \mathbf{r}_2)$ and $\mathbf{r}_r = \mathbf{r}_2 - \mathbf{r}_1$, respectively. The center-of-mass momentum of such system is *not conserved* – it is governed by the force depending on the relative coordinate, $\mathbf{F}_c(\mathbf{r}_r) \propto (\partial/\partial \mathbf{r}_r)[\varphi(-\mathbf{r}_r) - \varphi(\mathbf{r}_r)]$, where $\varphi(\mathbf{r}_r)$ [$\neq \varphi(-\mathbf{r}_r)$] is the net ("particle+wake") potential produced by each grain. When the system performs finite motion where $\mathbf{r}_r(t)$ and $\mathbf{r}_c(t)$ are correlated (for instance, due to resonances), one can construct such a loop for the center-of-mass coordinate that $\oint \mathbf{F}_c d\mathbf{r}_c \neq 0$, i.e., the total kinetic energy can grow due to wake-mediated interactions.

This is why wake-mediated interactions play a destructive role in (weakly damped) Newtonian dynamics. For instance, when the gas pressure and therefore the damping rate is low enough, strings (like those shown in Fig. 4.5b) spontaneously break apart. In the next section we show an interesting counterexample (see Fig. 4.7): Nonconservative hydrodynamic interactions in microfluidic systems described by (highly dissipative) Brownian dynamics result in self-organization of water droplets embedded in flowing oil and in the formation of stable chains which can sustain undamped phonons.

The influence of wakes is considered to be negligible when particles are levitating in relatively homogeneous regions of the bulk plasma, where the ion drift velocity is significantly decreased (in comparison with the sheath and striation regions, see Sec. 5.1) and the particle potential is rather isotropic.[4] In general, the potential depends on several parameters, such as the thermal ion Mach number M_T, ion collisionality ℓ_{in}/λ, and the ion scattering parameter β_{di} (see Secs. 2.2.2 and 2.3). Practically, the wake-mediated interactions have no noticeable influence on the particle dynamics when M_T is substantially smaller than unity and hence the deviation from a spherically symmetric interaction is small [see Eq. (2.7)]. Experiments in a homogeneous plasma environment (which is normally accessible under microgravity, see Sec. 5.1) provide very good conditions to study generic processes occurring in 3D liquids and solids – crystallization and melting, lane formation, onset of turbulence, etc.

Remarkably, the influence of wakes can be completely eliminated even in the presence of strong electric fields. This is usually possible in experiments with particle monolayers. Such 2D complex plasmas are very convenient model systems to study generic mechanisms of melting and freezing which operate in any (classical) system with a given (Hamiltonian) interaction between particles (see Sec. 7.3.1).

Thus, in order to study generic mechanisms of melting one has to disable *plasma-specific* mechanisms associated with non-Hamiltonian wake-mediated interactions. As we pointed out above, wake-mediated interactions make the particle dynamics nonconservative and can cause the energy

---

[4]We note, however, that the wake-mediated interactions can emerge even in the absence of an external field: Perturbations of ion fluxes induced by neighboring particles result in the shadowing interactions (see Sec. 2.2.3) which can have the effect similar to that of the hydrodynamic interactions in colloidal dispersions (see next section).

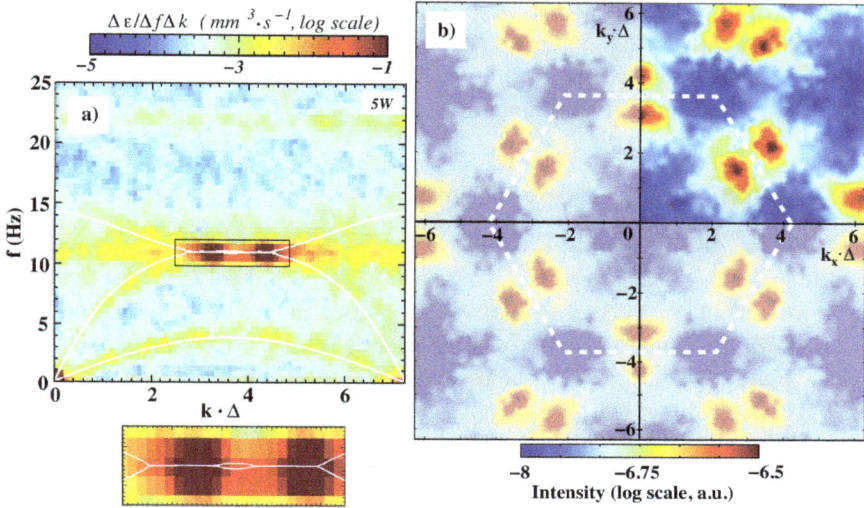

Fig. 4.6   Fluctuation spectra for 2D plasma crystal at the onset of the mode-coupling instability triggered by the wake-mediated interactions. The spectra were measured in the experiment performed with particles of 9.15 μm diameter. The logarithmic color coding is used for the "fluctuation intensity" $\Delta\epsilon/\Delta k\Delta f$ (energy density, for unit mass, per unit wave number and frequency): Yellow ridges represent stable wave modes, the unstable hybrid mode are the dark-red "hot spots". (a) Dispersion relations at $\theta = 30°$, solid lines are theoretical curves for three DL wave modes: out-of-plane transverse, in-plane longitudinal, and in-plane transverse (from top to bottom). The wave number k is normalized by the inverse interparticle distance Δ^{-1}, the shown range of k corresponds to the first two Brillouin zones (branches are symmetric with respect to the border between the two zones). Below the fluctuation spectra the magnified hybrid mode in the unstable regime (marked by a rectangular box) is shown. (b) Fluctuation spectra in the **k** plane integrated over frequency. Dashed hexagon indicates the border of the first Brillouin zone, red spots represent the unstable hybrid mode. From Couëdel *et al.* (2010).

growth in the case of inner resonances. Ivlev and Morfill (2001) showed that for a 2D plasma crystal, the resonance between DL wave modes can trigger the mode-coupling instability which, in turn, can cause the crystal melting. The unique fingerprint characterizing the instability is the crossing of the longitudinal in-plane and transverse out-of-plane DL branches, which results in the (wake-induced) resonance coupling between them and the emergence of an unstable *hybrid mode* operating in a certain proximity of the crossing (Ivlev *et al.*, 2003; Couëdel *et al.*, 2010, 2011). This is illustrated in Fig. 4.6: The intersection occurs in the vicinity of the first Brillouin zone boundary, resulting in the anomalous energy release (dark-red "hot spots"). Thus, the influence of wake-mediated interactions on the particle dynamics in 2D plasma crystals is completely eliminated by avoiding the mode intersection. Practically, this is achieved by increasing the strength of the vertical confinement.

4.3.2   *Hydrodynamic interactions*

In colloidal dispersions, the term "hydrodynamic interactions" refers to nonreciprocal forces induced by particles in the liquid solvent (see Sec. 3.4 for details). These forces can play a very important role in the presence of shear flows [but can also affect the kinetics of relaxation towards quiescent states (Dhont, 2003)]. Nevertheless, the influence of solvent hydrodynamics beyond trivial advection is often completely neglected in many theories, even if the aim is to describe dense colloidal dispersions. For certain situations (e.g., quiescent systems) this approximation is partially motivated by physical intuition. In most cases, however, the omission of solvent hydrodynamics is an undesirable but unavoidable compromise made in order to achieve tractable closed expressions (Brader, 2010). Let us elaborate on this point.

In the presence of the hydrodynamic interactions the probability density flux, Eq. (4.4), entering the Smoluchowski equation is generalized to the following form:

$$\mathbf{J}_i = \mathbf{v}_i \Psi + \sum_j \mathsf{M}_{ij} \cdot \left(\mathbf{F}_j \Psi - k_\mathrm{B} T \frac{\partial \Psi}{\partial \mathbf{r}_j} \right).$$

Here, for completeness we also added the advective flux due to the hydrodynamic velocity $\mathbf{v}_i$ which is a superposition of the the solvent velocity $\mathbf{v}(t, \mathbf{r}_i)$ and fluctuation velocity induced by other particles. Now the motion of particle i is related to the force on particle j via (3×3) mobility tensor M_{ij} which, in its simplest form, is determined by Eqs. (3.10) and (3.11) (and which is related to the diffusion tensor via $\mathsf{D}_{ij} = k_\mathrm{B} T \mathsf{M}_{ij}$). The corresponding Langevin equation is also modified (Edwards and Doi, 1986),

$$\dot{\mathbf{r}}_i = \sum_j \mathsf{M}_{ij} \cdot (\mathbf{F}_j + \mathbf{L}_j) + k_\mathrm{B} T \sum_j \frac{\partial \mathsf{M}_{ij}}{\partial \mathbf{r}_j},$$

(here we consider the overdamped limit). Furthermore, the hydrodynamic interactions naturally modify normalization of the Langevin random force $\mathbf{L}_i$. Now the elements $L_{i,\alpha}$ ($\alpha = x, y, z$) are related to the corresponding elements of the mobility tensor,

$$\langle \mathbf{L}_i(t) \rangle = 0, \qquad \langle L_{i,\alpha}(t + t') L_{j,\beta}(t) \rangle = 2 M_{ij,\alpha\beta}^{-1} k_\mathrm{B} T \delta(t').$$

The first common approximation is to replace the mobility/diffusion tensor by a scalar [viz., $\mathbf{D}_{ij} = D_0\delta_{ij}$], so that $\mathbf{J}_i$ is reduced to Eq. (4.4)], thus neglecting the influence of the configuration of the N colloidal particles on the mobility of a given particle. The second approximation is to neglect the particle-induced fluctuation of the advective flow. For low colloidal densities this may be reasonable at $Pe \ll 1$ but can be expected to break down at $Pe \gtrsim 1$, as hydrodynamics becomes increasingly important in determining the particle trajectories (Brader, 2010). For dense suspensions close to a glass transition the role of hydrodynamics is not quite clear. For certain situations (e.g., glasses close to yield) the relevant value of Pe is very small and suggests that hydrodynamic interactions should not be of primary importance.

Another effect arising from the hydrodynamic interactions is the existence of short-range lubrication forces (Kim and Karilla, 1991) which reduce the mobility when the surfaces of two particles approach contact. These forces play an important role in strong flow and are responsible for driving cluster formation and shear thickening (Bender and Wagner, 1996).

The hydrodynamic interactions caused by strong fluid flow can also result in formation of chain-like structures similar to those observed in complex plasmas. On the other hand, the underlying physical mechanism operating in fully damped systems is very different. To illustrate these similarities and differences, let us consider the experiment performed with an *emulsion*, where so-called microfluidic crystals (formed by the introduction of droplets of immiscible fluid into a liquid-filled channel) were observed (Beatus *et al.*, 2006), as shown in Fig. 4.7. These operate at very low Reynolds numbers and hence viscous dissipation dominates inertial effects (like in colloidal dispersions). In particular, one can expect that phonons in microfluidic crystals are fully damped (see Sec. 4.2).

Surprisingly, it turns out that microfluidic systems sustain phonons which are not damped and exhibit unusual dispersion relations: Figure 4.7 demonstrates that unlike harmonic crystals where each mode has two branches and is invariant with respect to the $\mathbf{k}$-reversal, in the hydrodynamic crystal this symmetry is broken by the flow field. Each mode – longitudinal and transverse – has a single branch with $\omega(-k) = -\omega(k)$, yet these modes have the same form but opposite signs.

Damping plays a crucial role here, which can be nicely demonstrated by making the comparison with complex plasmas. The main contribution to the hydrodynamic interactions at small Reynolds numbers is provided by the dipolar term in the expansion of the flow potential $\varphi(\mathbf{r})$ over Re. The

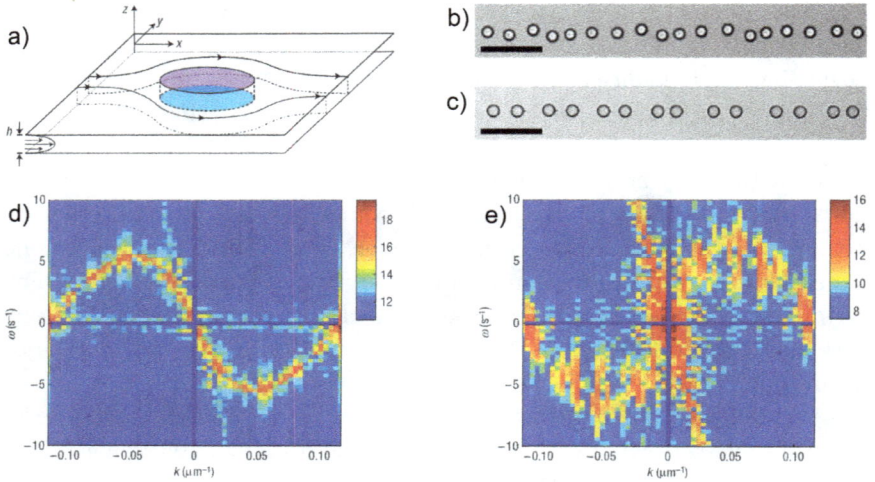

Fig. 4.7   Microfluidic strings. (a) A droplet is a rigid disc between two plates, oil flows along the x-direction. (b,c) Images of transversal (b) and longitudinal (c) acoustic waves. (d,e) Intensity plots of the logarithm of the power spectrum of longitudinal (d) and transversal (e) waves showing the dispersion relations in the (ω, k) plane. The oblique straight line $\omega = -k u_{\rm d}$ is due to the fact that the video camera was co-moving with droplets at the velocity $-u_{\rm d}$. From Beatus *et al.* (2006).

interaction force is proportional to the gradient of the potential. In complex plasmas, where the small parameter is the ion Mach number M_T and the wake potential is described by Eq. (2.7), the leading ($\propto M_T$) interaction term has similar form. However, while the dynamics of individual particles in complex plasmas is governed by Newtonian equations of motion, $\ddot{\mathbf{r}}_n \propto \sum_{m \neq n} \nabla\varphi(\mathbf{r}_n - \mathbf{r}_m)$, in a fully damped microfluidic system the dynamics is Brownian, $\dot{\mathbf{r}}_n \propto \sum_{m \neq n} \nabla\varphi(\mathbf{r}_n - \mathbf{r}_m)$. This has a very different impact on the stability.

If we consider a homogeneous 1D string with dipolar interactions, then the longitudinal (x) and transverse (y) perturbations in the Brownian regime are described by $\dot{x}_n = 2C\sum_m m^{-4}(x_{n+m} - x_{n-m})$ and $\dot{y}_n = -C\sum_m m^{-4}(y_{n+m} - y_{n-m})$ (where C is a positive constant), whereas for the Newtonian regime we have similar equations with the second-order time derivatives.[5]   In the former case these equations yield dispersion

---

[5] For the microfluidic experiment shown in Fig. 4.7 the constants and exponents under the sum are different, because the interactions are described by a 2D dipole.

relations $\omega(k)$ of a skew sine-like form which are obviously *stable*. In the Newtonian regime, however, the l.h.s. is proportional to ω^2 (instead of $i\omega$), which makes the transverse mode absolutely unstable.

Thus, particles with nonreciprocal interactions induced by the surrounding fluid flow gain energy from the flow. If the damping is too weak (like in complex plasmas) and the energy gain cannot be compensated by dissipation, the system becomes unstable. On the contrary, in fully damped systems (like colloidal dispersions) these two processes balance each other. This can result in various types of *self-organization* and the formation of dissipative structures similar to the string formation shown in Fig. 4.7. In a broader context, the damping plays a constructive role for the self-organization (Prigogine, 1980), and the resulting dissipative structures can be considered as the manifestation of a non-equilibrium phase transition [with other examples being, e.g., the formation of convection (Bénard) or turbulent (Taylor) vortices (Cross and Hohenberg, 1993), or lane formation, see Sec. 10.5.1].

4.3.3   *Variable charges*

There is another interesting phenomenon which affects the interparticle interactions and – under specific conditions – might also trigger instabilities. This is associated with the fact that both in complex plasmas and colloidal dispersions the individual particle charges fluctuate randomly with time around some equilibrium value which, in turn, is generally a function of the spatial coordinates (Fortov *et al.*, 2005). Let us briefly discuss the influence of charge variations on the particle dynamics.

The case when the charge is a function of the coordinates, $Q = Q(\mathbf{r})$ represents the simplest class of non-Hamiltonian dynamics (Zhakhovskii *et al.*, 1997; Zhdanov *et al.*, 2005): The force $Q\mathbf{E}$ acting on a particle in a potential electric field $\mathbf{E}(\mathbf{r}) = -\nabla\varphi(\mathbf{r})$ *cannot* be expressed in terms of a gradient of a scalar function, because $\nabla\times(Q\nabla\varphi) \equiv \nabla Q\times\nabla\varphi$ is not equal to zero in the general case. The dynamics is Hamiltonian only when the charge gradient is collinear with the electric field (in this case, the force depends on a single longitudinal coordinate and therefore it can always be written as a derivative of some scalar function over the coordinate). Note that the nonconservative electrostatic force $Q\mathbf{E}$ is in a certain sense analogous to the buoyancy force $\propto T\mathbf{g}$ entering the Boussinesq equations which drives free fluid convection [see first Eq. (4.7)]: The local charge $Q(\mathbf{r})$ plays the role of the local fluid temperature $T(\mathbf{r})$, and the electric field substitutes the

Complex Plasmas and Colloidal Dispersions

gravity **g**. This force can trigger the "charge-driven" convection in complex plasmas (Zhakhovskii *et al.*, 1997; Vaulina *et al.*, 2000) – an interesting example of self-organization which can be experimentally observed under specific conditions.

Now let us consider random charge fluctuations assuming a constant (i.e., independent of **r**) mean value Q (Vaulina *et al.*, 1999; Ivlev *et al.*, 2000). Interactions in this case, although they can be described by a (time-dependent) Hamiltonian, do not conserve energy. Using the stochastic properties of the charge fluctuations [see Eqs. (2.2) and (2.3)] it was shown that due to resulting random variations of the particle eigenfrequency (e.g., confinement frequency $\Omega_{\mathrm{conf}} \propto Q$) the charge fluctuations can trigger the parametric instability of particle oscillations (Ivlev *et al.*, 2000). Then the mean energy grows exponentially with time – the instability condition is $2\nu_{dn} \lesssim \sigma_Q^2 \Omega_{\mathrm{conf}}^2 / \Omega_{\mathrm{ch}}$. However, even in complex plasmas the charge-fluctuation instability is only possible at very low pressures (far below ~ 1 Pa) when the damping is extremely weak. In colloidal dispersions, such instability is, of course, always inhibited.

4.4 Summary

In this chapter we have summarized the key similarities and complementarities of complex plasmas and colloidal dispersions. These ideas underpin the particle-resolved studies described in Chap. 6-11. In a nutshell, colloids and plasmas are alike because both consist of mesoscopic particles suspended in a fluid medium. The interactions between the particles define the thermodynamics of the system, in an analogous way to atoms.

As we explore in Chap. 6 and 7, these similarities are most evident when the interparticle interactions in the two systems are the same. Under such conditions, many basic properties – phase behavior, spatial correlations – are *indistinguishable*. However, even in this case the essential complementarity between the two systems is also clear: particle dynamics and dynamical correlations are entirely different. This leads, for example, to strongly correlated complex plasmas sustaining phonon spectra analogous to atomic systems – since microparticles in a plasma obey (quasi)Newtonian dynamics. Conversely, the particle dynamics in colloidal systems is fully damped, leading to the effect of "phonon diffusion".

There is another very interesting similarity in terms of specific nonreciprocal interactions induced by the surrounding fluids. These interactions, termed hydrodynamic in colloids, are rather akin to the wakes-mediated interactions in plasmas, although of course the density of the fluid medium plays an important role in the strength of the coupling here.

One should note a greater degree of control over interparticle interactions in the case of colloids, with a resulting increase in complexity in terms of phase behavior. However, we stress that the key generic features of the richness already observed in colloidal systems should be accessible to the field of complex plasmas going forward – relevant examples are presented in Chap. 6 and 11.

Chapter 5

Overview of Experimental Methods

5.1 Complex Plasmas

In the beginning of the 1990s, when plasma crystals had just been discovered, practically all experiments were carried out using low-temperature radio-frequency (rf) discharges. In the following years direct current (dc) discharges were also employed, providing new possibilities, such as studies of large 3D particle clouds in ground-based experiments. In this section we introduce the two most important types of gas-discharge plasma – the rf (capacitively coupled) and dc discharges – which are still used in most complex plasma experiments (Fortov and Morfill, 2010). Other types of complex plasmas which can also be created in laboratory conditions are briefly mentioned as well.

5.1.1 *RF discharges*

Most complex plasma experiments have been and still are performed in capacitively coupled rf discharges. There are several reasons for this (Raizer, 1991; Lieberman and Lichtenberg, 1994): (i) the capacitive sheath near the electrode produces a strong negative self-bias (with magnitude of 10–100 V), which can be effectively used to confine and levitate negatively charged microparticles; (ii) the rf frequency is so high that neither dust particles nor ions can respond to the ac electric field, and therefore both species tend to acquire the temperature of the neutral component; (iii) the electron temperature is of the order of few eV, which ensures a rather high value of the charge on the dust particle.

Usually, the smaller the particle size is, the more layers in the vertical direction can be formed in ground-based experiments. When the particle size is below ~ 1 μm it is even possible to fill the bulk region of discharge with particles, because gravity in this case is no longer the dominant force operating in the system. On the other hand, for particle-resolved studies the required particle size is above 1 μm. Gravity restricts such a system to a few layers in the vertical direction. To perform experiments in the bulk with fully 3D complex plasma systems, it is necessary to use additional levitating forces, e.g., gas flow or thermophoresis – or to remove gravity. The latter is possible on parabolic flights, sounding rockets, or in experiments onboard the International Space Station (ISS). For such experiments a special symmetric design of the rf discharge is required, otherwise the homogeneity gained through microgravity will be destroyed.

Below we describe two major types of rf-discharge setups which are employed for complex plasma experiments on the ground and under microgravity conditions.

5.1.1.1   *Ground-based experiments: GEC reference cell*

The modified Gaseous Electronics Conference (GEC) rf-reference cell (Hargis *et al.*, 1994) shown in Fig. 5.1 is a setup which is particularly well suited for laboratory studies of 2D complex plasmas (see Fig. 2.3 illustrating the plasma environment in this case). The electrode system consists of a driven rf electrode in the bottom of the apparatus and a grounded counter electrode at the top. If the latter is removed, the metal vacuum chamber surrounding the electrode system (which provides the vacuum conditions for the low-pressure discharge) acts as the counter-electrode. The plasma can be ignited in inert gases (typically argon and krypton) at pressures in the range from ~ 0.1 Pa to ~ 1000 Pa (depending on the gas). Molecular gases can be used as well, but the advantage of inert gases is that they do not form negative ions, which substantially simplifies the analysis of plasma chemistry.

The electrode is driven at a rf frequency of 13.56 MHz. The resulting ac electric field provides efficient acceleration and heating of electrons to an average temperature of a few eV. This ensures weak ionization of neutral gas – in the central (bulk) region of the discharge the typical plasma density is $10^8 - 10^{10}$ cm^{-3}. On the other hand, (relatively) heavy ions are not affected by the rf field and remain at room temperature due to frequent collisions with neutral atoms (typical ionization fraction n_i/n_n is in the range of $10^{-8} - 10^{-6}$).

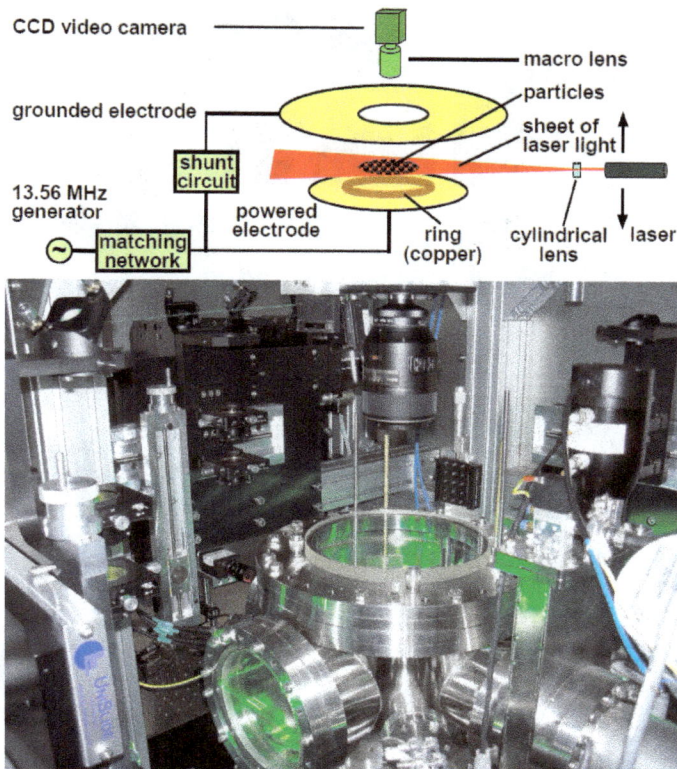

Fig. 5.1   Sketch shows typical rf electrode system with lower driven electrode and the grounded upper-ring electrode.   The microparticles are injected by a dispenser (not shown) and levitate in the (vertical) sheath electric field, horizontally they are trapped by a parabolic potential due to a ring placed on the electrode. The microparticles are illuminated by a laser beam expanded into a sheet parallel to the electrode, and the scattered light is observed at 90° by a video camera. The image below shows a typical assembly of the GEC rf-Reference Cell. From Fortov and Morfill (2010).

The sheath above the driven electrode (see Sec.   2.4) is very impor-tant for complex plasmas under gravity conditions.   This exerts an up-ward electrostatic force acting on negatively charged microparticles, which can balance gravity and allow particle levitation in a narrow region in the sheath. Since particles used in basic complex plasma experiments normally have an almost monodisperse size distribution, they easily form monolayer structures (the smaller the charge-to-mass ratio, the lower the leviation height). The monolayer is illuminated by a sheet of laser light parallel to

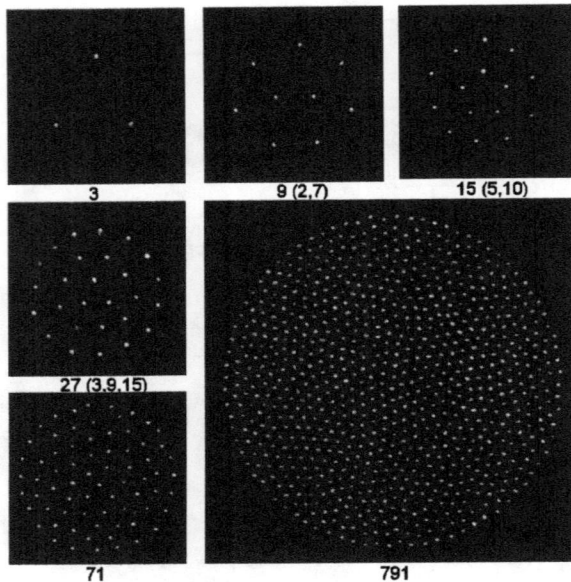

Fig. 5.2   Images showing 2D clusters of different numbers of particles. From Juan *et al.* (1998).

the electrode, as shown in Fig. 5.1. The light scattered by individual microparticles is recorded at 90° by a video camera, allowing fully resolved dynamic studies of 2D liquid and crystalline systems.

Experiments with single particles levitating in the sheath are important for measuring some basic properties, like the microparticle charge or the plasma wake (see Secs. 2.1 and 2.2.2). If the number of particles is increased step by step, one can investigate the formation of clusters (Juan *et al.*, 1998), small crystalline structures dominated by surfaces, and observe the formation of configurations with "magic" numbers of particles, as illustrated in Fig. 5.2. By adding more and more particles to a 2D cluster, the transition to a "infinite" monolayer (where boundary effects play no important role) can be eventually reached.

The plasma chamber can be further modified by adding manipulation devices. Two major kinds of manipulation are possible. With laser beams one can act on the microparticles from outside, without disturbing the plasma itself. Static laser beams are used to create steady-state shear flow, while movable beams (driven by scanning mirrors) introduce local heating (see, e.g., Secs. 7.3.1 and 10.3). A plasma-disturbing manipulation is a wire

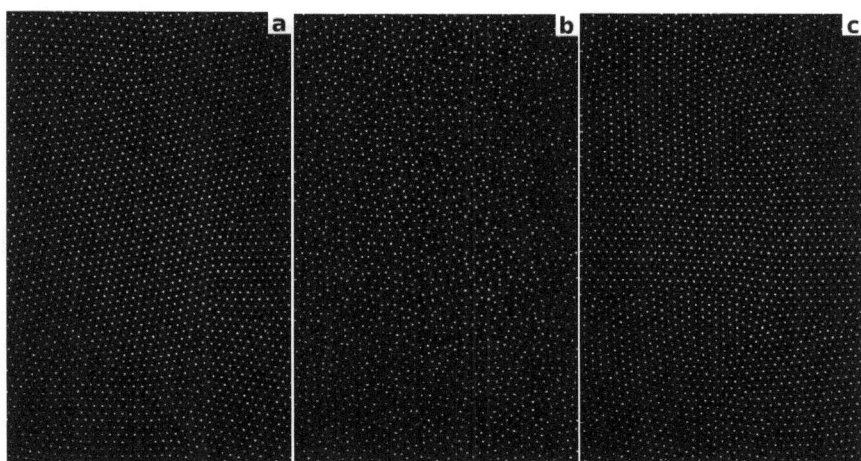

Fig. 5.3   Melting and recrystallization in 2D complex plasmas. Original images show (a) initial crystalline state before the melting, (b) liquid state shortly after the melting (1.6 s after the pulse), and (c) metastable state at later stage of recrystallization (7.0 s after the pulse). Two (vertical) parallel wires causing the melting (and simultaneously providing the horizontal confinement) are beyond the image frame. From Knapek *et al.* (2007).

or a wire system (with controllable bias) introduced into the plasma. The monolayer (which is usually at a similar height with the wire) can be excited through short voltage pulses or an alternating signal. The latter kind of manipulation is very efficient to study the process of non-equilibrium recrystallization (Knapek *et al.*, 2007) illustrated in Fig. 5.3: A short negative electric pulse applied to two parallel wires causes a disturbance, pushing the particles away from both wires to the center and leading to the melting of a 2D crystal (Fig. 5.3b). After the pulse, the particle velocities gradually decrease due to a weak neutral gas friction and recrystallization starts (Fig. 5.3c). The recrystallization kinetics can be investigated in terms of the measured kinetic temperature and structural properties (see Sec. 7.3.1 for details).

Usually, the electric force in the sheath above the lower electrode provides complete compensation for gravity. In addition, the thermophoretic force (exerted in the presence of temperature gradient, see Sec. 2.4) can significantly contribute to the force balance. The thermophoretic force can mimic quasi-"zero g" conditions in complex plasmas, thus allowing us to study 3D complex plasmas in ground-based laboratories. For instance, in

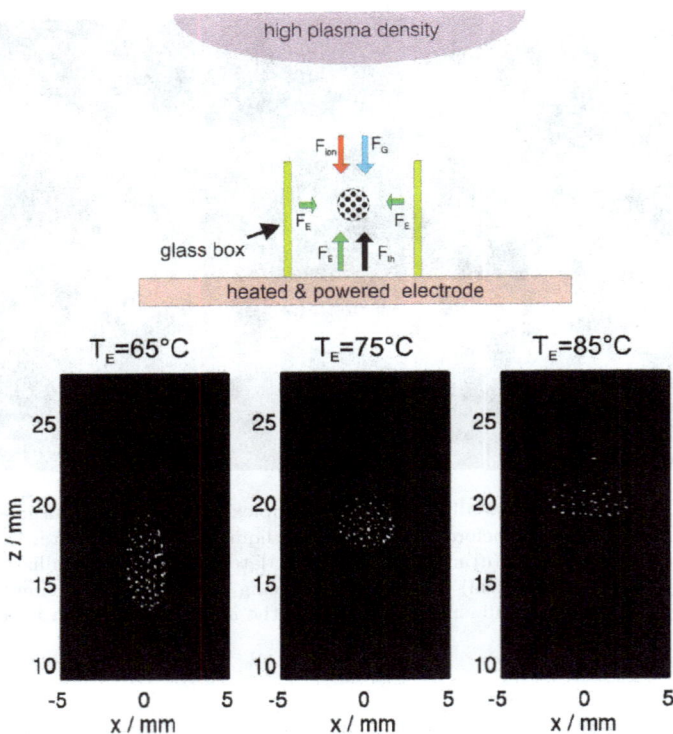

Fig. 5.4   3D particle clusters in complex plasmas. Top panel shows the sketch of the assembly for the formation of Yukawa (Coulomb) clusters by using thermophoresis. Different forces acting on the microparticles inside the confining glass box are indicated: gravity (F_g), ion drag (F_{ion}), electrostatic (F_E), and thermophoretic (F_{th}). Bottom panel demonstrates the vertical cross sections of clusters obtained at different temperature gradients. From Arp *et al.* (2005).

a combination with special confining system on the lower electrode, thermophoresis makes it possible to produce the so-called Yukawa (or Coulomb) balls – 3D spheroidal clusters of microparticles (see Fig. 7.22). By positioning a cubic glass box of a certain hight on the lower rf electrode and tuning the electrode temperature, it is possible to create a proper potential well for particles levitating inside the box (Arp *et al.*, 2005). A sketch of the apparatus and the balance of forces acting on particles in the trap are shown in Fig. 5.4 along with examples of the resulting particle clouds (for different heating temperatures). Using such a system, many interesting static and dynamic studies of "nanoclusters" can be performed.

5.1.1.2 *Microgravity experiments: PK-3 Plus setup*

Unlike "regular" plasma species – electrons and ions, microparticles are strongly affected by gravity (the same decoupling occurs in colloidal dispersions, see Sec. 3.5.2. and 3.5.3). As we pointed out above, a dc electric field in the rf sheath is usually employed to compensate for gravity, which provides favorable conditions to study 2D systems. However, in order to perform precision measurements with large isotropic 3D systems, microgravity conditions are absolutely necessary (Morfill *et al.*, 1999; Nefedov *et al.*, 2003).

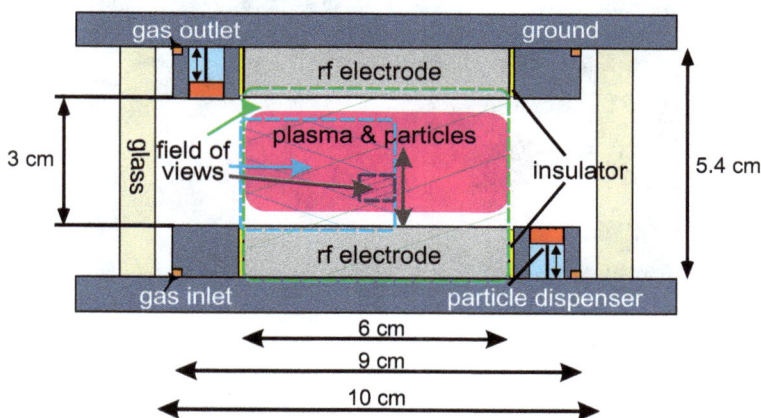

Fig. 5.5   The sketch showing a side view of the PK-3 Plus plasma chamber. From Thomas *et al.* (2008).

Under microgravity conditions, microparticles are pushed away from the sheath, towards the center of a chamber. In this case, a well-balanced symmetrically driven rf electrode system can provide a homogeneous plasma distribution with identical sheaths on both electrodes, which is required for a homogeneous distribution of microparticles. A cross-sectional schematic of the newest microgravity setup, the PK-3 Plus chamber ("PK" stands for "Plasma-Kristall"), is shown in Fig. 5.5 (Thomas *et al.*, 2008). The vacuum chamber is a glass cuvette of a cuboid form. The thermal homogeneity is very important in this case, because a temperature gradient of only 1 K/cm would give rise to a thermophoretic force on microparticles equivalent to $\sim 10^{-1}$ g, which would obviously destroy the microgravity conditions. Therefore the electrodes of the PK-3 Plus chamber are thermally coupled to the metal flanges. An insulator ring of high thermal conductivity prevents a temperature gradient between the electrodes and the structure. The metal parts of the chamber – the electrodes, flanges and connectors between the upper and lower flange – are manufactured from aluminum (which has high thermal conductivity), to further prevent a temperature drop across the chamber.

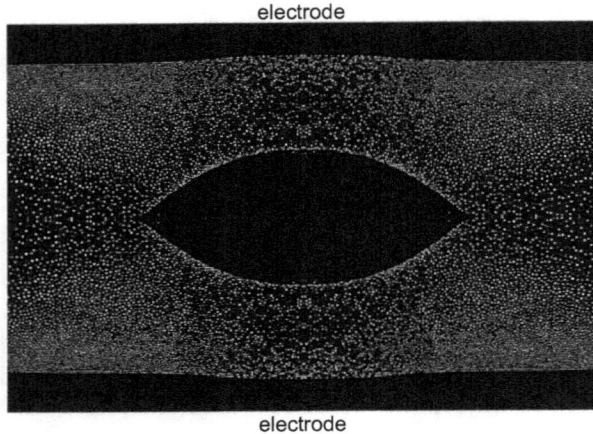

Fig. 5.6   Distribution of microparticles (3.4 μm in diameter) between the two electrodes under microgravity conditions. Particles are dispersed throughout the experimental volume, forming large 3D complex plasmas with the void in the center (see text). From Fortov and Morfill (2010).

The optical particle detection system consists of laser diodes with cylindrical optics, to produce a thin laser sheet (of $\simeq$ 150 μm thickness) perpendicular to the electrode surface, and video cameras observing the scattered light at 90° with different resolutions (different fields of view indicated in Fig. 5.5). The cameras and lasers are mounted on a horizontal translation stage allowing the depth scan, interference filters at the laser wavelength are used to filter out the plasma glow.

The formation of large homogeneous 3D systems can be inhibited by the void – the microparticle-free region in the center of the discharge shown in Fig. 5.6. The void is formed when the ion drag force, which pushes the particles outwards, overcomes the electric force in some vicinity of the discharge center. Under certain conditions the void can be closed, which is very important for many dedicated experiments (e.g., crystallization or demixing of binary fluids). With the PK-3 Plus setup, there are three presently known methods to close the void: (i) by reducing rf discharge power [as in the preceding PK setup (Lipaev *et al.*, 2007)]; (ii) by using a symmetrical gas flow; (iii) by low-frequency modulation of the electrode dc bias (Thomas *et al.*, 2008). The last method is also used to initiate a phase transition from an isotropic fluid to the so-called string-fluid phase (see Sec. 11.1).

5.1.2   *DC discharges*

5.1.2.1   *Ground-based experiments*

A dc gas discharge can also be conveniently used to study strongly coupled complex plasmas (Fortov *et al.*, 1996b; Lipaev *et al.*, 1997; Nefedov *et al.*, 2000). A sketch of a typical experimental setup is shown in Fig. 5.7. A discharge is usually generated in a vertically positioned cylindrical tube. Like rf discharges, here microparticles introduced into the plasma are illuminated by a vertical laser sheet and their individual positions are recorded from a side with a video camera. Typical experiments are performed for a gas pressure in the range of 10–600 Pa and discharge current of 0.1–10 mA.

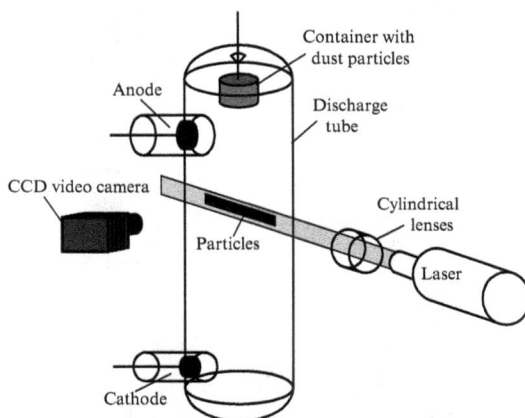

Fig. 5.7   Sketch of a dc experimental setup for studying strongly coupled complex plasmas. From Fortov and Morfill (2010).

Ordered structures of particles are observed in the regions where the axial electric forces are strong enough to balance gravity. The levitation conditions can be met in standing striations (formed in the positive column of the glow discharge), in an electric double layer (transition region from the narrow cathode part of the positive column to the wide anode part), in specially designed multi-electrode systems (having several electrodes at different potentials), etc.

Most experiments have been performed in standing striations of glow discharges – regions where plasma parameters exhibit spatial periodicity in the axial direction, with a characteristic length scale of about a few cm (Fortov *et al.*, 2004a). The concentration of electrons, their energy distribution, and the electric field inside striations are highly nonuniform. The electric field is relatively strong at the striation head (up to 10–15 V/cm, in a region occupying 25–30 % of the total striation length) and relatively weak (~ 1 V/cm) outside this region. The maximum value of the electron concentration is shifted relative to the field maximum towards the anode. The striation is essentially a 2D object, with typical center-to-wall potential difference (at the striation head) of about 20–30 V, which is quite sufficient to levitate microparticles.

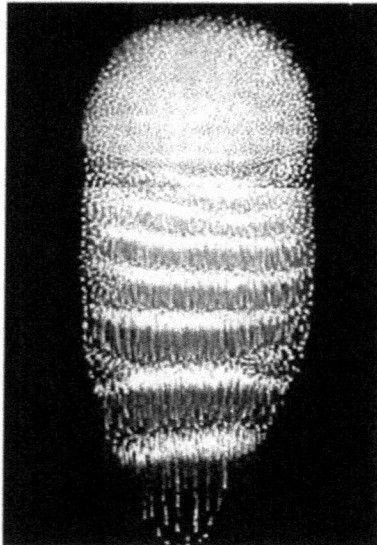

Fig. 5.8   Self-excited dust acoustic waves in a dc gas discharge at $p \simeq 30$ Pa neutral gas pressure. The wave frequency is $\omega \simeq 60$ s^{-1}, wave number $k \simeq 60$ cm^{-1} (the corresponding phase velocity is $\omega/k \simeq 1$ cm/s). From Fortov *et al.* (2000).

Experiments in dc discharges allow us to observe a variety of wave phenomena in complex plasmas. Using electrostatic manipulation techniques one can excite stable dust-acoustic waves (Barkan *et al.*, 1995; Thompson *et al.*, 1997; Merlino *et al.*, 1998). Under certain conditions inside striations, waves can be triggered due to the ion-streaming instability (Fortov *et al.*,

2000), as shown in Fig. 5.8. Also, solitons and shocks can be generated by implementing pulsed gas flow (Fortov *et al.*, 2004b; Molotkov *et al.*, 2004).

Furthermore, like complex plasmas in rf discharges, the transition from crystalline structures to fluids can be observed in dc discharges. This is achieved either by lowering the pressure or increasing the discharge current (Lipaev *et al.*, 1997). During this phase transition the interparticle distance remains approximately constant while the screening length decreases considerably, so that the screening parameter κ_p increases (see Fig. 7.2).

5.1.2.2 *Microgravity experiments: PK-4 setup*

The results obtained in ground-based experiments with dc discharges (Fortov *et al.*, 1996b; Lipaev *et al.*, 1997) were used as the basis for the first experiments to study dc complex plasma under microgravity conditions, which were performed onboard the Mir space station (Nefedov *et al.*, 2002; Fortov *et al.*, 2003). Note that under microgravity conditions the electric field of the positive column causes axial drift of negatively charged microparticles (in the direction of anode; in rf discharges, dc component of the electric field is practically absent near the center). In order to eliminate the drift, the additional electrode under the floating potential was used in the experiments.

The experiments on the Mir station confirmed that dc discharges can be used for studying complex plasmas under microgravity conditions. Therefore, a completely new concept for an experimental setup had been proposed – the PK-4 facility – which is planned for operation onboard the ISS starting in 2014 (Thoma *et al.*, 2007).

A schematic of the PK-4 experiment is presented in Fig. 5.9. The gas-discharge chamber is a 40 cm long U-shaped glass tube[1] with an inner diameter of 3 cm, the plasma is generated in neon gas at pressures between 10 Pa and 100 Pa. The setup is equipped with four particle dispensers, so that different combinations of monodisperse spherical microparticles of size between 1.5 μm and 15 μm can be used. In principle, nanoparticles produced in the discharge chamber as well as externally injected rod-like particles can also be studied. The particles are transported to the center of the discharge tube (with the help of the electric field of the dc discharge or by the gas flow), where they are illuminated by a laser sheet (17 mm$\times$0.15 mm) and record with video cameras at 120 frames per sec-

---

[1]Such a shape of the chamber was determined by a number of technical and scientific requirements, in particular by the restricted size of space setups.

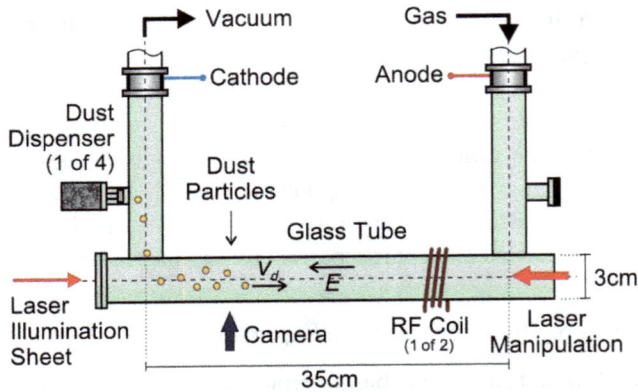

Fig. 5.9   Sketch of the PK-4 setup. From Thoma *et al.* (2007).

ond. When desired, rf coils can also be used to create a combined dc and rf discharge. The discharge can be sustained either in a pure dc regime or in the so-called "polarity switching" mode – when the discharge polarity is changing between the anode and cathode at a frequency of up to 1 kHz.

Unlike the PK-3 Plus (which is based on a rf discharge in a compact plasma chamber), the PK-4 setup is ideally suited to produce homogeneous elongated 3D clouds of microparticles, which is essential for various high-precision studies, e.g., experiments with shear flow (Sec. 10.1) or electrorheological plasmas (Sec. 11.1).

5.1.3   *Other types of complex plasmas*

Several alternative methods have been also proposed to generate complex plasmas, which were implemented in some experiments.

Thermal complex plasmas. In atmospheric-pressure plasmas, all species – including small liquid or solid particles – are often in local equilibrium and hence have equal temperatures (Sodha and Guha, 1971; Yakubov and Khrapak, 1989; Fortov *et al.*, 2006). An experimental study of the formation of ordered structures in such plasmas was carried out at temperatures of 1700–2200 K [see review by Fortov *et al.* (2004a) and references therein]. In principle, rather large dimensions of the available plasma volume along with uniform density distribution (typical volume of ~ 10 cm^3 and particle density $\sim 10^7$ cm^{-3}) help in eliminating boundary effects, which makes thermal plasmas favorable for equilibrium studies. However, issues related to temperature gradients and convection remain very severe in these systems.

Complex plasmas at cryogenic temperatures. Traditionally, the coupling parameter of electrons and ions is increased by cooling to temperatures ~ 1 mK. It is possible to create condensed crystalline ionic systems in cryogenic gas discharges (Fortov *et al.*, 2006), by employing laser cooling of atomic ions in non-neutral plasmas confined in Paul rf traps, storage rings, and Penning-type traps (Gilbert *et al.*, 1988). Complex plasmas at cryogenic temperatures were studied first by Fortov *et al.* (2002), where structures of polydisperse particles in rf and dc discharges cooled to liquid nitrogen temperatures were experimentally observed. Structural and dynamic properties of ordered microparticle structures are critically dependent on the temperature of heavy plasma species (atoms/molecules and ions) and, hence, on the discharge temperature – by decreasing it down to cryogenic level one can substantially reduce the plasma screening length. As a result, microparticles in cryogenic plasmas can form very dense crystalline structures (Fortov *et al.*, 2002; Asinovskii *et al.*, 2006).

Complex plasmas induced by UV radiation. One of the dominating dust charging mechanisms in space is photoemission. Under the influence of intense UV fluxes, microparticles can acquire positive electric charges of the order of $10^2 - 10^5$ e and form crystalline, liquid, or gaseous structures (Fortov *et al.*, 1998; Rosenberg *et al.*, 1996, 1999; Vaulina *et al.*, 2001). Detailed results of experimental studies of microparticles charged by photoemission under microgravity conditions (the charging was induced by solar radiation onboard the Mir space station) were presented by Fortov *et al.* (1998) and Vaulina *et al.* (2001, 2002b).

5.2   Colloidal Dispersions

Many colloidal dispersions occur naturally, for example – in biological systems where multiple components are suspended in an aqueous phase. However, experimental conditions for particle-resolved studies, namely control over interactions between particles, refractive index, and density matching of colloids and solvent, require the preparation of purpose-designed colloids. There are extensive reviews on colloid preparation, e.g., by Lyklema (2005), and therefore here we give merely the briefest overview to illustrate the main model systems used in experiments. Note that since the same kinds of particles are used in complex plasmas as well as in colloidal dispersions, the synthesis techniques are relevant to both fields.

Particle-resolved studies of colloids have their origin in a biological material: More than a century ago Jean Perrin (1913) used gum agar to demonstrate sedimentation-diffusion equilibrium and thus to measure Boltzmann's constant. Tens of thousands of particles were counted in this experiment by hand! Thankfully, computational techniques since have freed Ph.D. students from such laborious tasks. Below we summarize methods of 2D and 3D microscopy currently used for the particle tracking.

5.2.1   *Preparation of colloidal dispersions*

Colloids (as well as dust particles) are usually synthesized by polymerization of small molecules. These monomers may be miscible with the continuous phase in which the reaction is conducted (dispersion polymerization), or the monomers may themselves form droplets if they are immiscible with the continuous phase. Such droplets constitute an emulsion (which is itself a suspension of one liquid – the monomers, in another – the solvent), resulting in emulsion polymerization.

In both dispersion and emulsion polymerization, the reaction is started by an initiator and often (but not always) occurs at elevated temperatures (e.g., 80° C). In emulsion polymerization, the emulsion droplets become the colloids upon completion of the reaction. In the case of dispersion polymerization, on the other hand, while the monomers are miscible with the solvent, during the growth from oligomers to polymers they gradually become immiscible. These then nucleate small droplets, coalesce, and grow further through addition of unreacted monomers. At later stages, the second nucleation period can occur. Upon completion of the synthesis, colloids resulting from such secondary nucleation events are much smaller, leading to a bimodal size distribution.

In any case, both types of synthesis lead to a distribution of sizes – the *polydispersity*. This is typically quoted as the standard deviation in diameter normalized by the diameter, $\delta\sigma/\sigma$, and is usually about 4–5 % for "monodisperse" polymethyl methacrylate (PMMA) particles. Other components, such as fluorescent dyes, oligomers for steric stabilization (see Sec. 3.2), shells for "core-shell" particles, can require further reaction steps before or after the synthesis.

Once prepared, the colloids are then dispersed in a suitable solvent via agitation (if mixed up from dry) or by repeated centrifugation and solvent replacement (if the particles are to be kept wet). Density and index matching usually require that a mixture of two or more solvents are used. A salt may be added to this mix (to control the screening of the electrostatic interactions between colloids, see Sec. 3.2) along with other components, such as polymer (to induce depletion interactions, see Sec. 3.3).

5.2.2   *Optical/confocal microscopy*

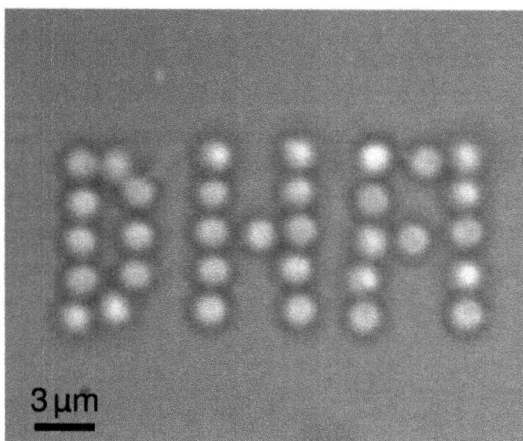

Fig. 5.10   Conventional optical microscopy imaging of colloids. A typical optical micrograph of 2 μm silica particles in an aqueous suspension is shown. "DHA" refers to dynamic holographic assembler, a sophisticated holographic optical tweezer technique for colloid manipulation. From Benito *et al.* (2008).

We assume the reader is familiar with optical microscopy. Conventional bright-field imaging is generally employed for (quasi) 2D systems. As Fig. 5.10 shows, there is ample contrast (without refractive index matching) to produce an image from which particle coordinates may be extracted.

Despite being proposed a long time ago by Minsky (1957), the potential of confocal microscopy was not realized until the 1980s. In the meantime, laser light sources were developed which compensate for low-level illumination inherent to the technique, along with image-processing software and the hardware required to run it. Confocal microscopy has two key advantages over conventional microscopy: Improvement in resolution and rejection of out-of-focus blur, enabling 3D imaging of thick specimens (Sheppard and Shotton, 1997; Ribbe, 1997).

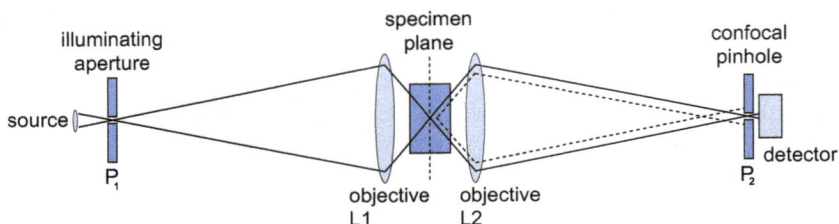

Fig. 5.11   Principle of confocal microscopy: Transmission mode. Light is focused into a point in the sample by the condenser lens. The confocal pinhole rejects all light except that from the point in focus (dotted lines show reflection of light from out-of-focus regions). From Sheppard and Shotton (1997).

Image formation in the confocal microscope differs from conventional optical microscopy, because it uses both a point detector and source. Figure 5.11 shows a schematic of such an instrument, in which light is transmitted through the sample. The figure shows that only a single (diffraction-limited) point of light from the focal plane reaches the detector: The confocal pinhole P_2 discriminates against light from all other points of the focal plane, and from out-of-focus planes.

The condition of a single point in focus requires that the specimen (or, more usually, the probe beam and pinhole) are scanned to produce an image. The fact that only a single point contributes to the image at any instant means that the overall intensity is very low. The intensity of each point (and hence the rate of scanning) can be increased with the use of a laser light source, the confocal laser scanning microscope (CLSM). The use of multiple pinholes in a Nipkov disc massively enhances the light throughput, enabling scanning rates in the plane of up to 10^3 frames per second (scanning rates on conventional CLSMs are of order 1-10 frames per second). If the specimen is scanned in all three directions, a 3D image may be constructed, which is not directly possible with wide-field microscopy – this requires image restoration via deconvolution (Pawley, 2006).

We see from Fig. 5.11 that if the confocal pinhole is removed, then all light transmitted by the entire field of view reaches the detector. In this case we have a conventional wide-field microscope without the ability to reject out-of-focus light.

In practice, the transmission mode of Fig. 5.11 is seldom used. *Epi-illumination* is utilized instead (where the condenser and objective are the same lens), as shown in Fig. 5.12. Although this does not in principle affect the confocal optics, alignment of various components is far easier, as

23, 2012   9:2

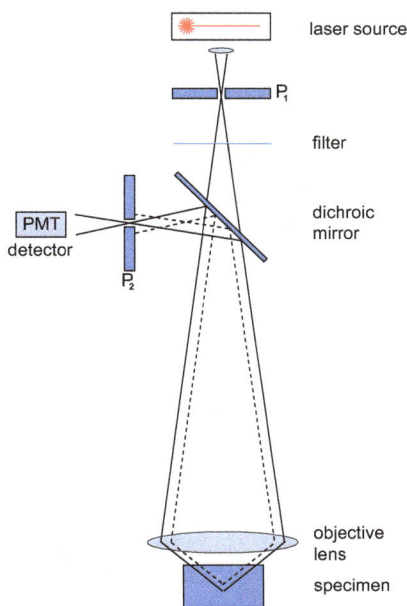

Fig. 5.12   *Epi*-illumination in confocal mode. The objective and condenser lenses are the same, and the dichroic mirror allows the illuminating beam to pass through, while reflecting the returning beam to the detector. Light is focused into a point in the sample by the condenser lens. From Sheppard and Shotton (1997).

L_1 and L_2 are replaced by a single lens. The work discussed here uses confocal microscopy in epi-fluorescence mode. Colloids are labeled with a fluorescent dye, whose molecules are excited by a light source of appropriate wavelength, and then decay emitting a photon of longer wavelength (Pawley, 2006). A filter is used to reject light that is simply reflected, whereas fluorescent light is passed through to the detector. The resulting image is then a map of the fluorescent dye in the specimen, i.e., of the colloids (if necessary, it is also possible to label the solvent rather than the colloids). Removing the filter allows reflected light to reach the detector, which is thus termed "reflection mode". In this case, contrast is generated by refractive index mismatch in the sample.

5.2.2.1   *Application to colloids*

The first documented use of confocal microscopy to study colloids is that of Yoshida *et al.* (1991), who looked at lattice spacing in crystals of charged colloids. They analyzed their images in 2D to obtain the in-plane coordinates of the colloids in the crystalline layers as a function of the distance to the sample cell wall, forming the basis of the real-space analysis technique. However, confocal microscopy is distinguished as a 3D technique and hence full 3D resolution of coordinates is desirable. This was first achieved by van Blaaderen and Wiltzius (1995) and van Blaaderen *et al.* (1992) who studied the 3D structure of colloidal glasses and crystals at the single-particle level. Owing to the high packing fractions used ($\phi \simeq 0.6$), particles were often touching (at least, the typical surface separation was below the resolution of the microscope). To circumvent the problems this might cause for particle tracking, van Blaaderen and Wiltzius used core-shell colloids shown in Fig. 5.13, where only the core was fluorescently labeled. Subsequently, real-space analysis was employed to follow particles in time, to provide direct evidence of dynamic heterogeneity in colloidal supercooled liquids in 2D (Kegel and van Blaaderen, 2001) and 3D (Weeks *et al.*, 2001). These works used particle tracking algorithms similar to those of Crocker and Grier (1995), which are discussed in the next section.

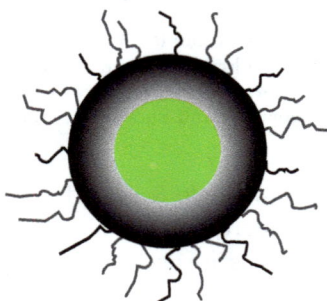

Fig. 5.13   Core-shell PMMA colloid suitable for confocal microscopy. The core is labeled with a fluorescent dye, the shell is undyed PMMA. Steric stabilization is provided by the polyhydroxyl stearic acid stabilizer hairs which are not drawn to scale. The stabilizer layer is typically about 10 nm, the colloids are typically 2 μm in diameter.

5.3   Real Space Analysis

Here we summarize methods of particle tracking in space and time, which are commonly used in both complex plasma and colloidal dispersion experiments. In conclusion, we also make a few remarks on recent developments in the particle tracking technique.

5.3.1   *Spatial particle tracking*

Most particle tracking algorithms are similar to the schematics illustrated in Fig. 5.14 (Royall *et al.*, 2003). Ideally, the brightest pixel of the digitized particle image would lie at the center of the object and provide a reasonable approximation for the coordinate (Fig. 5.14a). Taking the first moment of the pixel intensities within the dashed circle (the centroid) leads to a better approximation for the particle coordinates,

$$\mathbf{R} = \frac{\sum_i I_i \mathbf{r}_i}{\sum_i I_i},$$

where $\mathbf{r}_i$ is the coordinate and I_i is the intensity (above the background level) of the ith pixel (Crocker and Grier, 1995).

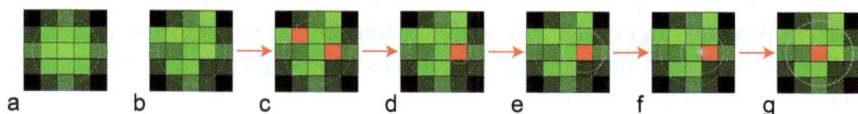

Fig. 5.14   Particle tracking in 2D. (a) Ideal image with brightest pixel in center. (b-g) In practice, noise leads to multiple local maxima. By iteratively taking the centroid of an increasing circle, the identified center is "pulled" towards the middle of the object.

In practice, noisy data lead to multiple local maxima (bright pixels) which typically do not lie at the center of the object (Fig. 5.14b). One may then choose one local maximum which lies within the region associated with a particle and exclude others, for example based on their intensity. Taking a centroid around this brightest maximum, one can iteratively move towards the center of the object (Royall *et al.*, 2003). Noise is typically tackled by applying a Gaussian blur to the image, or some similar form of filtering [e.g., moments can be calculated by fitting intensity profiles with quadratic polynomials (Ivanov and Melzer, 2007)]. Note, however, that such filtering typically does not remove multiple maxima, nor maxima located away from the particle center. Furthermore, in order to avoid distorted images a high frame rate is required, such that particle displacement during each frame is sufficiently small.[2]

---

[2]For further details of particle tracking, the interested reader is referred to Eric Weeks' web site http://www.physics.emory.edu/~weeks/idl/.

5.3.2   *Tracking in time*

Complex plasmas and colloidal dispersions are prized for their experimentally accessible dynamics. This brings a new dimension to real-space analysis, as the tracking of each particle as a function of time allows, for example, the determination of transport properties (Ohtsuka *et al.*, 2008), relaxation times (Kegel and van Blaaderen, 2001; Weeks *et al.*, 2001) and lattice fluctuations in crystals (Royall *et al.*, 2003, 2006; Reinke *et al.*, 2007).

Particle tracking in time is a non-trivial problem. Unlike computer simulation, where each particle is labeled, here one seeks to identify trajectories between frames sampled at different times. Since the identity of each particle is not known a priori, a maximum-likelihood approach is used. For stochastic systems, it can be shown that performing the minimization

$$\sum_{i,j} (\mathbf{r}_i - \mathbf{r}_j)^2 \to \min,$$

(where i and j are permutations of the coordinates tracked in consecutive frames) provides an effective strategy for identifying trajectories of individual particles. This algorithm is broadly employed for particle tracking in time (Kegel and van Blaaderen, 2001; Weeks *et al.*, 2001).

5.3.3   *Non-equilibrium systems*

The methods so far discussed assume that the system is quiescent. Often this is not the case, for example when the system is under shear (Besseling *et al.*, 2007) or is otherwise driven [e.g., by an electric field (Vissers *et al.*, 2011a,b)]. In these situations, more sophisticated algorithms are required. For instance, in the case of affine shear one knows *a priori* the velocity in each plane. Thus, it is possible to subtract the local velocity as a function of the transverse coordinate z [in the direction of shear (Besseling *et al.*, 2008)]. Having removed the movement originating from the shear, one may now proceed as before in the case of quiescent systems.

Non-affine shear, in particular phenomena such as shear banding, present a more challenging situation. Here, however, one may track 2D coordinates in each plane and thus develop a measure of the mean velocity as a function of z. Once this non-affine velocity is subtracted, one may again proceed as discussed above. Alternatively, one can use cross-correlation of Fourier-transformed 2D images as an accurate means by which drift or local velocity is removed (Derks *et al.*, 2004). Such methodology may also be applied to colloids moving under electric fields (Vissers *et al.*, 2011a,b), where Poiseuille flow leads to a parabolic velocity profile.

5.3.4   *Recent developments*

Complex plasmas.   While the characteristic diffusion time $\sim \sigma^2/D_0$ in overdamped colloidal dispersions is rather large (of the order of seconds), typical dynamic timescales in complex plasmas $\sim \Delta/v_T$ are 1-2 orders of magnitude shorter.   This imposes a very serious technical challenge for time-resolved 3D particle tracking in complex plasmas.

There have been several techniques proposed so far for 3D tracking. In a few experiments (Annaratone *et al.*, 2004; Antonova *et al.*, 2006), a cloud was illuminated by a thick laser beam with a color gradient due to mixed light from two lasers operating at different wavelengths. The depth measure given by the gradient was determined by using two video cameras. Other methods employ a parallel shifting of the illuminating (narrow) laser sheet, either by using a pair of synchronized rapidly oscillating mirrors (Samsonov *et al.*, 2008) or by moving a translation stage (Pieper *et al.*, 1996). While the latter technique is too slow to follow particles in time (unless they form a stable crystal), the other methods are rather complicated from the experimental point of view. In addition, they require significant computational efforts to reconstruct instantaneous 3D positions and velocities.

A very different approach to obtain time-resolved 3D data is to use so-called digital in-line holography – a method which provides encoding of 3D information in a single holographic image. This approach has been successfully used in fluid dynamics, and recently was also employed to measure 3D particle coordinates in complex plasmas, demonstrating quite encouraging results (Kroll *et al.*, 2008, 2010). At the moment, the major drawbacks of this method are the need to use very large particles ($\gtrsim 10$ μm, so that special hollow spheres must be employed to provide levitation in experiments on ground) and rather small working volumes.

Yet another approach to obtain 3D dynamic information is to apply stereoscopic particle image velocimetry (PIV) technique (Thomas, 1999; Thomas *et al.*, 2004), where the velocity vectors in a narrow slice are deduced by flashing two ns laser pulses (with a typical interval in a ms range) and recording the scattered light with two synchronized video cameras. Such technique was recently extended to the so-called tomographic PIV (Williams, 2011), where the thickness of the illuminated volume is significantly increased in comparison with the stereoscopic PIV. We note, however, that the PIV technique does not provide particle-resolved information – all characteristics are averaged over small groups of particles, yielding a coarse-grained picture.

Colloidal dispersions. While the algorithms introduced to the field by Crocker and Grier (1995) have undoubtedly formed the backbone of most work, it is important to note more sophisticated approaches. Among the pioneers of the field, van Blaaderen and Wiltzius (1995) sought to improve the accuracy of coordinates tracked in the axial direction by fitting a Gaussian to the integrated intensities in each plane that constituted a given colloid. More recently, Jenkins and Egelhaaf (2008) pushed the limits of the technique when they sought to identify contacts between colloids in a dense sediment. Their approach delivered ultra-high precision coordinate location. They achieved this by determining an empirical image of a colloid (similar to a point-spread function for a colloid). The original image was then reconstructed by placing a point-spread function on each coordinate determined by conventional means (see Sec. 5.3.1). The coordinates on which the point-spread functions are centered are then refined to give an optimal agreement between the reconstructed image and the original image. The advantage of this approach is that no *a priori* assumption is made about the microscope optics. This is very helpful, since the image of a colloid depends on the individual synthesis.

Other more exotic developments include that of Lu *et al.* (2007), who were able to follow events in real time (typically, due to the computational efforts involved, this is done off-line). This is important, since the region of interest can often drift out of the field of view (which is usually quite small due to large file sizes that time-resolved 3D imaging generates). Lu *et al.* (2007) made a number of impressive innovations which enabled the analysis to be carried out in real time. This tracking then allowed feedback to be applied to the stage, such that the region of interest could be followed should it drift out of the field of view.

We close this section by noting some of the most important current developments in the field of colloid physics, namely the progress in tracking of anisotropic particles (Glotzer and Solomon, 2007). Thus far we have implicitly been discussing spheres. Since a sphere has only three degrees of freedom, simply determining the coordinates is sufficient to fully describe spherical colloids. Ever-present polydispersity is typically neglected, except in the simple case of binary systems where two-color imaging enables the two species to be differentiated (Leunissen *et al.*, 2005; Royall *et al.*, 2007c). Mukhija and Solomon (2007) have made important developments in colloidal rods, and also in the imaging of such systems. Their approach enabled not only the center-of-mass coordinate to be found, but also orientation and rotational diffusion to be measured.

Chapter 6

Simple Liquids

Of the three basic states of matter, liquids require far more sophisticated theoretical models than gases or solids. The former may be treated to a first approximation as an ideal gas, the latter by a harmonic bead-spring model. No such straightforward analytical approaches exist for liquids which combine the disorder of gases and the strong correlations of solids. In other words, there is no obvious smallness parameter to be implemented, which makes the description of the liquid state rather challenging (Hansen and MacDonald, 2006).

In this chapter we focus on simple liquids (of one species), where classical particles interact via a spherically symmetric potential. Before presenting the main results from particle-resolved studies, we briefly summarize the basic structural and dynamical properties of such liquids.

6.1   Liquid Structure

Structure is one of the principal characteristics of the liquid state. Packing effects lead to a succession of shells around each particle, which gradually lose correlations with increasing distance from the central particle. The simplest model which captures this behavior are hard spheres. Experiments began with Bernal's ball bearings, which provided important insights into the liquid structure (Bernal, 1959). Unlike colloids and complex plasmas, ball bearings have negligible thermal energy.

Systems with hard-sphere or other repulsive interactions (such as the one-component Yukawa system) cannot have a liquid-vapor phase transition, but rather form highly correlated fluids at high densities. Nevertheless, many relevant properties of liquid structure and dynamics can still be captured when only repulsion is taken into account (Weeks *et al.*, 1971), which suggests that packing effects generally dominate. Therefore, we start this chapter with a general consideration of strongly coupled fluid systems, and then move on to the role of attractive interactions driving liquid-vapor condensation (in Sec. 6.3).

The development of particle-resolved studies enables a unique opportunity to measure liquid structure directly, in real space. To illustrate this point, let us consider N particles occupying the volume V: The instantaneous microscopic density is $n(\mathbf{r}, t) = \sum_i^N \delta[\mathbf{r} - \mathbf{r}_i(t)]$, while the mean microscopic density in equilibrium is determined by $n^{(1)}(\mathbf{r}) = \langle \sum_i^N \delta(\mathbf{r} - \mathbf{r}_i) \rangle$ (where $\langle \ldots \rangle$ is the canonical ensemble-average, equivalent to the time-average in equilibrium). The basic structural measure of the liquid state is the pair correlation (radial distribution) function $g(\mathbf{r})$ which is defined as (Hansen and MacDonald, 2006; March and Tosi, 2002)

$$g(\mathbf{r}) = \frac{1}{nN} \left\langle \sum_{i \neq j}^N \delta\left(\mathbf{r} - \mathbf{r}_i + \mathbf{r}_j\right) \right\rangle,$$

(where $n = N/V$ is the mean macroscopic number density). For isotropic liquids, correlations depend only on the absolute value of the radial distance. Typical examples of $g(r)$ obtained by direct measurement of particle positions in liquid colloidal dispersions and complex plasmas are presented in Fig. 6.1. It has been shown that there is a unique one-to-one correspondence between $g(r)$ and the effective pair potential $V(r)$ for a homogenous system of particles at a given density and temperature (Henderson, 1974; Chayes and Chayes, 1984).

Another important measure of pair correlations is the direct correlation function $c(r)$ which is introduced via the Ornstein-Zernike (OZ) relation (March and Tosi, 2002; Barker and Henderson, 1976),

$$h(r) = c(r) + n \int d\mathbf{r}' c(|\mathbf{r} - \mathbf{r}'|) h(r').$$

The OZ relation reflects the fact that in the framework of pair correlations the "total" correlation between two points separated by distance r [and represented by $h(r) = g(r) - 1$] is a sum of "direct" correlation [described by $c(r)$] and "indirect" correlations propagated via increasingly large numbers of intermediate particles (Hansen and MacDonald, 2006). The direct correlation function depends on $V(r)$ through a *closure relation*. There have been different closures proposed for different classes of interaction potentials (Hansen and MacDonald, 2006; Barker and Henderson, 1976).

From the one-to-one correspondence between $g(r)$ and $V(r)$ one can conclude that if two correlation functions are the same then the corresponding

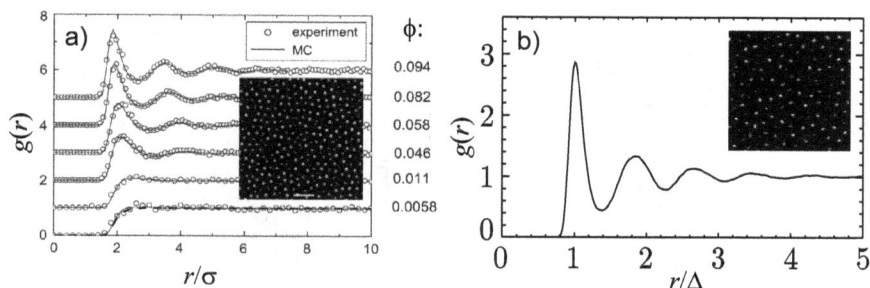

Fig. 6.1   Liquid structure obtained from particle-resolved studies. The radial distribution functions $g(r)$ calculated from 3D particle positions in charge-stabilized colloids [a), from Royall *et al.* (2003)] and complex plasmas [b), courtesy of P. Huber]. The results in (a) are for different packing fractions ϕ (as indicated, r is normalized by the colloid radius σ); in (b), the (estimated) screened coupling parameter is $\Gamma^{(s)} \simeq 0.94\Gamma_m^{(s)}$ [see Eqs (7.1) and (7.2), r is normalized by the mean interparticle distance Δ]. The insets show instantaneous particle positions (thin slices) representing fluid states used to calculate $g(r)$ ($\phi \simeq 0.094$ for colloids).

effective pair interactions are identical. For example, Fig. 6.1a shows that $g(r)$ obtained from Monte Carlo simulations of a Yukawa system (with the interaction parameters $\epsilon_Y/k_BT = 140$ and $\kappa\sigma = 5.0$) are indistinguishable from experimental data. However, since the pair correlations in dense systems are dominated by the hard core [Weeks *et al.* (1971)], the accuracy with which the interactions can be extracted from $g(r)$ (measured with finite errors) diminishes at higher density.

6.2   Liquid Dynamics

Like their structure, the dynamics of liquids are fundamentally different to those of gases (where there are occasional collisions) and crystalline solids (where atoms primarily oscillate around their lattice positions).

The basic measure of liquid dynamics is the mean squared displacement of a single particle versus time, $\delta r^2(t) = \langle|\mathbf{r}(t) - \mathbf{r}(0)|^2\rangle$ – the dependence which is approximated by the scaling $\delta r^2 \propto t^\alpha$, where the exponent $\alpha > 0$ is generally a slowly varying function of time. The diffusion is referred to as "normal" when $\alpha \simeq 1$, and is called "subdiffusion" for $\alpha < 1$ or "superdiffusion" for $\alpha > 1$. The dynamics of colloids is Brownian on all meaningful timescales due to strong viscous damping caused by the solvent (see discussion in Sec. 4.2). Therefore, the diffusion in colloidal liquids is "normal" both at very short times (corresponding to a "free" motion within

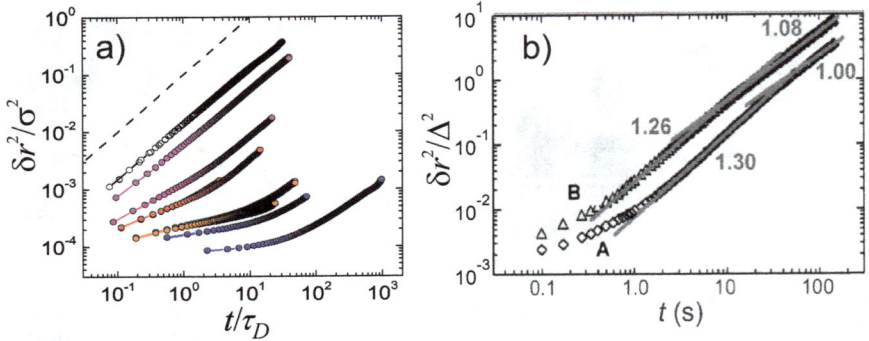

Fig. 6.2   Mean squared displacement of particles $\delta r^2(t)$ measured in experiments with 3D colloidal liquids (a) and 2D liquid complex plasmas (b) [from Lai and I (2002)]. In (a), different curves correspond to different values of packing fraction ϕ (increasing from top to bottom), distance and time are normalized by the particle diameter σ and Brownian diffusive timescale τ_D, respectively, the dashed line denotes the slope $\alpha = 1$. In (b), two curves represent two experiments performed at different temperatures, numbers indicate values of the effective diffusion exponent α, distance is normalized by the mean interparticle separation Δ.

the shell of neighbors – "cage") and at asymptotically long times, while the intermediate stage (due to in-cage motion) is usually subdiffusive. A typical example of $\delta r^2(t)$ is shown in Fig. 6.2a.[1] Liquid complex plasmas, on the other hand, often reveal a sequence of super- and subdiffusive regimes: At timescales below the damping time, their dynamics is more akin to atoms and, hence, the "free" motion within the cage is ballistic ($\alpha \simeq 2$). This normally crosses over to subdiffusion (in-cage oscillations) followed by a gradual transition to normal diffusion at asymptotically long times (Lai and I, 2002; Nunomura *et al.*, 2006). Figure 6.2b illustrates the diffusive behavior of liquid complex plasmas (the initial ballistic phase is not shown here). One can see that the eventual transition to normal diffusion at long timescales is mediated by a noticeable superdiffusive stage.

This distinction in the diffusive behavior observed in strongly coupled colloidal dispersions and complex plasmas is associated with different dynamic regimes (see Sec. 4.2). Recent computer simulations by Ott and Bonitz (2009) demonstrated that in weakly damped complex plasmas the transient stage (associated with finite damping) can be rather long and is usually characterized by effective superdiffusive behavior, as shown in

---

[1]In supercooled liquids the intermediate plateau becomes very extended, so that the long-time diffusion regime eventually disappears when glassy state is reached (see Sec. 9.1.1).

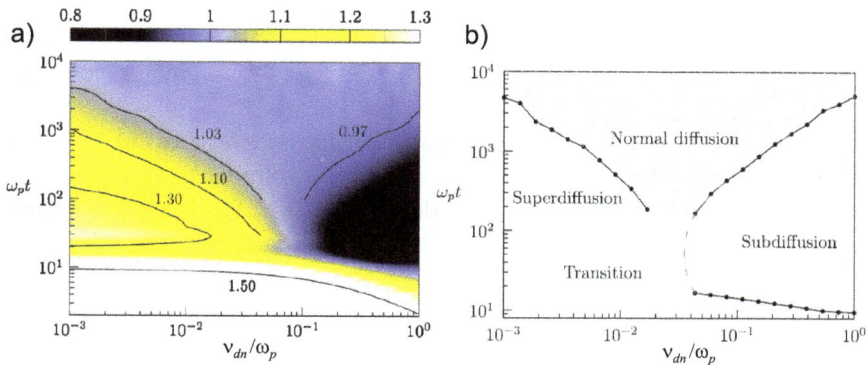

Fig. 6.3   Effective diffusion exponent α in 2D complex plasmas. Shown are contour plots of α in the plane $(\omega_p t, \nu_{dn}/\omega_p)$, where time t and friction coefficient ν_{dn} are normalized with the plasma frequency ω_p. Results are for the plasma screening parameter $\kappa_p = 3$ and (Coulomb) coupling parameter $\Gamma = 0.16\Gamma_m$ [where Γ_m is the value at the melting line, see Eqs. (7.1) and (7.2)]. (a) Gray-scale representation with contour lines for diffusion exponent $\alpha = 1.50, 1.30, 1.10, 1.03, 0.97$. (b) Schematic representation with labeled regions of super-, sub-, and normal diffusion. From Ott and Bonitz (2009).

Fig. 6.3.[2] This example is very instructive in demonstrating the role of short-time particle dynamics.

We note that several complex plasma experiments [e.g., see references in the paper by Ott and Bonitz (2009)] demonstrated persisting superdiffusive long-time behavior. However, one should bear in mind that so far such behavior has been observed either in relatively small and inhomogeneous systems [e.g., experiment by Ratynskaia *et al.* (2006)], or in systems with noticeable large-scale flow [e.g., experiment by Liu and Goree (2008)]. One should also mention that the long-time behavior of colloidal dispersions is always diffusive Dhont (2003).

While averaged quantities, such as the mean squared displacement, can be inferred from scattering data (van Megen *et al.*, 1998), observations at the individual-particle level are highly desirable in order to get more detailed structural and dynamic information. For instance, particle-resolved studies allow us to directly obtain the van Hove correlation function $G(\mathbf{r}, t)$ (Hansen and MacDonald, 2006) defined as[3]

---

[2]The normalization here is by the plasma frequency $\omega_p = \sqrt{2Q^2/ma_{WS}^3}$, which is expressed via the Wigner-Seitz radius $a_{WS} = (\pi n_{2D})^{-1/2}$.

[3]The product of two δ-functions is reduced in $G(\mathbf{r}, t)$ to single δ-function by utilizing the translational invariance in space and integrating over one coordinate.

$$\frac{1}{n}\langle n(\mathbf{r},t)n(\mathbf{0},0)\rangle = \frac{1}{N}\left\langle \sum_{i,j}^{N} \delta\left[\mathbf{r} - \mathbf{r}_i(t) + \mathbf{r}_j(0)\right]\right\rangle \equiv G(\mathbf{r},t). \qquad (6.1)$$

The van Hove function describes density-density correlations in space and time, i.e., $G(\mathbf{r},t)d\mathbf{r}$ is the number of particles in a volume $d\mathbf{r}$ around a point $\mathbf{r}$ at time t, provided there was a particle at $\mathbf{r} = \mathbf{0}$ at $t = 0$. The van Hove function consists of "self" and "distinct" parts, $G(\mathbf{r},t) = G_\mathrm{s}(\mathbf{r},t) + G_\mathrm{d}(\mathbf{r},t)$,

$$G_\mathrm{s}(\mathbf{r},t) = \frac{1}{N}\left\langle \sum_{i}^{N} \delta\left[\mathbf{r} - \mathbf{r}_i(t) + \mathbf{r}_i(0)\right]\right\rangle,$$

$$G_\mathrm{d}(\mathbf{r},t) = \frac{1}{N}\left\langle \sum_{i\neq j}^{N} \delta\left[\mathbf{r} - \mathbf{r}_i(t) + \mathbf{r}_j(0)\right]\right\rangle,$$

which describe the evolution of single-particle distribution and pair correlations, respectively, so that $G_\mathrm{s}(\mathbf{r},0) = \delta(\mathbf{r})$ and $G_\mathrm{d}(\mathbf{r},0) = ng(\mathbf{r})$.

The density-density correlation function in $\mathbf{k}$-space is obtained from Eq. (6.1) by the Fourier-transformation of the van Hove function,

$$\frac{1}{n}\langle n_\mathbf{k}(t)n_{-\mathbf{k}}(0)\rangle = \int d\mathbf{r}\, G(\mathbf{r},t)e^{-i\mathbf{k}\cdot\mathbf{r}} \equiv F(\mathbf{k},t), \qquad (6.2)$$

[where $n_\mathbf{k}(t) = \sum_i e^{-i\mathbf{k}\cdot\mathbf{r}_i(t)}$ is the instantaneous particle density in Fourier space]. Equation (6.2) introduces the co-called intermediate scattering function (ISF), $F(\mathbf{k},t)$, which is an important characteristics of the liquid dynamics (see Sec. 9.1.1). The ISF can be deduced from scattering experiments (Hansen and MacDonald, 2006; March and Tosi, 2002): The static structure factor, $S(\mathbf{k}) = F(\mathbf{k},0)$, is measured in (X-ray or neutron) diffraction experiments, and is related to the pair correlation function via $S(\mathbf{k}) = 1 + n\int d\mathbf{r}\,[g(\mathbf{r}) - 1]e^{-i\mathbf{k}\cdot\mathbf{r}}$ (unity is subtracted under the integral to facilitate convergence, and the δ-function at $\mathbf{k} = \mathbf{0}$ is omitted). Furthermore, the power spectrum of $F(\mathbf{k},t)$ yields the dynamic structure factor,

$$S(\mathbf{k},\omega) = \frac{1}{2\pi}\int dt\, F(\mathbf{k},t)e^{i\omega t}, \qquad (6.3)$$

which is proportional to the cross section measured in inelastic neutron scattering experiments.

We note that results from particle-resolved experiments on these and related quantities remain rare. Early work in 2D systems by Marcus *et al.*

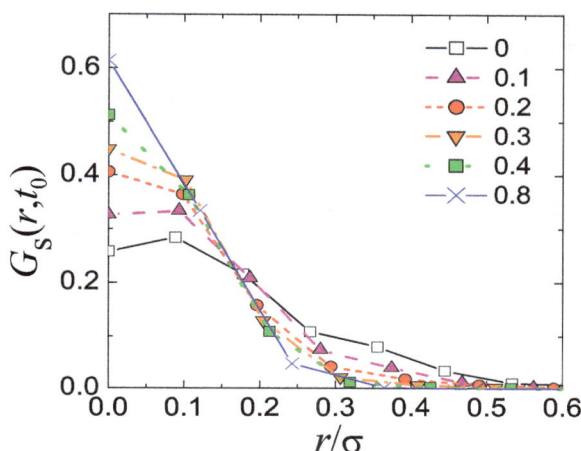

Fig. 6.4   Probability distribution of the particle displacement in colloidal dispersions. Shown is the van Hove self-correlation function $G_s(r, t_0)$ calculated at time $t_0/\tau_D \sim 0.1$. State points refer to the amount of polymer (in g/l) added to induce attractions. From Ohtsuka *et al.* (2008).

(1996) reported measurements of the ISF and van Hove functions. Similar 3D quantities shown in Fig. 6.4 were obtained for a system with attractive interactions caused by depletion forces (Ohtsuka *et al.*, 2008). One can see that attraction between colloids at higher polymer concentrations result in enhanced clustering, leading to much stronger spatial correlations in the particle motion and, generally, to a significant slowdown in dynamics.

6.3   Liquid-Vapor Phase Transition

Colloid-polymer mixtures are a suitable model system to study the liquid-vapor transition at the individual-particle level (Royall *et al.*, 2007a). The effective interaction between colloids in this case is the AOV pair potential described by Eq. (3.8). The attractive part of the AOV potential can be tuned by changing the polymer concentration, which makes it possible to broadly explore the phase diagram.[4]

---

[4]Colloid-polymer mixtures can be considered as a special case of binary mixtures discussed in detail in Sec. 8.3.

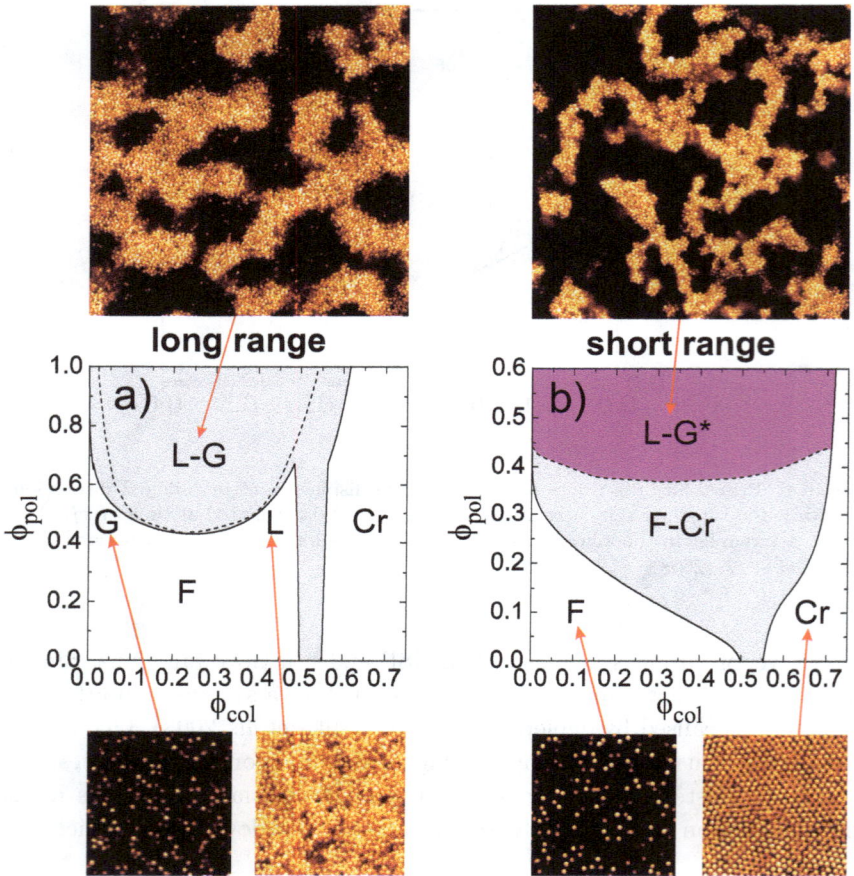

Fig. 6.5   Phase diagrams of colloid-polymer mixtures. The effective interactions between colloids are described by the AOV potential [Eq. (3.8)], the diagrams are plotted in the plane spanned by the volume fractions of polymers, ϕ_{pol}, and colloids, ϕ_{col}. Left panel shows the phase diagram for relatively long-ranged attraction: At lower ϕ_{col} the fluid (F) system exhibits phase separation into liquid (L) and gas (G) upon quenching through the coexistence (solid) line, for higher ϕ_{col} the system is crystalline (Cr). Right panel represents the case of short-ranged attraction. In this case the liquid-gas phase separation is metastable (L-G*) and dynamical arrest of the colloid-rich phase results in gel formation.

Figure 6.5 demonstrates two characteristic regimes of the phase diagram for colloid-polymer mixtures. The left panel represents a system with a relatively large polymer-colloid size ratio of $q = 0.45$, resulting in "long-ranged" attraction. It is known that for such mixtures (viz., for $q \gtrsim 0.3$)

the colloidal gas and liquid are thermodynamically stable (Lekkerkerker *et al.*, 1992). Quenching into the spinodal decomposition region in Fig. 6.5a (above the dashed line) results in the characteristic patterns (depicted in the frame above) where both single-particle and mesoscopic coarsening length scales are seen. On the other hand, for sufficiently short-ranged interactions (obtained by reducing the relative polymer size) the liquid-gas phase separation is no longer thermodynamically stable. This is because reducing the attraction range leads to a denser liquid (Elliot and Hu, 1999), so that eventually the density reaches the critical value corresponding to the hard-sphere freezing and the liquid becomes metastable to crystallization. This situation is presented in Fig. 6.5b (corresponding to $q = 0.18$), where the "liquid" density is sufficiently high for dynamical arrest to occurs (see Chapter 9) and, hence, for a gel to be formed (as shown in the frame above).

Particle-resolved studies of the liquid-vapor phase transition with colloid-polymer mixtures were hampered for some time by the need to use sufficiently large colloids (so that they can be observed individually). In turn, this requires rather large polymers, to ensure that the attraction range is sufficient to avoid gelation. A match was found with 1.3 μm colloids and 3.1×10^7 molecular weight polymers, where q was about 0.4 (Royall *et al.*, 2007a). This revealed the equilibrium liquid-vapor interface at the single-particle level, as shown in Fig. 6.6.

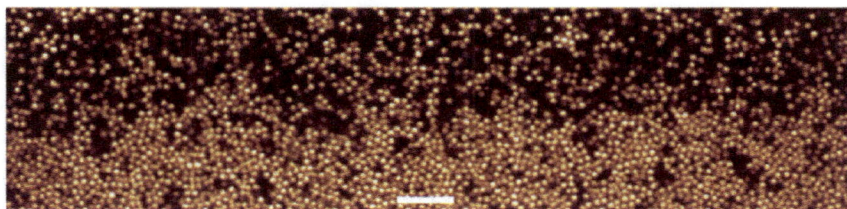

Fig. 6.6   The particle-resolved liquid-vapor interface observed in colloid-polymer mixture. From Royall *et al.* (2007a).

The tunability of the interactions in colloid-polymer mixtures enabled a wide range of effective temperatures to be accessed – far more than in molecular systems. Since the effective temperature is set by the inverse of the polymer concentration (see Fig. 6.5) and the latter can be freely adjusted, the range of effective temperatures is limited only by deviations from the simple depletion picture at high polymer concentration.

Complex Plasmas and Colloidal Dispersions

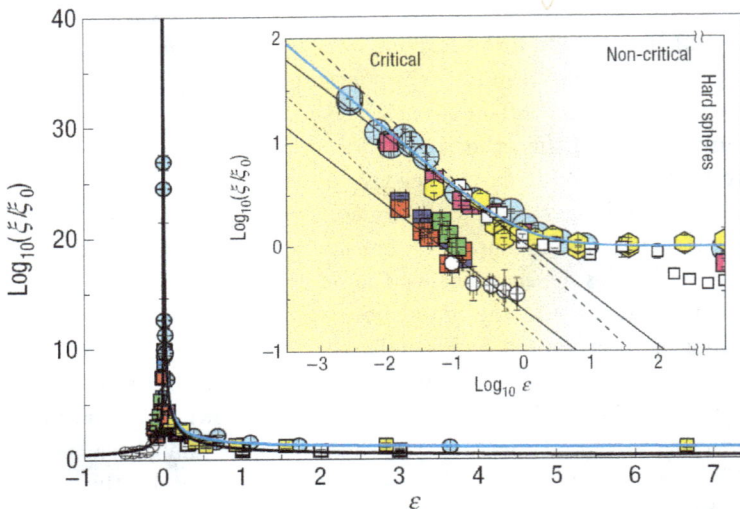

Fig. 6.7   Near- and far-critical behavior of the correlation length in colloid-polymer mixtures. Shown is the scaled correlation length ξ/ξ_0 versus the reduced effective temperature ε, obtained from experiments with colloids of two different sizes. The black solid line represents the critical exponent $\nu = 0.5$, the dashed line is for $\nu = 0.63$. The blue line is an empirical expression which describes $\xi(\varepsilon)$ at *all* temperatures (both near to and far from criticality). The inset shows the same data on a log-log plot (two-phase data offset by -0.5 for clarity). From Royall *et al.* (2007a).

The wide range of accessible effective temperatures led to a very interesting observation (Royall *et al.*, 2007a): Figure 6.7 shows that critical scaling which was only thought to be valid very close to the critical point (so that the structural correlation length ξ is much larger than the discreteness scale σ) in fact holds throughout a very large region of the phase diagram, until ξ decreases down to $\sim \sigma$. This dependence is described by the critical scaling relation $\xi = \xi_0 \varepsilon^{-\nu}$, where ε is the reduced effective temperature and ν is the critical exponent. Within the accuracy of the experimental data, the fitting is equally well described with $\nu = 0.63$ (3D Ising universality class; note that binary mixtures with short-range interactions belong to the same class, see Sec. 8.3) and $\nu = 0.5$ (mean field).

In complex plasmas, there have been no dedicated experiments performed so far on the liquid-vapor phase transition. Theory predicts that an isotropic attraction between particles can be tuned by implementing external multiaxial fields (see Chapter 11). For the case when the vector of the applied field rotates in such a way that its directions are "isotropically

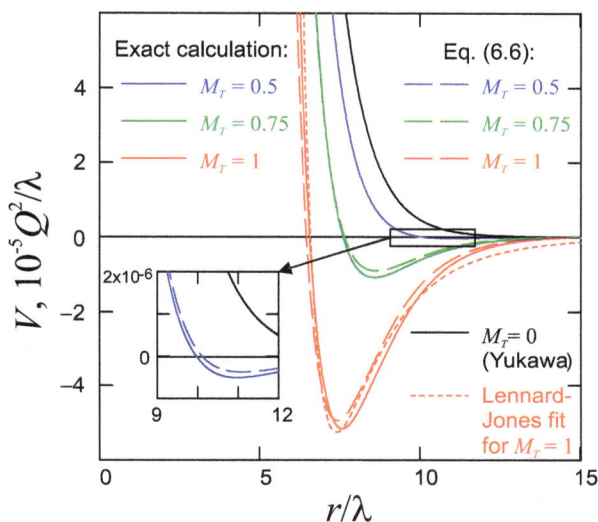

Fig. 6.8 Tunable isotropic interaction in complex plasmas. The potential energy $V(r)$ of pair interaction between particles of charge Q is obtained by applying an external ac field of a spherical polarization. For a subthermal induced flow (i.e., when the ion thermal Mach number $M_T \lesssim 1$) the potential always has an attractive part and is well described by Eq. (6.4). At $M_T \sim 1$, the Lennard-Jones form provides a good approximation for the potential well. The potential is shown in units of $10^{-5}Q^2/\lambda$, distance is normalized by the effective screening length λ. From Kompaneets *et al.* (2009).

distributed" in 3D and that its absolute value (and hence the thermal Mach number of ions, M_T) remains constant, Kompaneets *et al.* (2009) derived analytically the following approximate expression (valid for small M_T):

$$V(r) = \frac{Q^2}{\lambda}\left[1 + \frac{\pi M_T^2}{48}\left(\frac{16}{\pi} - 1 - \frac{r}{\lambda}\right)\frac{r}{\lambda}\right]\frac{e^{-r/\lambda}}{r/\lambda}. \tag{6.4}$$

Figure 6.8 shows that Eq. (6.4) provides remarkably good approximation to the exact potential calculated numerically for $M_T \leq 1$. Interestingly, for $M_T \sim 1$ the potential well can be approximated by the Lennard-Jones interaction.

The parameters of the critical point for the potential described by Eq. (6.4) are controlled by the Mach number. For instance, $M_T = 1$ corresponds to the critical temperature $k_B T_{cr} \simeq 7 \times 10^{-5}Q^2/\lambda$ and density $n_{cr} \simeq 10^{-3}\lambda^{-3}$ (Kompaneets *et al.*, 2009). For typical experimental conditions ($Q \sim 10^4\ e$ and $\lambda \simeq 50\ \mu m$) this yields $T_{cr} \sim 0.3$ eV, which is well above the room temperature. Thus, the liquid-vapor phase transition

108

Complex Plasmas and Colloidal Dispersions

and the corresponding critical behavior should, in principle, be observed in complex plasmas with tunable interactions.

6.4   Summary

This is the first chapter where we systematically discuss examples of particle-resolved studies. Already here the reader can see clear advantages of resolving classical systems at the single-particle level. By means of their real-space approach and application of the Ornstein-Zernike equation, colloidal dispersions allow us to determine one of the most significant quantities in a bulk classical system – the pair potential. A similar approach can also be employed in experiments with complex plasmas.

Systems with interparticle interactions characterized by the relatively long-range attraction form a classic example demonstrating the power of particle-resolved studies: An interface emerging upon gas-liquid phase separation is precisely the kind of problem best tackled in real space. Single-particle resolution enables the crossover from mesoscopic to microscopic behavior to be investigated, which can yield unexpected new physics, for example in the case of critical scaling. The first studies have been carried out with colloidal dispersions, while complex plasmas show great promise, not least in the possibility of studying the role of (quasi)Newtonian particle dynamics on interface fluctuations.

Chapter 7

Liquid-Solid Phase Transitions

Crystal growth is a very important branch of industry, with numerous applications ranging from semiconductors, substrates for high temperature superconductors, piezo sensors, ferroelectric memories, optical elements to nano-structures, quantum dots and organic systems. There are different facets to crystal growth – homogeneous nucleation, heterogeneous nucleation, epitaxial growth, molecular beam epitaxy, chemical vapor deposition, etc. Whilst techniques for visualization (and quality control) of crystal growth have improved greatly, the detailed kinetic understanding of dynamic growth processes is still far from complete. The same holds for nano- and microparticle contamination in production processes.

Measuring the (steady-state) crystalline structure in conventional (molecular) solids is well established and can be done using X-ray diffraction, scanning probe microscopy, coherent electron diffraction, etc. The advantage of crystals in this case is their stability and order, which compensates to some extent for the small (atomic) size resolution. From the crystal structure it is possible to classify material properties and the main interactions that lead to self-organization. Consequently, the determination and prediction of the bulk crystal structure, crystal surfaces (important for many chemical properties) and lately with increasing interest nano-crystal structures (which gives important clues about possible size-dependent new properties) is an important area of physics, linking material (mechanical) properties, electrical properties, superconductivity and chemical properties (Zhu *et al.*, 1987; Gleiter, 1989, 2000; Yokoyama *et al.*, 2005; Takagiwa *et al.*, 2006).

In the last ten years or so, there have been tremendous advances in the experimental investigation of dynamic effects in natural crystals, too. Here the fundamental processes on the atomic/molecular level, such as vibrations and rotations in molecules or bond breaking occur on timescales of femto- to picoseconds, and ultrafast optical spectroscopic techniques are needed for their resolution (Rose-Petruck *et al.*, 1999). Similarly, associated structural rearrangements of particles in perturbed crystal lattices (e.g., due to phonon propagation) can be resolved using sub-picosecond X-ray diffraction (Yazaki *et al.*, 2002) – truly remarkable achievements. These developments now open up the unique possibility to directly compare (and re-scale) measurements made by employing model systems and using natural systems and, hence, to gain a much deeper understanding of the kinetics of crystal growth.

Thus, the use of model systems that allow visualization in real space and time at the individual particle level is becoming more and more desirable. It is no surprise, therefore, that colloidal suspensions have been widely studied in order to learn more about the generic properties of self-organization [see, e.g., papers by Vlasov *et al.* (2001) and Gasser *et al.* (2001), and references therein]. The only essential limitation of colloids for this purpose is the damping by the suspension fluid, which makes it practically impossible to investigate dynamical effects (such as, e.g., latent heat).

With the discovery of plasma crystals, a new system became available for studying the fully resolved dynamics of self-organization processes. Research into 3D crystallization may benefit from this, and consequently a number of studies have been conducted, beginning with the investigation of basic crystal properties such as 3D crystal structure and acoustic modes (Zhdanov *et al.*, 2003; Zuzic *et al.*, 2000), liquid-solid phase transitions (Thomas and Morfill, 1996; Rubin-Zuzic *et al.*, 2006), etc.

In this chapter we summarize major advances achieved with complex plasmas and colloidal dispersions in particle-resolved studies of crystals and liquid-solid phase transitions. Given the fact that the crystallization/melting critically depends on the dimensionality, we first focus on "pure" 3D and 2D systems (starting with the equilibrium regime and proceeding to the most important non-equilibrium processes) and then discuss how the crystallization develops in confined systems (in particular, how the crossover from 3D to 2D melting scenarios occurs).

7.1   Equilibrium Phase Diagram

7.1.1   *3D systems*

We begin the consideration of the equilibrium phase diagram with a system of hard spheres (Gasser, 2009). Being the simplest model system showing a phase behavior analogous to that of atomic substances, it has been extensively studied in colloidal experiments. The crystallization of hard spheres was predicted theoretically (Kirkwood, 1981) and confirmed in numerical simulations (Alder and Wainwright, 1957; Wood and Jacobson, 1957) long time ago. The interaction between hard spheres does not have an energy scale and therefore the phase states of such systems are independent of temperature. The sole parameter controlling the phase equilibrium in this case is the volume fraction ϕ which determines the entropy, i.e., melting or freezing in a system of hard spheres is a purely *entropy-driven* phase transition (Frenkel, 1999). The definitive experiments confirming this were carried out by Pusey and van Megen (1986). As shown in Fig. 7.1, fluid and crystalline states coexist between the freezing point $\phi_f = 0.494$ and the melting point $\phi_s = 0.545$ (Hoover and Ree, 1968), and a pure face-centered-cubic (fcc) crystal is stable for $\phi > \phi_s$. While the configurational entropy naturally decreases upon the transition from a liquid to crystal, the overall entropy nevertheless grows due to significant increase of a free volume available for particles in a lattice (in comparison with that in a disordered liquid state). At higher volume fractions slow dynamics *gradually* set in (see Chapter 9), as indicated in Fig. 7.1 by the shading, and a glass transition is observed at $\phi_g \simeq 0.58$ (Pusey and van Megen, 1986).

Now let us discuss the opposite regime of soft interactions where the hard core plays no role. For this purpose, we naturally choose the Yukawa (Debye-Hückel) potential, which is the generic form of the interparticle interaction broadly used to describe both complex plasmas and charged colloidal dispersions. Pair Yukawa interactions of small ("point-like") particles are determined by two (energy and length) scales [see Eq. (2.4)], so that one can construct two similarity parameters characterizing the coupling strength. Here, we use the pair interaction energy at the mean interparticle distance normalized by the kinetic temperature – the *screened* coupling parameter $\Gamma^{(s)}$, and the interparticle distance normalized by the screening length – the plasma screening parameter κ_p,

$$\Gamma^{(s)} = \Gamma e^{-\kappa_p}, \qquad \kappa_p = \Delta/\lambda, \tag{7.1}$$

Complex Plasmas and Colloidal Dispersions

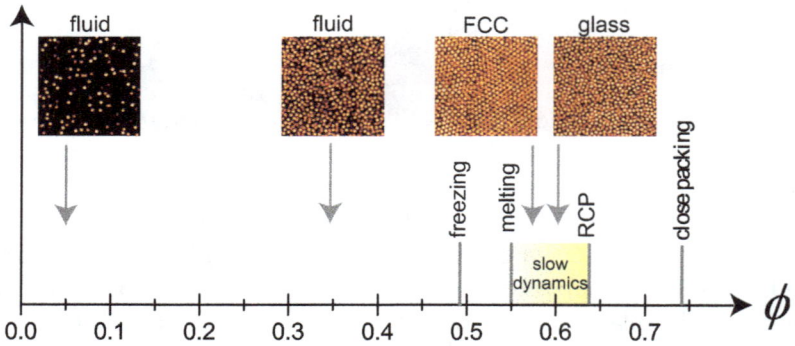

Fig. 7.1   Phase diagram of hard spheres. Monodisperse spheres freeze at a packing fraction $\phi = 0.494$ and melt at $\phi = 0.545$. A glass transition is observed at around $\phi_g = 0.58$. Random close packing (rcp) is indicated at $\phi \simeq 0.64$, and the maximum packing fraction (close packing) is at $\phi = 0.74$.

where $\Gamma = Q^2/k_B T \Delta$ is the coupling parameter of the bare Coulomb inter-action (see also Sec. 4.1.1). The system is usually called "strongly coupled" when $\Gamma^{(s)} \gtrsim 1$ and "weakly coupled" otherwise.

Figure 7.2 shows the phase diagram of a Yukawa system in the $(\Gamma^{(s)}, \kappa_p)$-plane. Above the melting line (indicated by solid curve) the system can form equilibrium face-centered cubic (fcc, at larger κ_p) and body-centered cubic (bcc, at smaller κ_p) solid phases. The bcc-fcc transition line tends to $\kappa_p \simeq 1.72$ in zero-temperature limit ($\Gamma^{(s)} \to \infty$) and approaches the melting line at $\Gamma^{(s)} \simeq 3.47$ and $\kappa_p \simeq 6.90$ (triple point) (Hamaguchi et al., 1997). The melting line can be approximated by the following simple dependence (Vaulina et al., 2002a):

$$\Gamma_m^{(s)} \simeq \frac{106}{1 + \kappa_p + \frac{1}{2}\kappa_p^2}, \tag{7.2}$$

which yields a remarkably good agreement with the results of numerical simulations (Hamaguchi et al., 1997; Vaulina et al., 2002a; Kremer et al., 1986; Robbins et al., 1988; Meijer and Frenkel, 1991; Stevens and Robbins, 1993) at $\kappa_p \lesssim 10$. The phase states shown in Fig. 7.2 were calculated from parameters measured in experiments with complex plasmas (Khrapak et al., 2011) and colloidal dispersions (Royall et al., 2003, 2006), under the *assumption* of the Yukawa interaction.

There are different phenomenological criteria for crystallization (melt-ing), which are usually based on various mean characteristics and often are practically independent of the exact form of the interparticle interaction. The best known is the Lindemann criterion (Lindemann, 1910), according

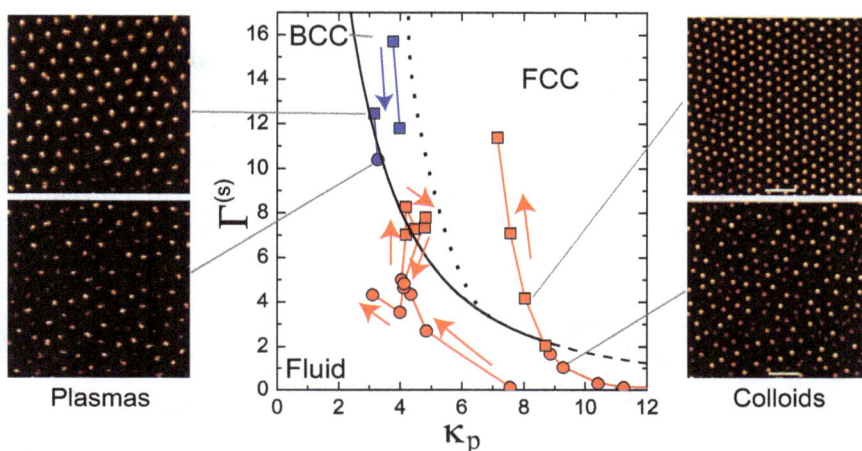

Fig. 7.2 Phase diagram of Yukawa systems. The fluid-solid phase boundary (solid line) is the analytic approximation [Eq. (7.2), transition to the dashed line at large κ_p denotes that its validity is limited by $\kappa_p \lesssim 10$]. The approximate position of the bcc-fcc crystal boundary is indicated by the dotted line. Narrow coexistence regions are not shown here. Symbols represent various crystallization/melting experiments in complex plasmas [from Khrapak *et al.* (2011), blue] and colloidal dispersions [from Royall *et al.* (2003) and Royall *et al.* (2006), red]: Squares and bullets indicate, respectively, solid and liquid phases (as observed), arrows show the direction in which parameters varied during the experiments. Characteristic snapshots of observed liquid and crystalline phases are also presented.

to which melting occurs when the ratio of the root-mean-square particle displacement to the mean interparticle distance reaches a critical value of $\simeq 0.07 - 0.1$ (Löwen, 1994b; Saija *et al.*, 2006). Another criterion was suggested by Hansen and Verlet (1969) who observed that in 3D hard-sphere systems the first maximum of the static structure factor is equal to $\simeq 2.85$ at the melting curve [for inverse-power-law interaction potentials $\propto r^{-n}$ this value varies in the range from $\simeq 2.6$ for $n = 1$ to $\simeq 3.0$ for $n = 12$ (Hansen and Schiff, 1973)]. There also exists a criterion for the pair correlation function proposed by Raveche *et al.* (1974): For inverse-power-law interactions, the critical ratio of the first (nonzero) minimum to the first maximum lies in the range from $\simeq 0.1$ (for $n = 1$) to $\simeq 0.26$ (for hard spheres) at the melting curve.[1] A simple dynamic crystallization criterion, similar to some extent to the Lindemann criterion, was proposed by Löwen *et al.* (1993). According to this criterion, crystallization occurs when the

---

[1] There were also attempts to relate melting to a critical magnitude of the first peak of $g(r)$. However, no systematic studies (e.g., for different interactions) have been reported in this direction.

Complex Plasmas and Colloidal Dispersions

long-time diffusion constant (normalized to its short-time counterpart) falls
below a critical value of $\simeq 0.1$. This dynamic criterion holds both for 2D
and 3D systems (Löwen, 1996). A similar criterion can also be obtained for
finite damping, and then the critical value depends on the dissipation ratio
ν/Ω_E. In the Newtonian dynamic regime, the critical diffusion constant
(normalized to $\Omega_E \Delta^2$) has a distinct asymptote of $\simeq 3.2 \times 10^{-3}$ (Vaulina
et al., 2002a; Ohta and Hamaguchi, 2000).

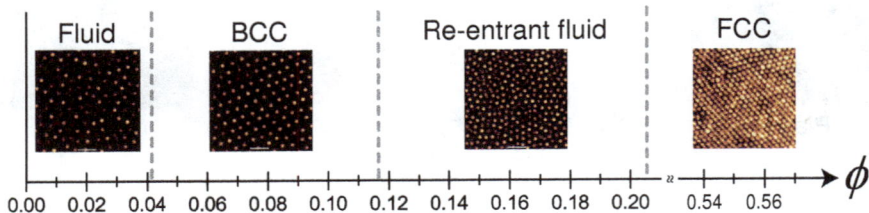

Fig. 7.3   Reentrant melting in a system of charged colloids. The phase diagram is
deduced from the experiment where particles interacted via Yukawa potential, but the
charge rapidly decreased with the packing fraction ϕ. From Royall *et al.* (2006).

Finally, one should make an important note about crystallization in
dense systems. Generally we expect that as density is increased, a system tends to exhibit more ordering – the classic example here are the hard
spheres. Very few anomalies exist in this case, with water being a well-
known example. Royall *et al.* (2006) showed that charged colloids can
also exhibit an order-disorder transition as a function of increasing density.
While this is not expected for a system with Yukawa interactions, in this
experiment the colloid charge was a strongly decreasing function of packing
fraction. So much did the charge decrease that the coupling became sufficiently weak to observe reentrant melting, as shown in Fig. 7.3. Of course,
at even higher densities the hard cores caused freezing around the hard
sphere transition (see Fig. 7.1). Royall *et al.* (2006) attributed the reentrant melting to charge depletion ("scavenging" due to quasineutrality of
charged systems). A very similar mechanism operates also in complex plasmas (Sec. 2.1.1), where the charge depletions is expected to cause reentrant
melting too (Khrapak *et al.*, 2010b).

7.1.2   *2D systems*

The melting processes of 2D and 3D crystals are very different, which becomes evident even on the level of simple phenomenological criteria. For

instance, the Lindemann criterion in its "classical" form is not applicable to 2D systems because of the divergence of the mean squared displacement δr^2 in the thermodynamic limit [in accordance with the Landau-Peierls theorem, δr^2 diverges logarithmically with the system size (Landau and Lifshitz, 1978)]. Therefore, Bedanov *et al.* (1985) proposed a modified criterion for the *relative* mean squared displacement δr_r^2 calculated for a pair of neighbors. The modified Lindemann criterion (where the normalization is also by the mean interparticle distance) yields critical values of $\simeq 0.16 - 0.2$ at the melting line (Zheng and Earnshaw, 1998). As regards structural phenomenological criteria (e.g., the Hansen-Verlet criterion), to our knowledge they have not been systematically tested so far for 2D systems.

One of the central questions in understanding 2D phase transitions concerns the critical parameters that determine which melting scenario will be realized in a particular experiment. Essentially, there are two competing scenarios providing quite different pictures of the mechanisms by which defects in a hexagonal lattice develop. The first one is based on the Kosterlitz-Thouless-Halperin-Nelson-Young (KTHNY) theory (Nelson, 2002; Kosterlitz and Thouless, 1973; Halperin and Nelson, 1978; Nelson and Halperin, 1979; Young, 1979). This scenario involves two phase transitions, with an intermediate, so called "hexatic phase" in between: One transition is due to the unbinding of (weakly correlated) dislocation pairs, and the other is due to the dissociation of dislocations into disclination pairs. The relevant question here is whether particular predictions of the KTHNY theory (viz., the appropriate scaling behavior of the correlation functions associated with the crystalline and hexatic phases, the sequence of two second-order phase transitions in the thermodynamic limit) are indeed observed in the experiments. A possible alternative scenario is the theory of grain-boundary induced melting (Chui, 1982, 1983), where the hexatic phase is preempted by a first-order phase transition (similar to that operating in 3D systems, see below).

These issues have been discussed extensively over the last decades [see, e.g., reviews by Strandburg (1988) and by Alba-Simionesco *et al.* (2006)]. The jury is still out on the nature of the freezing transition for many experiments and numerical simulations, especially for those performed with particles interacting via short-ranged potentials. Therefore, the critical role of the interaction range as well as of the system homogeneity and size requires further careful investigation.

To demonstrate major distinctions from the melting process occurring at grain boundaries (i.e., at lines formed by strongly correlated dislocations, see Sec. 7.3.1), we now focus on the equilibrium phase behavior of those systems which exhibit the KTHNY melting scenario. We briefly recall that the traditional way to distinguish phases in KTHNY theory derives from the shape of correlation functions for two order parameters (Strandburg, 1988) – bond-orientational, $\psi_{6j} = N_{nn}^{-1} \sum_k e^{i6\theta_{jk}}$, and translational, $\psi_{Gj} = e^{iG \cdot r_j}$. The order parameters are defined for each particle j located at position r_j, θ_{jk} is the angle of the bond between particle j and its neighbor k, G is a primary reciprocal lattice vector, and N_{nn} is the number of nearest neighbors. The corresponding correlation functions in space and time are $g_\alpha(r) = \langle \psi_{\alpha i}(r_i) \psi_{\alpha j}(r_j) \rangle$ and $g_\alpha(t) = \langle \psi_{\alpha i}(t') \psi_{\alpha i}(t'+t) \rangle$, where $r = |r_i - r_j|$ and α denotes the bond-orientational (6) or translational (G) order.

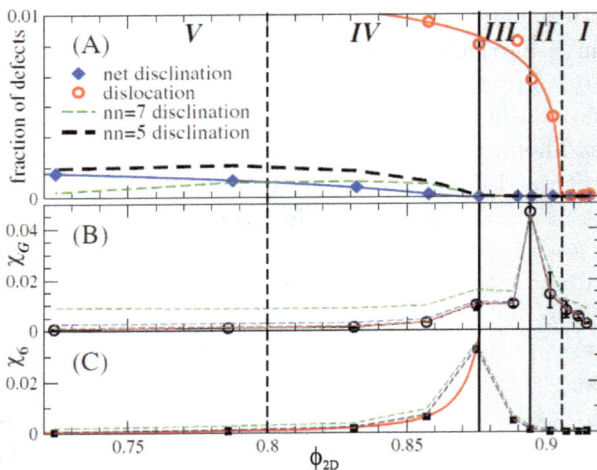

Fig. 7.4   Phase diagram of a hard-sphere-like 2D system. Shown is the concentration of dislocations (and also disclinations, a) as well as translational, χ_G (b), and bond-orientational, χ_6 (c), susceptibilities as functions of the areal fraction ϕ_{2D}. Vertical solid lines partition crystal (regions I and II), hexatic (region III), and liquid (regions IV and V) phases as determined from susceptibilities in (b) and (c). From Han *et al.* (2008).

We illustrate KTHNY melting with two experiments [for other examples, see, e.g., papers by Murray and van Winkle (1987) and Marcus and Rice (1997)]. In one experiment by Han *et al.* (2008), the melting was studied in a system of thermosensitive colloids. The pair potential between particles in this colloidal suspension is short-ranged and repulsive, and the effective sphere diameter σ (and hence the areal fraction $\phi_{2D} = \frac{\pi}{4} \sigma^2 n_{2D}$) can be tuned by varying temperature. This allowed the authors to follow the spatiotemporal evolution of the same sample area through the entire sequence of transitions. In order to avoid ambiguities from finite-size effects, the divergence of translational and bond-orientational susceptibilities (defined as $\chi_\alpha \propto \langle \psi_\alpha^2 \rangle - \langle \psi_\alpha \rangle^2$, where $\psi_\alpha = N^{-1} \sum_i^N \psi_{\alpha i}$ is the order parameter averaged over the system) was used as indicator to determine the phase transition points. With this approach, five regimes were assigned to the phase diagram shown in Fig. 7.4. A number of KTHNY predictions were quantitatively confirmed, especially near the hexatic-liquid transition, but the order of the two phase transitions was not resolved.

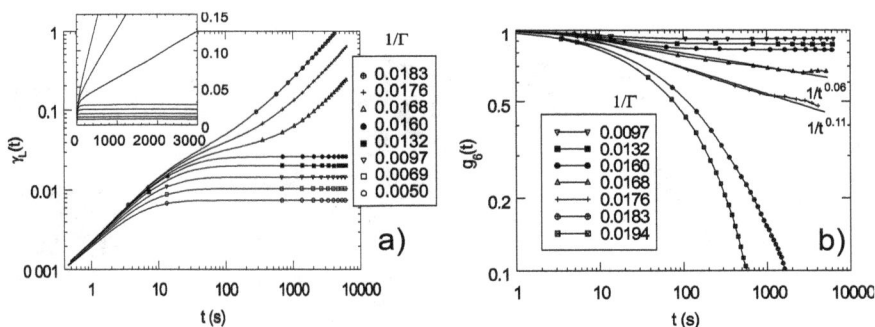

Fig. 7.5   Melting in 2D colloids. (a) Dynamic Lindemann parameter, $\gamma_L(t)$, for different values of the scaled temperature Γ^{-1}. The long-time limit of $\gamma_L(t)$ is bounded in the crystalline phase and diverges in the liquid phase. The inset shows the data in a linear plot to illustrate the change in the behavior of $\gamma_L(t)$. (b) The bond-orientational correlation function in time, $g_6(t)$. From Zahn and Maret (2000).

The other experiment by Zahn *et al.* (1999) and Zahn and Maret (2000) was performed with super-paramagnetic spherical colloids which were confined by gravity to a horizontal flat water/air interface. The interparticle interaction was controlled via a vertical magnetic field B which induced a magnetic moment χB, with χ being the effective magnetic susceptibility per particle. The repulsive dipole-dipole potential dominated the interaction and the coupling parameter is $\Gamma = (\chi B)^2 (\pi n_{2D})^{3/2} / k_B T$. Thus, changing the magnetic field strength allows external tuning of the coupling parameter and the study of phase transitions in a controlled way. Figure 7.5a shows the temporal evolution of the 2D modified Lindemann parameter, $\gamma_L(t) \propto \sqrt{\delta r_r^2(t)}$, calculated using the relative neighbor-neighbor displacement. One can see a clear bifurcation between bound to unbound behavior occurring at $\Gamma \simeq 62.5$. Figure 7.5b shows the bond-orientational correlation function, $g_6(t)$, from which three distinct regimes can be identified:

Complex Plasmas and Colloidal Dispersions

The crystalline regime at $\Gamma \gtrsim 62.5$, where $g_6 =$ const; an intermediate regime at $56.8 \lesssim \Gamma \lesssim 59.5$, where $g_6(t)$ decays as a power law; the isotropic liquid regime at $\Gamma \lesssim 54.6$, where $g_6(t)$ decays exponentially. Note that in the intermediate regime $g_6(r)$ also exhibits a power-law decay (Zahn *et al.*, 1999). These findings clearly indicate the hexatic phase and hence support the two-stage melting also for systems with a r^{-3} interaction potential.

7.2 Liquid-Solid Interfaces and Kinetics of 3D Transitions

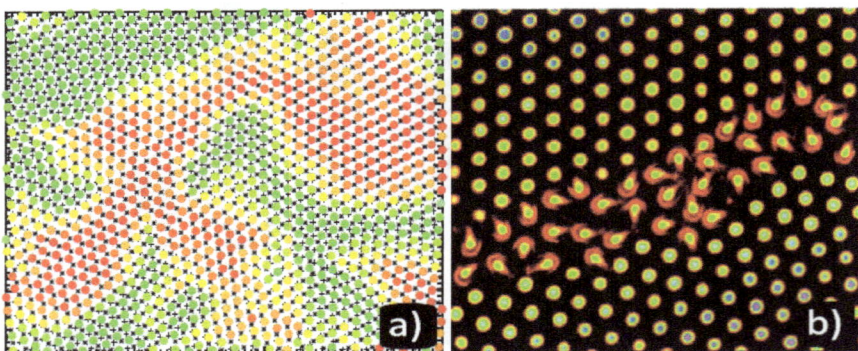

Fig. 7.6 (a) Domain structure of a 3D plasma crystal. Three consecutive lattice planes are shown, each particle in the middle plane is color-coded in accordance with the local order (red corresponds to the fcc lattice cell and green to hcp), particle in two adjacent planes are indicated by crosses and stars. From Zuzic *et al.* (2000). (b) Domain interface in a crystalline colloid. Near such interfaces (grain boundaries) the crystal is premelted – the particles move rapidly and show liquid-like diffusion (red represents the most movements, violet is for the least). Courtesy of A. Alsayed and A. G. Yodh.

To substantiate the importance of complementary studies, we compare the crystallization experiments performed with complex plasmas and colloidal dispersions. Figure 7.6a shows different lattice structures found locally (in a single cell around each particle) in a 3D plasma crystal, color-coded onto a single lattice plane (Zuzic *et al.*, 2000). We observe the co-existence of the (presumed) ground state (fcc) and a metastable state (hexagonal close-packed, hcp), separated by smooth (but well defined) domain borders [which are also seen between domains of the same structure but different lattice orientation, analogous to nanostructured solids (Gleiter, 1989, 2000)]. Such borders appear to be similar to the grain boundaries observed in colloids (Alsayed *et al.*, 2005; Pusey, 2005; Pusey

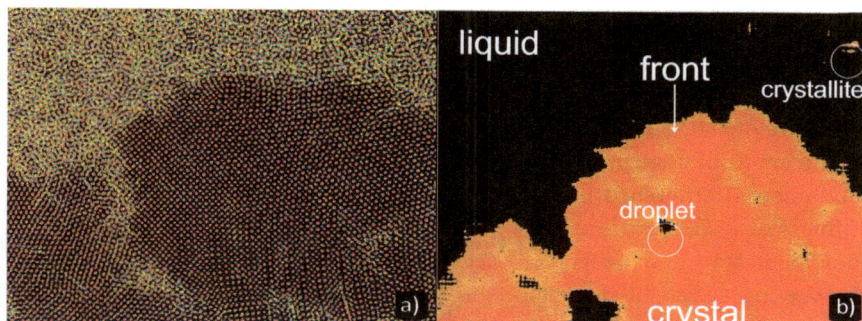

Fig. 7.7   Crystallization front in a 3D complex plasma. Figure (a) illustrates the front of heterogeneous nucleation propagating upwards. It shows a superposition of 10 consecutive video frames (about 0.7 s), particle positions are color-coded from green to red, i.e., "caged" particles appear redder, "fluid" are multicolored. (b) The local order for figure (a), where red implies high crystalline order, black denotes the fluid phase, and yellow indicates transitional regions. Along with the crystallization front, droplets and crystallites are seen that may grow and then dissolve again. From Rubin-Zuzic *et al.* (2006).

and van Megen, 1986; Gasser *et al.*, 2001) – a beautiful example is presented in Fig. 7.6b.

As regards the crystallization kinetics, one can observe both homogeneous and heterogeneous nucleation, and which pathway is realized in the experiment depends strongly on the boundary conditions: In the bulk region, where boundaries play no role, the system exhibits homogeneous nucleation. In this case, one normally observes coexistence of crystalline domains of different structure and orientation, as shown in Fig. 7.6. However, closer to the boundaries crystallization often develops in the form of a front propagating from boundaries inwards. This process in complex plasmas is illustrated in Fig. 7.7.

Let us discuss these and other prominent features of 3D crystallization kinetics (as observed in particle-resolved experiments with complex plasmas and colloidal dispersions) in some more detail.

7.2.1   Liquid-solid interface in equilibrium

Solidification or melting processes occur via a liquid-solid (fluid-crystal) interface which separates the disordered and ordered phases. Understanding thermodynamic and structural properties of this interface on a microscopic scale is pivotal to control and optimize crystal nucleation and the emerging

microstructure of the material. In equilibrium, creating a crystal-fluid interface results in a free-energy penalty characterized by the so-called interfacial
tension (per unit area). Unlike the liquid-gas or fluid-fluid coexistence, the
structure of the solid-fluid interface depends on its orientation with respect
to the crystalline structure (Woodruff, 1980). This anisotropy identifies a
fundamental difference between the interfacial tension and the interfacial
stiffness of a solid surface.

Calculating and measuring crystal-fluid interfacial tensions is a very
challenging task. In principle, the capillary fluctuations of the interface
give access to the stiffness. Therefore, particle-resolved studies have been
very useful to determine the thermal undulations of the interface and extracting the stiffness and the tension out of the data. Recent experimental studies of a hard-sphere solid-fluid interface (Hernandez-Guzman and
Weeks, 2009; Ramsteiner *et al.*, 2010; van Duc *et al.*, 2011) suggest that the
interfacial tension is about $0.6k_BT/\sigma^2$, which is comparable to computer
simulation data (Davidchak and Laird, 2000; Mu *et al.*, 2005; Amini and
Laird, 2006; Zykova-Timan *et al.*, 2009) and predictions of the density functional theory (Curtin, 1987; Marr and Gast, 1993; Ohnesorge *et al.*, 1994).
Another experimental route to estimate the surface tension is measure the
homogeneous crystal nucleation rate, which can then be fitted to classical
nucleation theory (Palberg, 1999). Note, however, that such approaches
assume a spherical nucleus, which is often not the case (see next section).

7.2.2   *Homogeneous versus heterogeneous nucleation*

One of the fascinating and still puzzling questions of statistical physics is
crystal nucleation out of an undercooled melt. As we pointed out above,
there is a basic distinction between homogeneous and heterogeneous nucleation: While homogeneous nucleation occurs out of a bulk liquid, initiation
of heterogeneous nucleation requires pre-existing inhomogeneity (like a wall,
external grain, or seed). Heterogeneous nucleation is normally characterized by much higher rate and therefore can mask homogenous nucleation
completely (Turnbull, 1950).

Homogeneous nucleation proceeds via critical nuclei needed to initiate
subsequent crystal growth. These nuclei come from fluctuations of the
metastable initial state. The simplest description of homogeneous nucleation is provided by the so-called classical nucleation theory (Landau and
Lifshitz, 1978). Within this theory, there are two basic contributions to the
Gibbs free energy of a crystalline nucleus: First, the undercooling favors

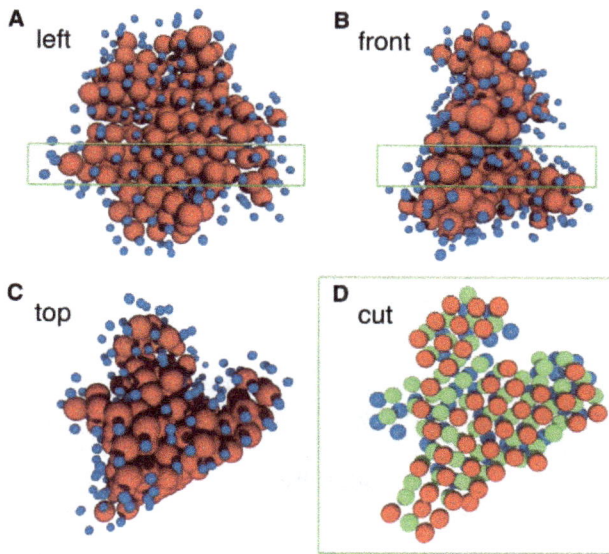

Fig. 7.8   Homogeneous nucleation in experiments with colloidal dispersions. (a-c) Snapshots of a crystallite of post-critical size in a system of charged hard spheres with the volume fraction $\phi = 0.47$, shown from three different directions. The crystallite consists of 206 crystal-like particles represented by red spheres (drawn to scale), while blue spheres (reduced in size for clarity) depict 243 particles which share at least one crystal-like "bond" with red spheres but are not identified as crystal-like. (d) A cut with a thickness of three particle layers through the crystallite, illustrating the hexagonal layer structure; blue, red, and green spheres represent particles in the consecutive layers. The cut was taken from the region indicated by the green boxes in (a) and (b). The particle size in (d) is reduced in order to improve the visibility of the second and third layers. From Gasser *et al.* (2001).

larger nuclei, and the corresponding (bulk) free energy gain is proportional to the volume of the nucleus and the undercooling magnitude. Second, there is a free energy penalty as a solid-fluid interface is created around the nucleus. The latter (interfacial) contribution scales with the area of the nucleus. This competition yields a maximum in the free energy as a function of the nucleus size, which determines the critical size.[2] Then the nucleation rate is given by $J_0 \exp(-\Delta G/k_{\mathrm{B}}T)$, where ΔG is the resulting free energy barrier and J_0 is a kinetic pre-factor which defines the inverse nucleation timescale.

---

[2]Normally, a spherical shape of the nucleus is assumed in classical nucleation theory, such that its size and the corresponding free energy barrier can be worked out analytically.

The actual size, shape and structure of the critical nucleus can be calculated by computer simulation techniques (Auer and Frenkel, 2004) or be measured in real-space experiments (Gasser *et al.*, 2001). It turns out that the nucleus shape can be very different from spherical and the interface can be quite fuzzy – a real-space snapshot for hard-sphere-like colloids is presented in Fig. 7.8. Although the general trends of nucleation in such systems are captured by classical nucleation theory, quantitative discrepancies (by many orders of magnitude in the nucleation rate) between experiments and simulations (Auer and Frenkel, 2004) remain unresolved. Clearly, more work is needed to address this issue.

We briefly mention another mechanism of crystal nucleation which can operate in systems with short-ranged attractions. Such systems exhibit a metastable critical point (see Fig. 6.5b). It has been supposed that density fluctuations close to the critical point can provide a pathway by which nucleation proceeds much more quickly – the density fluctuations act to lower the free energy barrier to nucleation (Ten Wolde and Frenkel, 1997). This mechanism has been confirmed in experiments with attractive colloids by Savage and Dinsmore (2009) who mapped the energy landscape of nucleation and indeed found a lowering in the energy barrier (which they associated with the metastable critical point). We note that such an accurate sampling of the energy landscape in a particle-resolved experiment (where, unlike simulation, no biasing is employed to facilitate rare events) is a remarkable achievement.

Heterogeneous nucleation of crystals, on the other hand, can be induced at inhomogeneities and has also been studied extensively for hard-sphere colloids. Various kinds of inhomogeneities have been considered: (i) a larger particle introduced as a "model grain" (de Villeneuve *et al.*, 2005; Cacciuto *et al.*, 2004), (ii) a "seed" imposed to the system (van Teeffelen *et al.*, 2008; Hermes *et al.*, 2011), or (iii) a smooth or a structured wall which triggers the nucleation (Heni and Löwen, 2000; Hoogenboom *et al.*, 2003a,b; Cacciuto and Frenkel, 2005). The structure and size of the seed cluster can be controlled at wish, by fixing colloidal particles with laser tweezers or/and offering a prescribed structure to the undercooled or overcompressed colloidal fluid. Using different seeds, the nucleation rate and the resulting crystalline structure can be broadly tuned and, thus, a variety of crystalline structures can be grown. A similar idea is used in so-called epitaxy (Hoogenboom *et al.*, 2003a): A structured wall (favoring a crystal) is offered as a template to the colloidal solution. With the help of gravity, large monocrystal can then be grown by subsequent formation of crystalline layers.

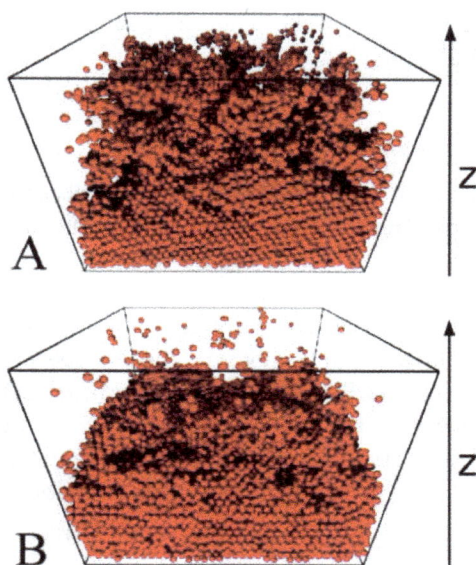

Fig. 7.9   Heterogeneous nucleation in colloidal dispersions. Shown are snapshots from (a) confocal microscopy and (b) numerical simulations of a system of hard spheres with the volume fraction $\phi = 0.52$. In this reconstruction, only crystalline particles are shown and the simulation box is reduced in the z direction. From Sandomirski *et al.* (2011).

Real-space experiments in this case allow us to identify crystalline parti-cles and therefore serve to explore the mechanisms governing heterogeneous crystal nucleation.

The resulting crystallization front has been followed using particle-resolved experiments with hard-sphere colloids (Dullens *et al.*, 2006; San-domirski *et al.*, 2011). A real-space picture of growing solid-fluid interface is presented in Fig. 7.9a. Similar to the experiments with complex plasmas shown in Fig. 7.7 (where the detailed structure of the front and different crystal domains with different structure and/or orientation were observed), here the growth velocities and a depletion zone of particles above the ad-vancing interface have been also directly measured by confocal microscopy (Sandomirski *et al.*, 2011). Unlike the crystallization in complex plasmas (where the front was 2–3 layers wide), the width of the interfacial region for hard-sphere colloids extends over 6–10 layers. Quantitative agreement for the interface structure and dynamics has been achieved by comparing

Complex Plasmas and Colloidal Dispersions

experiments with Brownian dynamics computer simulations of hard-sphere systems (Sandomirski *et al.*, 2011). The corresponding simulation snapshot is shown in Fig. 7.9b.

7.2.3   *Kinetics of phaselets*

An interesting feature of heterogeneous nucleation observed in experiments with complex plasmas (Rubin-Zuzic *et al.*, 2006) is the presence of *phaselets* – small droplets in the crystal phase and small crystallites in the fluid phase. Traces of these droplets can be clearly seen in Fig. 7.7b.

Figure 7.10 summarizes the measured characteristics of phaselets. There are two general features worth noting: (i) The size spectra of both droplets and crystallites are compatible with power laws (Fig. 7.10, right column). This suggests that within the observable size range (~ 10 to $\sim 10^3$ particles) there is no characteristic length scale that either determines formation or dissolution. (ii) The larger crystallites and droplets tend to live longer

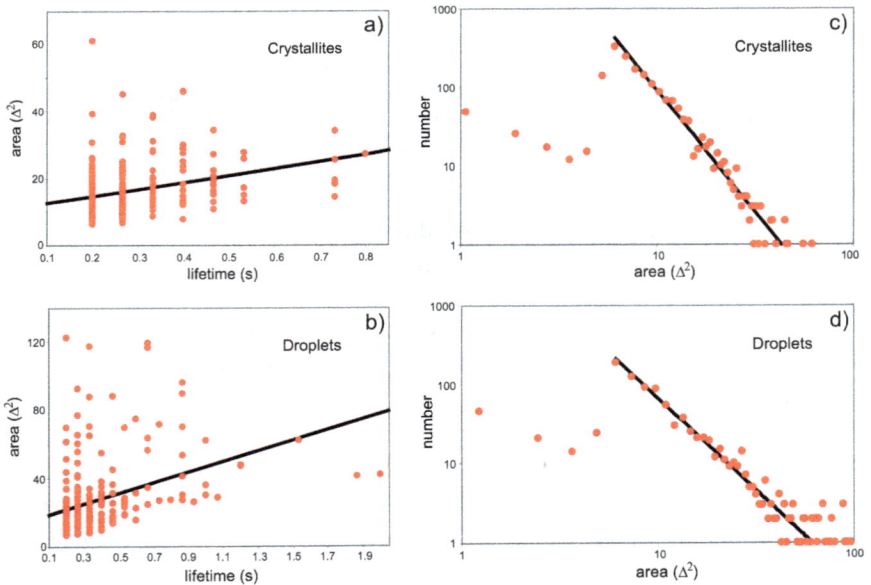

Fig. 7.10   Characteristics of crystallites and droplets in the experiment shown in Fig. 7.7. Area of crystallites (a) and droplets (b) measured in units of a single particle cell (squared interparticle distance Δ^2) versus their lifetimes, and histograms showing number of crystallites (c) and droplets (d) versus their areas. From Rubin-Zuzic *et al.* (2006).

(Fig. 7.10, left column). By "life time" we mean the growth + dissolution phases, so that this result is not too surprising. There is, however, a
substantial spread in the individual life times.

The crystallites above the front are apparently the traces of (relatively
slowly developing) homogeneous nucleation discussed above. As for the
droplets observed in the crystal regime, the mechanism responsible for their
formation should be quite different, because thermodynamically, both the
bulk and the surface contributions cause the free energy to increase. It
is possible that after the initial solidification, a gradual relaxation from a
metastable to a ground state (for example, from hcp to bcc or fcc structure,
as one can see in Fig. 7.6) occurs downstream from the crystallization front.
This is naturally accompanied by a release of latent heat. Then the droplets
could be a local manifestation of this relaxation. The larger the droplet,
the longer it takes to dissipate the released heat, and the longer its lifetime.
Clearly, this effect can only be seen in complex plasmas, since in colloidal
dispersions the latent heat is completely damped out.

7.2.4   *Grain-boundary melting*

Another particular feature seen in Fig. 7.7 is a narrow (a few lattice constants in extent) premelted region in the crystalline regime (perpendicular
to the front) where particles exhibit enhanced mobility signifying *grain-
boundary melting*. The effect of grain boundaries was first demonstrated in
a number of experiments using colloidal suspensions (Alsayed *et al.*, 2005;
Gasser *et al.*, 2001). The example shown in Fig. 7.6b is from the experiment
carried out in a suspension of thermosensitive colloids (microgel particles
whose diameter and hence the packing fraction depends on temperature,
see Sec. 3.2.3). It is noteworthy that grain boundaries also form in conventional solids, where they separate domains (grains) of locally ordered
regimes (Gleiter, 2000).

Thermodynamically, grain boundaries are different (both in energy and
entropy) from the homogeneous crystal regimes. When such a grainy crystal
is heated and approaches its melting point, the grain boundaries may play a
special role – they can act as "seeds" of pre-melting regions (Alsayed *et al.*,
2005; Pusey, 2005).

The fact that grain-boundary melting can be also observed in complex
plasmas is significant for (at least) three reasons:

- the measurement "slice" shown in Fig. 7.7 was obtained in a large
  (millions of particles) complex plasma assembly, many interparti-

cle spacings away from the boundaries. Hence the measurements
support the conclusion that grain-boundary melting is not (necessarily) associated with system boundaries.

- the particle interaction is primarily electrostatic. This implies that
  the same process – grain-boundary melting – occurs in different
  systems with very different forms of binary particle interactions.
  In other words, the process is generic and does not depend on any
  peculiarities or special features of the system.
- the strongly coupled complex plasma system is almost undamped.
  This implies that energy transport is (to a large extent) governed by
  phonons in the crystalline phase and dust-acoustic (sound) waves in
  the fluid regime, making the heat transport important – in contrast
  to fully damped colloids.

Summarizing the examples discussed in this section, we can conclude
that a ubiquitous and still poorly understood process – like crystallization
and melting – needs different inputs, different constraints, generalization
from different sources and new approaches, so that the principal mechanisms can be identified and combined to a fundamental kinetic theory.

7.3   Kinetic Processes in 2D Crystals

2D model systems are ideal for studying non-equilibrium processes occurring in liquids and solids, because (regardless the timescales involved) the
complete information about all particles can be directly obtained at each
moment of time. Let us consider some examples.

7.3.1   *Scaling laws of 2D crystallization*

There are several mechanisms that can result in the melting of 2D plasma
crystals. These mechanisms can generally be divided into two categories
– *generic* (i.e., those operating in any classical system with a given pair
interaction between particles, provided the interaction can be described by
a Hamiltonian) and *plasma-specific* (which are associated with the energy
exchange between microparticles and ambient plasma and can be considered
as a result of the system openness). Fortunately, the plasma-specific melting
mechanisms (resulting in drastic increase of the particle kinetic temperature
above the temperature of neutral gas) can be easily disabled by proper
choice of experimental conditions, as discussed in Sec. 4.3.

To illustrate the generic melting processes operating in 2D plasma crystals, below we discuss the paper by Nosenko *et al.* (2009) summarizing results from four experimental series (Knapek *et al.*, 2007; Nosenko and Goree, 2004; Nosenko *et al.*, 2006, 2008). Each of these series was composed of several experiments performed at different pressures, employing qualitatively different heating methods and microparticles of different sizes. Before the heating was applied, particles in a monolayer self-organized in an ordered triangular lattice, as shown in Fig. 7.11a. The lattice always contained defects revealed by a Voronoi diagram (and defined as lattice sites where particles have a number of nearest neighbors other than six). Defects usually clustered together to form domain boundaries. In three experimental series by Nosenko and Goree (2004) and Nosenko *et al.* (2006, 2008) the particles were kept under steady heating conditions, whereas in the series by Knapek *et al.* (2007) the evolution during a rapid recrystallization was observed:

- Series I *"uniform heating"*: A laser-heating method was used. The kinetic temperature of particles, T, grew with the laser power applied; the concentration of defects, c, was on average uniform in the heated rectangular region.
- Series II *"temperature gradient"*: The same heating method was used. Unlike series I, T and hence c had a spatial gradient outside of the heated region (due to the heat transport, see Sec. 10.3).
- Series III *"shear flow"*: A planar (Couette) shear flow was created, causing shear-induced melting of the monolayer. Similar to series II, T and c varied in the direction perpendicular to the flow.
- Series IV *"recrystallization"*: A monolayer was melted by applying a short electric pulse (see Fig. 5.3), and then the system was left to cool down naturally due to the friction. The evolution of T and c was examined during the stage of recrystallization.

Figure 7.11 allows us to identify a qualitative difference between *steady-state* experiments I-III and the *unsteady* experiment IV, by comparing the results on the uniform heating (I, left column) and the recrystallization (IV, right column) in different "temperature regimes": At high temperatures illustrated in Figs. 7.11 b,c monolayers contain a substantial fraction of 5/7-fold defects (dislocations), a few 4/8 defects can be seen as well. The number of defects in experiment I is noticeably larger than that in experiment IV. The situation changes at lower temperatures shown in Figs. 7.11 d,e – now the number of defects in experiment I is smaller than that in

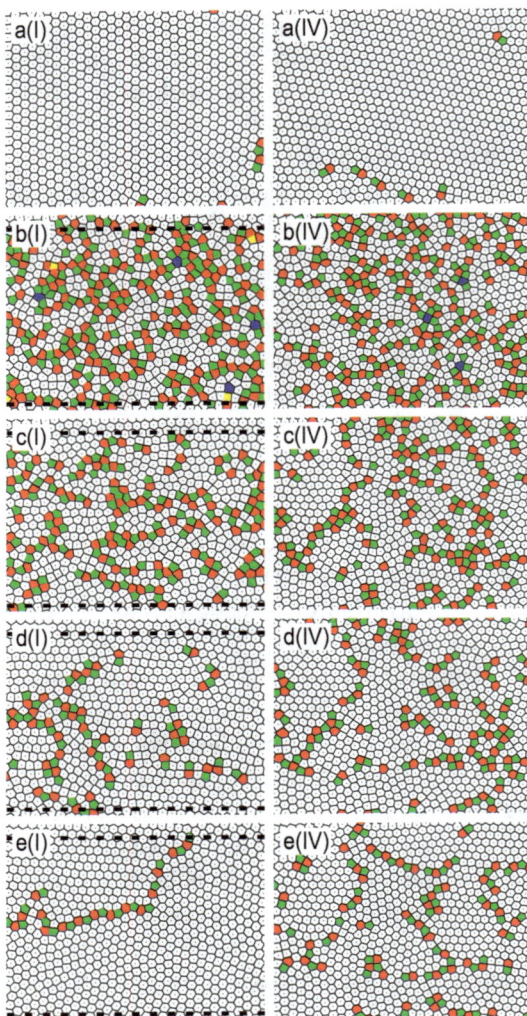

Fig. 7.11   Maps of defects in 2D complex plasmas. The left and right columns illustrate the difference between, respectively, the uniform heating experiment I (horizontal dashed lines indicate the heating region) and the recrystallization experiment IV. The dots are individual particles located inside their Voronoi cell: green/red colors represent 5/7-fold defects, yellow/blue are for 4/8-fold defects. Different "temperature regimes" are shown: (a) is the initial equilibrium state, before the external heating was applied; (b) and (c) are the "high-temperature" regimes, where steady-state experiments I-III reveal higher concentration of defects; (d) and (e) demonstrate the reversed trend at lower temperatures, where the recrystallization is characterized by a higher number of defects. The temperature gradually decreases from (b) to (e). From Nosenko *et al.* (2009).

experiment IV. At the same time, at lower temperatures the dislocations have a common tendency to cluster into strings separating domains with local hexagonal order.

The quantitative analysis of experiments I-IV is summarized in Fig. 7.12. The number of 5/7-fold defects $N_{5,7}$ was measured and their concentration $c_{5,7} = N_{5,7}/N$ was plotted as a function of temperature [$c_7(T)$ is nearly identical to $c_5(T)$ and is not shown here]. In order to compare the experiments performed at different conditions, the temperature was measured in units of the inverse coupling parameter $\Gamma^{-1} = k_B T \Delta / Q^2$.

A remarkable result obtained from the analysis is the principal difference in the behavior of $c(T)$ seen in Fig. 7.12 for steady-state experiments I-III and the recrystallization experiment IV. In the latter case, the concentration exhibits a relatively slow power-law scaling over two decades of temperature, $c(T) \propto T^\alpha$, whereas experiments I-III demonstrate the Arrhenius dependence, $c(T) \propto e^{-E_a/k_B T}$. The least-squares fit of $c(T)$ yields the activation energy $E_a = 1 - 2$ eV of 5/7-fold defects for experiments I-III (see footnote on p. 3 for the relation between eV and $k_B T$), and the temperature exponent $\alpha \simeq 0.37$ for experiment IV.

It is generally believed that the value of the defect core energy E_c plays a critical role in the realization of the melting scenario (Strandburg, 1988; Alba-Simionesco *et al.*, 2006). The KTHNY mechanism discussed in Sec. 7.1.2 should operate when $E_c \gtrsim 2.8 k_B T_{hex}$, otherwise the grain-boundary-induced melting should occur (see also discussion in Sec. 7.4.1). In the equilibrium case, the activation energy in the Arrhenius law for the defect concentration is equal to twice the core energy (Strandburg, 1988; Tobochnik and Chester, 1982). We obtain $E_c = \frac{1}{2} E_a = 0.5 - 1.3$ eV, whereas the estimates for the hexatic transition temperature yield $T_{hex} = 6 - 9$ eV. Hence, E_c is $10 - 20$ times smaller than the critical value $2.8 k_B T_{hex}$, which strongly suggests that in the discussed experiments the grain-boundary-induced melting scenario is realized.[3]

The conclusion that the KTHNY melting mechanism is "preempted" by a first-order phase transition can also be drawn from the analysis of the bond-orientational correlation function $g_6(r)$ (see inset in Fig. 7.12). Indeed, the cases with the power-law scaling exponent $\eta_6 \leq 0.25$ (shown above the red dotted line) correspond to a nearly perfect crystal with singular grain boundaries illustrated in Figs. 7.11a(I),e(I). At the same

---

[3]Here, we leave aside the fundamental issue of "renormalization" of the core energy in the presence of other dislocations, as well as the definition of the corresponding core radius in the elastic limit, see review by Strandburg (1988).

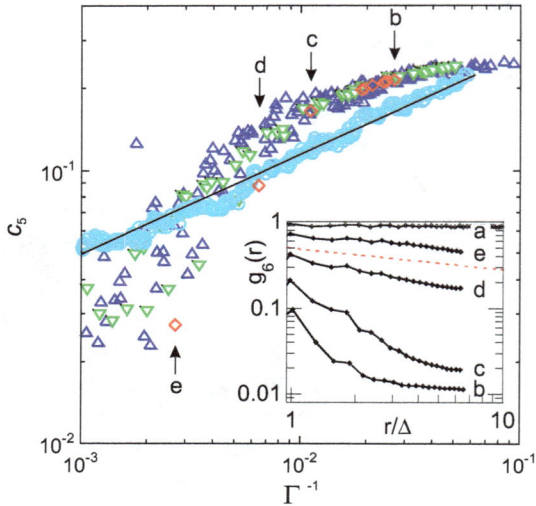

Fig. 7.12   Concentration of defects versus temperature. The figures show the 5-fold defect concentration c_5, the symbols correspond to different experimental series: ◇ for uniform heating (I), ▽ for temperature gradient (II), △ for shear flow (III), and ○ for recrystallization (IV). (a) Log-log plots of $c_5(T)$ for all experiments. Vertical arrows point to different "temperature regimes" illustrated in Fig. 7.11b-e, normalized temperature is the inverse coupling parameter $\Gamma^{-1} = k_B T \Delta / Q^2$. The unsteady recrystallization series IV exhibits a power-law scaling $c(T) \propto T^\alpha$ (fit by a straight line). The inset shows the bond-orientational correlation function $g_6(r)$ for different temperature regimes in the experiment I (left column of Fig. 7.11). From Nosenko *et al.* (2009).

time, the cases with many (isolated or clustered) dislocations shown in Figs. 7.11b(I),c(I),d(I) have $\eta_6 > 0.25$. This clearly contradicts the prediction of the KTHNY theory on the scaling behavior of $g_6(r)$ (Strandburg, 1988).

Let us now focus on the principal difference revealed in the behavior of $c(T)$ for the experiments performed under steady and unsteady conditions. In all experimental series I-IV the damping rate ν_{dn} due to neutral gas friction was a factor of 10–30 smaller than the Einstein frequency Ω_E (which also characterizes the momentum/energy exchange in the mutual particle collisions). Therefore, it is not surprising at all that the measured velocity distributions of particles are well reproduced by the equilibrium Maxwellian form.

The structural equilibration, however, might take much longer than the equilibration of the velocity distribution. This is apparently the reason of the drastic difference in the behavior seen in Fig. 7.12 between series I-III and series IV: The former experiments were performed under steady conditions allowing structural relaxations during a few minutes, which is a factor of $10^3 - 10^4$ longer than the "atomistic" timescale Ω_E^{-1}. The Arrhenius law for $c(T)$ observed in these cases naturally suggests a (quasi) equilibrium regime of melting (Strandburg, 1988; Tobochnik and Chester, 1982). In contrast, the exponential cooling of the system occurring at timescales ν_{dn}^{-1} in the recrystallization experiment was too rapid for the structural equilibration.

One can expect that at the beginning of the recrystallization process, shortly after the system was suddenly melted, the balance of "source" and "sink" terms in the kinetic equation describing the evolution of defects is strongly shifted towards the source. Therefore, the observed concentration of defects at this high-temperature stage is smaller than the equilibrium Arrhenius value. As the temperature decreases, the balance is naturally shifted towards the sink, and at a certain point $c(T)$ should cross the equilibrium curve. The further cooling is accompanied by "freezing" of metastable states at lower T, and the resulting concentration of defects becomes systematically larger than the equilibrium value. One can suppose that a power-law (i.e., scale-free) dependence observed in the unsteady series IV might be a universal feature peculiar to non-equilibrium melting regimes (although the value of the exponent α is not necessarily universal, but might be a function of the dissipation ratio ν_{dn}/Ω_E). This hypothesis undoubtedly requires further thorough investigations.

7.3.2 *Dynamics of dislocations*

Even far above the melting line, dislocations are ubiquitous in both 2D and 3D crystals. Dislocations are essential for understanding such properties as plasticity, yield stress, susceptibility to fatigue, fracture, etc. Their generation and motion is of interest in materials science (Kittel, 1961), the study of earthquakes and snow avalanches (Kirchner *et al.*, 2002), colloidal crystals (Schall *et al.*, 2004), 2D foams (el Kader and Earnshaw, 1999), and various types of shear cracks (Abraham and Gao, 2000; Rosakis *et al.*, 1999).

In elastic theory, a dislocation's core is treated as a singularity in an otherwise continuous elastic material. Obviously, such a simplified approach is often too crude to capture essential quantitative characteristics of dislocations, whose scales are usually of the order of the lattice constant. In conventional solids dislocation dynamics is almost impossible to study experimentally at an atomistic level (Murayama *et al.*, 2002) because of the small distances between the atoms/molecules, high characteristic frequencies, and the lack of experimental techniques of visualizing the motion of individual atoms.

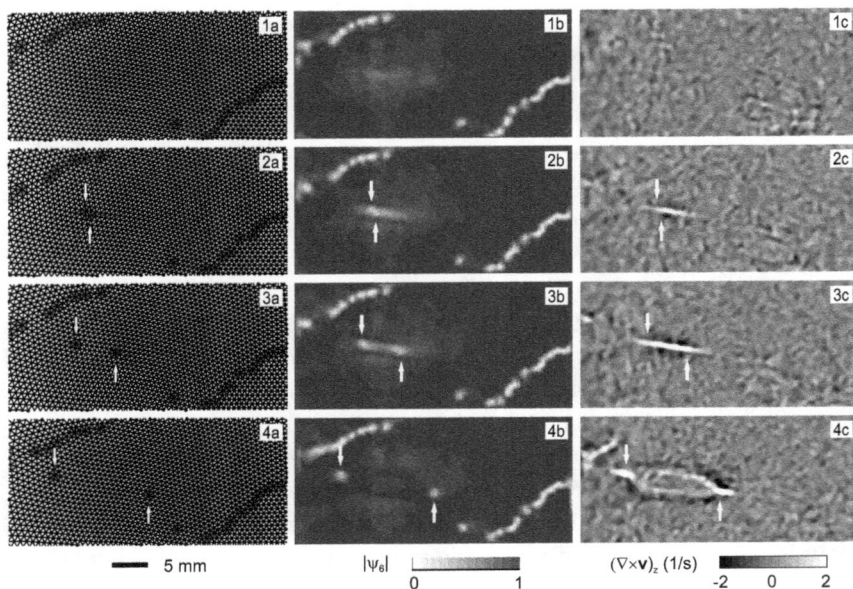

Fig. 7.13 Generation and dynamics of dislocation pairs in a 2D plasma crystal. Maps of (a) triangulation of the particle positions, (b) bond-orientational order parameter $|\psi_6|$, and (c) vorticity $|\nabla \times \mathbf{v}|$ are shown for four different instants of time: (1) 0.33, (2) 0.57, (3) 0.70, and (4) 1.00 s. A pair of dislocations is indicated by arrows. From Nosenko *et al.* (2007).

In contrast to molecular solids, complex plasmas turned out to be an exceptionally suitable model system for experimental study of the discrete structure and dynamics of dislocations. In the experiment by Nosenko *et al.* (2007) shown in Fig. 7.13, a 2D plasma crystal was heavily stressed due to inhomogeneous (parabolic) radial confinement. That was the reason for the strong variation of the number density across the crystal and, as a consequence, for the appearance of topological defects (indicated in Fig. 7.13a). Most of the defects formed linear chains that constitute domain boundaries in the crystal. During the course of the experiment, dislocations (i.e., isolated pairs of fivefold and sevenfold defects) were continuously generated due to the shear introduced by a slow rotation of the crystal. They then moved around and finally annihilated with each other or with domain boundaries.

In order to characterize dislocations at the discreteness limit, one first has to relate discrete and continuous measures of shear deformation. The most appropriate discrete measure (which is, at the same time, insensitive to uniform compressions, rotations, and translations, etc.) is the modulus of the bond-orientational order parameter $|\psi_6|$ shown in Fig. 7.13b. In the limit of weak simple shear, one can use the relation $|\psi_6| \simeq 1 - 9\gamma^2$, where γ is the shear strain; for weak pure shear, $|\psi_6| \simeq 1 - 2.25e^2$, where e denotes the elongation which is the measure of pure shear deformation (Nosenko *et al.*, 2007). The dislocation dynamics can be conveniently characterized in terms of 2D vorticity, $|\nabla \times \mathbf{v}|$, shown in Fig. 7.13c (where $\mathbf{v}$ is the particle velocity).

Figure 7.13b shows that the shear strain had a rather nonuniform distribution. It was higher (i.e., $|\psi_6|$ lower) in two kinds of locations. First, the shear strain was high in domain boundaries – the two nearly parallel bright stripes in Fig. 7.13b (or equivalently the chains of fivefold and sevenfold defects in Fig. 7.13a). Second, a "diffuse background" of shear strain appeared between the domain boundaries. The diffuse shear strain increased with time. When it locally exceeded a certain threshold, a pair of edge dislocations was created in that location, as one can see in the second row of Fig. 7.13; these dislocations appear as bright spots in (b) or as pairs of fivefold and sevenfold defects in (a), all indicated by arrows. Once a pair of dislocations was created, they moved apart rapidly (third and fourth rows). The Burgers vectors in such a pair were oppositely directed and equal in magnitude, so that the total Burgers vector was naturally conserved.

The creation of dislocation pairs is characterized by several distinct stages in the evolution of the shear strain: First, the shear strain builds up gradually in a certain location. Second, when the shear strain in this location exceeds a threshold, a pair of dislocations is born. Third, the shear stress is rapidly relaxed when the dislocations separate, and gradually drops to the background level. This cycle then starts over again, perhaps in a different location.

Dislocations that move supersonically create clear signatures – Mach cones that can be seen in Fig. 7.13c, fourth row. The Mach cones were composed of shear (but not compressional) waves, because they were excited by dislocations moving faster than the transverse acoustic velocity, C_t, but slower than the longitudinal velocity, C_l. The average speed of

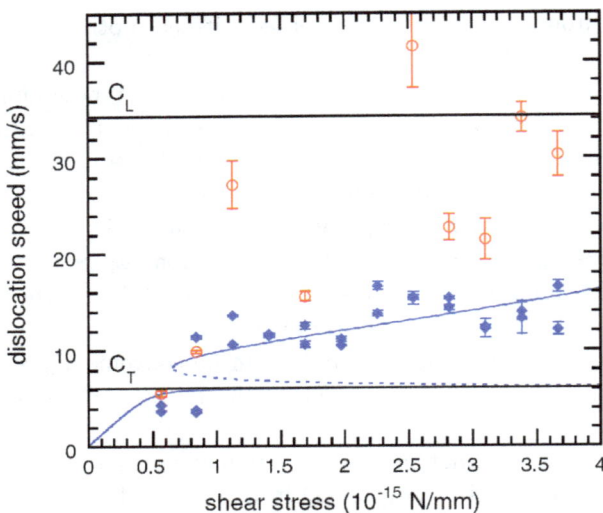

Fig. 7.14   Dislocation speed as a function of applied shear stress. Open circles are from high-speed motion during the initial relaxation, solid diamonds are from subsequent uniform motion. The curve is a fit of the theoretical prediction to the uniform motion data. Horizontal lines indicate the speeds of longitudinal and shear waves, C_l and C_t, respectively. From Nosenko *et al.* (2011).

supersonic dislocations in the experiment was about $2C_t$, which is in reasonable agreement with the results of particle-resolved simulations (Gumbsch and Gao, 1999) where gliding edge dislocations moving at the speed of $1.3C_t$ to $1.6C_t$ were observed. To the best of our knowledge, the results reported by Nosenko *et al.* (2007) provide the first experimental evidence that dislocations can move faster than C_t.

These findings were fully confirmed in recent experiments by Nosenko *et al.* (2011) where the the speed of dislocations was measured as a function of the shear stress (which was controlled by applying two counter-propagating laser beams of variable power to a homogeneous 2D crystal). Figure 7.14 shows the experimental results along with theoretical predictions by Rosakis (2001). The solid line is a least-squares fit of the theoretical stress-speed relation to the experimental data. In the velocity range between C_t and $\sqrt{2}C_t$ ($\simeq 8.6$ mm/s) the speed-stress curve has negative slope and therefore is unstable (shown by the dashed line). This implies that velocities of supersonic dislocations should exceed $\simeq 1.4C_t$ and, hence, explains the velocity gap seen in Fig. 7.14. Furthermore, theory predicts that supersonic motion is only possible above a critical stress corresponding to the critical point of the speed-stress curve. The predicted critical stress, $\simeq 0.6 \times 10^{-15}$ N/mm, is indeed very close to the measured value of $\simeq 0.8 \times 10^{-15}$ N/mm. Finally, in the regime above $\sqrt{2}C_t$ a monotonically increasing speed-stress relation is predicted, which is roughly the trend observed in the experiment.

7.4 Confined Systems

As we showed above, bulk freezing and melting processes in two and three spatial dimensions can be completely different. Confined systems allow us to follow a smooth transition between these regimes. Various types of confinement can be realized by imposing external fields and potential barriers for the particles. The most common types of confinement are slit confinement between two parallel plates, spherical (harmonic) confinements in trapping potentials, as well as periodic and aperiodic optical gratings. In principle, such confining fields can be realized both in colloids and complex plasmas.

7.4.1 *Crossover between 3D and 2D melting*

First, we shall briefly discuss the fundamental question of how the "intermediate" dimensionality (confinement) affects the crystallization/melting scenario. While 3D bulk crystals often melt via the grain boundaries (as illustrated in Figs 7.6 and 7.7, see also discussion in Sec. 7.2.4), in 2D crystals this process can develop either through grain boundaries (see Fig. 7.11 and discussion in Sec. 7.3.1) or in accordance with the KTHNY two-stage melting scenario, with the intermediate hexatic phase (see Sec. 7.1.2). The relevant question is therefore how the crossover between the melting in 3D and 2D systems occurs.

We illustrate the "dimensional crossover" with the recent experiments by Peng *et al.* (2010) which were performed by using thermosensitive colloids confined between two glass walls. The effective pair interaction in these experiments was quite close to the hard-sphere form, the volume fraction of colloids, ϕ, was controlled by slightly changing the ambient temperature. To avoid finite-size effects, a vertical vibration was applied, which effectively annealed the sample into very large domains (with $\sim 10^6$ particles per layer).

The results of the experiments are summarized in Fig. 7.15. It was found that "thick" ($\geq$ 5-layer) solids tend to melt at grain boundaries, by

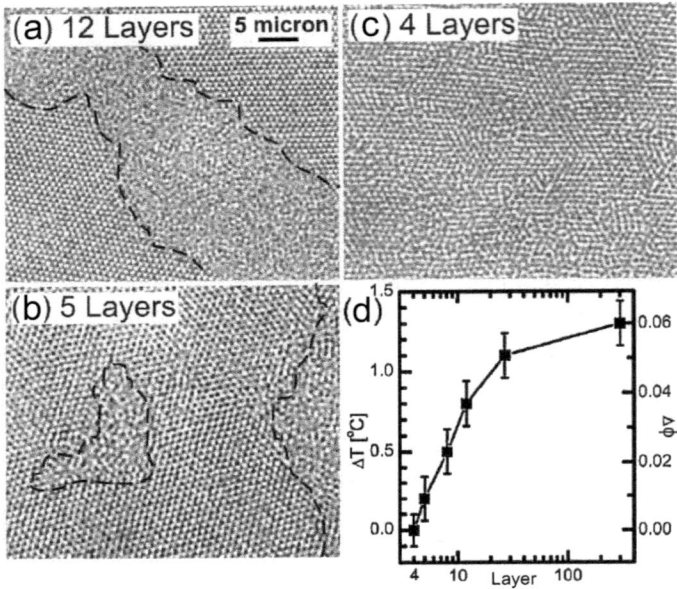

Fig. 7.15   Melting regimes of colloidal crystal films. (a) The melting develops from grain boundaries (indicated by dashed lines) in a solid whose thickness is about 12 layers, (b) a 5-layer solid melts from both grain boundaries and from within the crystalline domains (left loop indicates a liquid "lake", the dashed line on the right bounds a liquid "strip" which extends out of the image), and (c) in a 4-layer solid the melting starts at random positions within the sample. (d) The liquid-solid coexistence regime (expressed as the temperature window interval ΔT, or packing fraction interval $\Delta\phi$, for coexistence) decreases with the film thickness. From Peng *et al.* (2010).

generating liquid "strips" at grain boundaries (Figs. 7.15 a, b), while thin-film crystals (with ≤ 4 layers) abruptly melt at a critical volume fraction by generating many small defects and transient chainlike clusters at both grain boundaries and within crystalline domains (Fig. 7.15c). Furthermore, in the thickest films (about 20 layers), the coexistence regime was estimated to be in the range of ϕ from 53.9 % to 47.9 % (for bulk 3D systems of hard spheres, the solid-liquid coexistence regime is in the range from 54.5 % to 49.4 %). The liquid-solid coexistence regime decreased with film thickness and vanished at a critical thickness of 4 layers (Fig. 7.15d). Note that no reliable indications of the two-stage KTHNY melting scenario were found for thin films.

These results show that the classical bulk melting behavior accompanied by the solid-liquid coexistence can still be observed for rather thin films (comprised of ≥ 4 layers for hard-sphere interactions). On the other hand, the melting behavior of films consisting of 1–3 layers apparently depends on the interparticle interactions, and can tend either to the KTHNY scenario or be a one-step grain-boundary-melting process.

7.4.2   *Crystalline structures in slit confinement*

As one can conclude from the previous section, the slit geometry is particularly appealing since it interpolates between 2D and 3D bulk situations: If the plate distance is small the systems is practically confined to a monolayer, while for large slit widths the 3D bulk is approached. The third dimension (z) perpendicular to the slit is typically confined by an external one-particle potential $V_{\text{ext}}(z)$.

In order to realize a narrow slit confinement (both in complex plasmas and colloidal dispersions), the basic idea is to confine charged particles between two parallel plates (Pieranski *et al.*, 1983; van Winkle and Murray, 1986; Murray *et al.*, 1990b; Strandburg, 1992; Weiss *et al.*, 1995; Teng *et al.*, 2003; Chan *et al.*, 2004). The plates become charged, typically with the same sign of charge as the particles, so that the particle-wall interacting is repulsive (Murray *et al.*, 1990a; Murray and Grier, 1996; Teng *et al.*, 2003). Estimates of the repulsion strength show that it is typically considerably larger than the thermal energy $k_{\text{B}}T$. Therefore the system is confined almost completely to the middle of the slit and out-of-plane fluctuations can be kept sufficiently small. The number of layers in the system can be changed by tuning the wall repulsion, the particle density, or the wall distance.

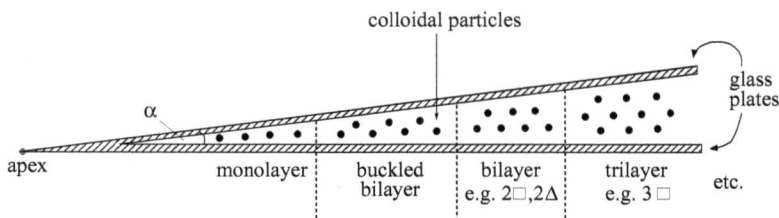

Fig. 7.16   Sketch of a wedge confinement of colloidal particles. For small opening angles α, there is a cascade of multilayered crystalline phases as a function of distance to the apex of the wedge.

If the two plates are not exactly parallel, forming a wedge, one can follow a whole cascade of solid-to-solid transitions along the wedge as shown

Complex Plasmas and Colloidal Dispersions

in Fig. 7.16. The latter configuration has been extensively studied in experiments with colloids (Pieranski *et al.*, 1983; van Winkle and Murray, 1986; Murray *et al.*, 1990b; Strandburg, 1992; Weiss *et al.*, 1995).

Depending on the nature of the confining boundary and the particles, different models for the potential well $V_{ext}(z)$ are realized. Below we focus on two generic types of confinement: The first type, confinement between *hard walls* typical to colloidal dispersions, simply involves geometric constraints and contains no energy scale. In this case, $V_{ext}(z)$ reads as

$$V_{ext}(z) = \begin{cases} 0 & \text{if} -L/2 \leq z \leq L/2, \\ \infty & \text{else,} \end{cases}$$

if the confining walls are located at $z = \pm L/2$, where L is the total (effective) slit width. The second type is the *parabolic* confinement (which is also typical to complex plasmas),

$$V_{ext}(z) = \frac{1}{2}\alpha z^2. \tag{7.3}$$

The confinement strength α can normally be controlled in experiments (in complex plasmas, it is related to the confinement eigenfrequency of a single particle via $\alpha = m\Omega^2_{conf}$). We note that the electrostatic confinement of charged particles can be screened. A reasonable approximation for the potential well in this situation is $V_{ext}(z) = V_0 \cosh \kappa z$ (Andelman, 1995). For small z, this obviously reduces to the parabolic form of Eq. (7.3).

7.4.2.1   *Hard spheres between hard walls*

Hard spheres confined between two hard slits constitute the simplest nontrivial model system. Temperature is irrelevant in this case, so that the system is characterized by the slit width L, particle diameter σ, and the mean interparticle distance. This results in two similarity parameters: A scaled (dimensionless) density,

$$\tilde{\rho} = n_{2D}\sigma^3/L,$$

where $n_{2D} = N/A$ is the number density per area, and a scaled slit width,

$$h = L/\sigma - 1,$$

so that $h = 0$ corresponds to a perfect monolayer. The quantity $\tilde{\rho}$ varies between zero and a close packing limit (recall that $\tilde{\rho} \equiv 6\phi/\pi$). By increasing h, the following cascade of close-packed configurations is obtained (Pieranski *et al.*, 1983; Schmidt and Löwen, 1996, 1997):

$$1\triangle \rightarrow B \rightarrow 2\square \rightarrow Rh \rightarrow 2\triangle,$$

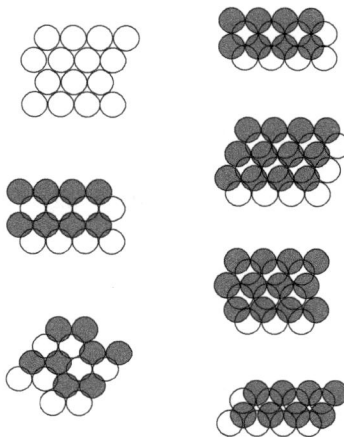

Fig. 7.17 Various close-packed structures. Left column: 1△, linear buckling, zigzag buckling (from top to bottom). Right column: 2□, linear rhombic, zigzag rhombic, 2△ (from top to bottom). Dark shaded and transparent spheres represent two consecutive layers. From Schmidt and Löwen (1997).

where 1△ is a triangular (or hexagonal) monolayer and B denotes a buckled phase of linear arrays which have a perpendicular zig-zag structure. The corresponding structures are sketched in Fig. 7.17. After the buckling, two intersecting square layers become stable, which is denoted by 2□. This configuration shears upon further increase of h into the rhombic phase Rh, until two staggered triangular layers, 2△, become the ground-state structure.

Beyond the bilayer regime which is limited by 2△, the cascade is (Oğuz *et al.*, 2009a; Ramiro-Manzano *et al.*, 2007),

$$2\triangle \rightarrow 2\text{hcp-like} \rightarrow 2\text{hcp}(100) \rightarrow 2\text{hcp-like} \rightarrow 2\text{P}_\square \rightarrow 3\square.$$

This involves a prism phase $\text{P}_\square$ and more complicated cut-outs and derivatives of the hcp phase followed by a complex series for higher number of layers (Neser *et al.*, 1997; Ramiro-Manzano *et al.*, 2007; Fortini and Dijkstra, 2006).

All the structures discussed above have been confirmed in experiments with charged colloids in the limit of high salt concentration (Neser *et al.*, 1997; Ramiro-Manzano *et al.*, 2007, 2006; Manzano *et al.*, 2009; Cohen *et al.*, 2004). This is illustrated in Fig. 7.18, where the volume fraction ϕ is shown as a function of $h + 1$ up to the point when a 4△ structure is reached.

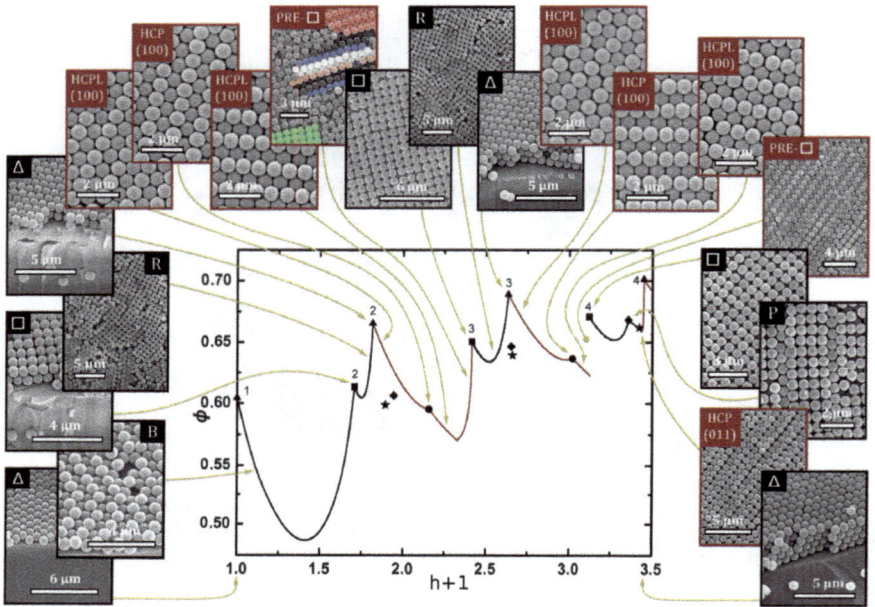

Fig. 7.18   Calculated volume fraction ϕ as a function of the reduced cell thickness value $h+1$. Solid lines correspond to the experimentally observed stable phases, 1 to 4 indicate the number of layers. The SEM images surrounding the graph show different structures: triangular ($\triangle$), square ($\square$), hcp(100), hcp(011), prismatic (P), buckling (B), rhombic (Rh), hcp-like (100), and pre-squared (pre-$\square$). From Ramiro-Manzano *et al.* (2007).

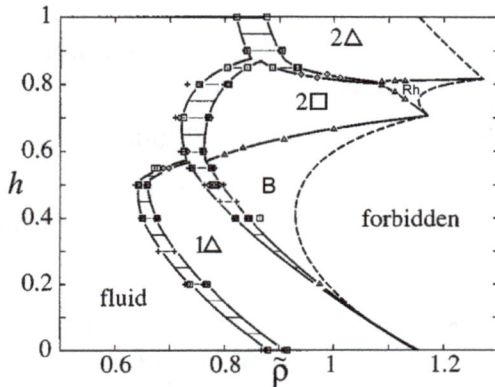

Fig. 7.19   Phase diagram of hard spheres of density $\tilde{\rho}$ confined between parallel hard walls with the reduced separation distance h. Solid lines are guides to the eye. Thin horizontal lines identify two-phase coexistence regions. From Schmidt and Löwen (1997).

In the regime of small h, the phase diagram in the $(\tilde{\rho}, h)$ plane is shown in Fig. 7.19. There is only a small stability regime for the rhombic phase. Exploring the full phase diagram beyond the $2\triangle$ phase away from the close-packing regime is still a fascinating problem where the first results (from simulations) have already been reported (Fortini and Dijkstra, 2006).

7.4.2.2 *Yukawa particles between hard walls*

At zero temperature, the stable crystalline structure for point-like Yukawa particles in a hard slit depends solely on the slit width L, screening length κ^{-1} ($\equiv \lambda$), and mean interparticle distance. Now, the density can be conveniently characterized by the number of particles in a cube of size L,

$$\eta_{\text{hw}} = n_{2\text{D}} L^2,$$

while the reduced slit width is

$$\tilde{L} = \kappa L.$$

The limit $\tilde{L} = 0$ corresponds to the unscreened Coulomb system (Goldoni and Peeters, 1998; Schweigert and Peeters, 1999a,b; Messina, 2009), while the hard-sphere case is obtained in the opposite limit $\tilde{L} \to \infty$. A Yukawa system with hard-wall confinement has been realized both in charged colloids (Ramiro-Manzano *et al.*, 2007; Fontecha *et al.*, 2007) and complex plasmas (Baumgartner *et al.*, 2009; Donko *et al.*, 2003).

For the bilayer regime, the phase diagram obtained by Messina and Löwen (2003, 2006) is shown in Fig. 7.20. The phases $1\triangle$, B, $2\square$, $2\triangle$, known from the hard-sphere case, are also present at finite screening. However, there is an additional rhombic phase IVA with a reentrant behavior, which is missing for the hard spheres. Experiments with charged colloidal dispersions by Ramiro-Manzano *et al.* (2007) and Fontecha *et al.* (2007) have confirmed the bilayer phase diagram (Fontecha *et al.*, 2005), which is also illustrated in Fig. 7.20. The additional rhombic phase IVa is confirmed experimentally.

The regime beyond bilayers was explored recently by Oğuz *et al.* (2009a), and a complex transition scenario from the $2\triangle$ to the $3\square$ layered phases was found. Apart from two prism phases $P_\triangle$ and $P_\square$ with triangular or square bases, a 2hcp-like and a 2hcp(100) phases are stable. Furthermore, there is a double-buckled structure which is reminiscent to a so-called "Belgian waffle iron" (BWI). Some of these additional phases (not all) have been confirmed experimentally by Ramiro-Manzano *et al.* (2007, 2006). In particular, the BWI phase still needs experimental verification.

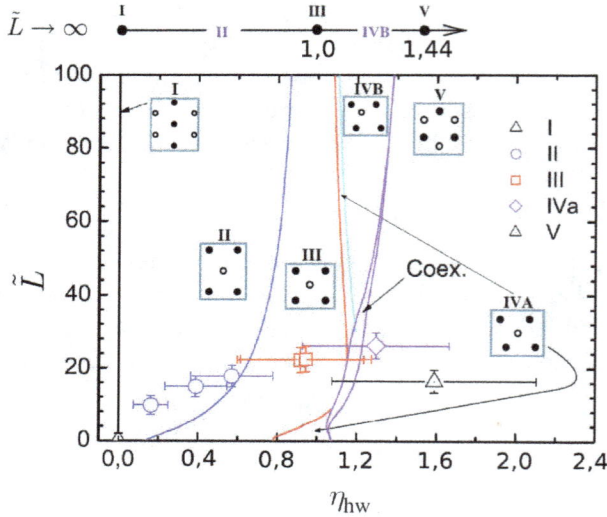

Fig. 7.20 Zero-temperature phase diagram of Yukawa particles of density η_{hw} confined between parallel hard walls with reduced separation distance $\tilde{L}$. The hard-sphere limit $\tilde{L} \to \infty$ is sketched on top. Different phases are illustrated by the insets showing the corresponding lattice structures, where the filled and open circles represent two consecutive layers. The coexistence domain of phases IV and V is also indicated. Symbols denote the positions of the experimentally observed phases. Despite the large error bars stemming from large measurement errors, a semi-quantitative overall agreement is observed. From Messina and Löwen (2003) and Fontecha *et al.* (2005).

Finally we mention that finite temperatures were also explored in Monte Carlo simulations by Grandner and Klapp (2008).

7.4.2.3 *Yukawa particles in parabolic confinement*

A harmonic potential well, Eq. (7.3), leads to stable structures which are different from those known for hard-wall confinement (Totsuji *et al.*, 1997; Totsuji and Barrat, 1988). Generally, a parabolic confinement tends to keep more particles in the central part of the slit. A special model for parabolic confinement was considered recently by Oğuz *et al.* (2009b), where the counterions were smeared out homogeneously across the slit.

For a parabolic confinement (which does not have a geometrical scale) the density is characterized by the parameter

$$\eta_{par} = n_{2D}/\kappa^2,$$

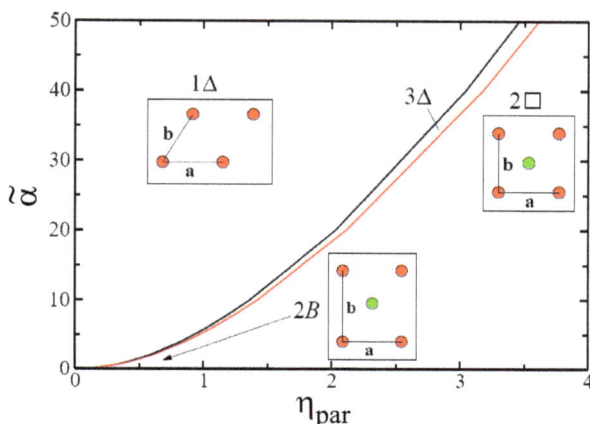

Fig. 7.21   Zero-temperature phase diagram of a Yukawa system in harmonic confinement. At a given confinement strength $\tilde{\alpha}$, the triangular monolayer is the ground-state
phase for low values of particle density η_{par}. Upon increasing η_{par} a second-order transition into the trilayer structure $3\triangle$ is observed. The further increase of η_{par} leads to
discontinuous transitions, either to the square phase $2\square$ (for $\tilde{\alpha} \gtrsim 0.2$) or to the rectangular buckling phase 2B (for $\tilde{\alpha} \lesssim 0.2$). The different phases with lattice vectors **a** and
b are sketched schematically in the insets, different particle colors represent consecutive
layers. From Oğuz *et al.* (2011).

while the normalized confinement strength is

$$\tilde{\alpha} = \alpha/Q^2\kappa^3.$$

For increasing density η_{par}, the cascade of stable phases is (Oğuz *et al.*,
2009b)

$$1\triangle \rightarrow 3\triangle \rightarrow 2\square \rightarrow 2\mathrm{Rh} \rightarrow 2\triangle \rightarrow 3\square \rightarrow 3\mathrm{Rh} \rightarrow$$
$$3\triangle \rightarrow 4\square \rightarrow 4\mathrm{Rh} \rightarrow 4\triangle \rightarrow 5\mathrm{Rh} \rightarrow 5\triangle \rightarrow 6\mathrm{Rh} \cdots.$$

Here, 3Rh–6Rh denote rhombic phases with 3–6 layers, respectively, with
the ABA–ABABAB stacking sequence, whereas the triangular phases $3\triangle$,
$4\triangle$ and $5\triangle$ have stacking sequence ABC, ABCA, and ABCAB, respectively.

Figure 7.21 shows the phase diagram for the beginning of this cascade
(Oğuz *et al.*, 2011). The buckling transition from a monolayer occurs into
a trilayer structure $3\triangle$, which is in clear contrast with the hard-wall case
(where buckling occurred into a bilayer with a rectangular unit cell).

In conclusion, let us make a short remark about the increased stability range of the $2\square$ phase. A bilayer of repulsive particles with two discrete possibilities of the vertical coordinate is *strictly equivalent* to a monolayer of binary particle mixture with negative nonadditivity (i.e., when the

cross-interactions are less repulsive than the interaction between the same species). The two species correspond to particles in the two layers (Das *et al.*, 2002). The negative nonadditivity leads to mixing of the two species, resulting in a nested 2□ phase where the degree of mixing is optimal.

7.4.3   *Other types of confinement*

Much related work has been done for a spherical parabolic confinement, which is the simplest model for traps created by optical tweezers confining colloidal particles as well as for thermophoretic traps in complex plasmas (see Fig. 5.4). In this case, the harmonic confining potential is

$$V(r) = \frac{1}{2}\alpha r^2,$$

with the trap centered at the origin and r denoting the radial distance. For a sufficiently large confining strength α and/or the number of trapped particles, there is a crossover to freezing within the confined cluster. For complex plasmas, this has been demonstrated for the so-called "Coulomb balls" (Arp *et al.*, 2004, 2005; Bonitz *et al.*, 2006, 2010). A vertical cross section of a particle configuration shown in Fig. 7.22 reveals a crystalline shell ("onion") structure around the potential minimum. There are "magic numbers" for which the shell structure is optimal (Peeters *et al.*, 1995; Apolinario *et al.*, 2006), and different dynamic eigenmodes of the cluster have been identified (Melzer *et al.*, 2001). A similar setup in two spatial dimensions can be realized for super-paramagnetic particles in a spherical environment (Bubeck *et al.*, 1999), as illustrated in Fig. 7.23. In principle, the number of particles trapped in clusters is relatively small, implying that "strict" thermodynamic limit cannot be reach for such systems. In practice, however, this number can be made sufficiently large to demonstrate clear freezing behavior.

Regarding other types of confinement, a static external potential which is periodic in one spatial direction can be realized for colloids by crossing laser beams and employing the tweezer effect (see Sec. 3.5.5). In the simplest form, the corresponding external potential reads

$$V(x) = V_0 \cos kx, \tag{7.4}$$

where V_0 is an energy amplitude and k is the wave number of the optical grating. The freezing behavior of colloidal particles in confinement (7.4) has been explored more than twenty years ago by Chowdhury *et al.* (1985), leading to the scenario of laser-induced freezing which exhibits interesting

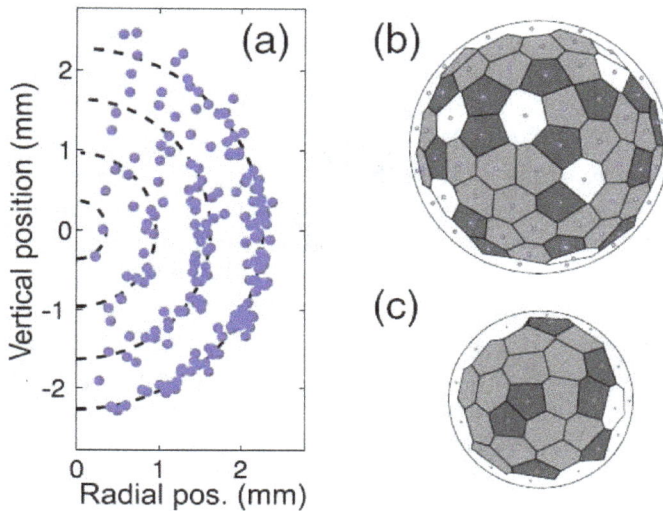

Fig. 7.22   Spheroidal cluster (Coulomb ball) consisting of 190 particles. (a) The shell structure is evident by projecting all particles onto a (vertical) plane. (b,c) Arrangements of the particles (small circles) in different shells, superimposed by the Voronoi cell analysis. Hexagons are shaded light gray and pentagons dark gray. From Arp *et al.* (2004).

Fig. 7.23   (a) Schematic side view of the experimental cell. A TEM mesh (M) serves as lateral confinements for super-paramagnetic particles. (b) Photograph (top view) of a single compartment which is occupied by a 2D colloidal system. From Bubeck *et al.* (1999).

reentrant effects (Chakrabarti *et al.*, 1995; Bechinger *et al.*, 2001; Strepp *et al.*, 2002): As the factor V_0 is continuously increasing, the system undergoes first a fluid-solid and then a solid-fluid transition. The corresponding situation for a liquid-vapor transition in such "substrate" confinement was also investigated in the context of colloid-polymer mixtures, and a split of the bulk critical point into two critical points separated by an intermediate structured phase was found (Götze *et al.*, 2003).

Gratings which are periodic in two spatial dimensions have also been studied. In particular, underlying triangular lattices lead to molecular crystals, where a "molecule" consists of a cluster (e.g., trimer) assembled in the same optical potential well (Brunner and Bechinger, 2002; Reichhardt and Olson, 2002). These molecular crystals show a melting and freezing behavior which is drastically different from bulk crystals of spherical particles. For instance, a two-step melting scenario was reported (Reichhardt and Olson, 2002), where the orientational order of the clusters is crucial.

By interfering five laser beams, even a quasi-crystalline decagonal substrate can be created (Mikhael *et al.*, 2008). A new pseudomorphic phase was observed in real-space experiments with colloids on these quasi-crystalline substrates, which shows both crystalline and quasi-crystalline structural properties. The structural elements of the pseudomorphic phase comprises an archimedean-like tiling with alternating rows of square and triangular tiles. Corresponding theoretical studies reveal a wealth of new phases to be observed on quasi-crystalline substrates, e.g., a quasi-crystalline phase with 20-fold bond order (Schmiedeberg and Stark, 2008).

7.5   Summary

The liquid-solid transition provides an avenue of research where the benefits of particle-resolved studies are hard to overstate. Hard spheres are quite well approximated by colloidal dispersions, yielding an excellent benchmark against which experiments and simulations/theory can be compared, not least due to the large amount of data for this system. Following our discussion of the role of interaction potential in phase behavior (see Chap. 6), it is natural to consider the crystal side of the phase boundary too, which has been done for Yukawa systems both in complex plasmas and colloid dispersions.

Perhaps it is in the non-equilibrium phenomena that particle-resolved studies truly come into their own. Put simply, the dynamics of freezing and melting are dominated by local nucleation events which are hard to access directly with conventional reciprocal-space experimental techniques. Real-space analysis allows direct visualization of such local phenomena, providing the key link to define local order parameters based on coordinate data. One day, such data may help us to develop a theory of the liquid-solid transition going beyond the crude assumptions of classical nucleation theory – which are known to be qualitatively incorrect perhaps in the majority of cases.

Among the key achievements have been direct observation of nucleation in 3D systems as well as investigation of 3D crystal melting from defects (grain boundaries). Remarkable progress has been also achieved in the particle-resolved studies of 2D crystal melting which can either evolve via the KTHNY two-stage scenario (as observed in some equilibrium colloidal experiments) or be induced at the grain boundaries (as suggested by available experiments with complex plasmas). Because of the uniqueness of particle-resolved studies, new phenomena can be tackled: For example, while the conventional Lindemann criterion is not directly applicable for 2D crystals (due to diverging fluctuations), particle-resolved studies offer the chance to compare displacements with neighboring particles. Furthermore, experiments with weakly damped complex plasmas (where the damping rate is small compared to the relevant dynamical eigenfrequency) allow us to directly study dynamics of defects in 2D and 3D crystals, and identify the role of latent heat in the melting/crystallization kinetics.

Chapter 8

Binary Mixtures

Both in nature and technological applications, fluid and solid systems are often mixtures of different particles (atoms). Each particle species is generally characterized by its interaction with other particles (e.g., interaction strengths and/or lengths are different), thus providing increasingly rich phenomenology – both in terms of structure and kinetics. Even for binary mixtures, the variety of phase states and structural correlations is very large in comparison with one-component systems. Below we present some truly fascinating examples from binary colloids and complex plasmas.

8.1 Static Liquid Structure

Static correlations in 2D liquids can be computed straightforwardly, using particle-resolved studies. Complementary to scattering experiments, where structural pair correlations are obtained in reciprocal space, pair correlation functions in real space can directly be accessed and compared to theory and simulations.

One important paradigmatic example is a 2D dipolar mixture, which can be realized for super-paramagnetic particles at a pending air-water interface exposed to an external uniaxial magnetic field. Such mixtures are known to be good glass formers (König et $al.$, 2005) and to exhibit complex crystal ground states (Assoud et $al.$, 2007). Therefore, their structural correlations have been studied in great detail (Kollmann et $al.$, 2002; Assoud et $al.$, 2009a,b; Ebert et $al.$, 2008). Figure 8.1 shows a comparison between experimental and simulation data for the three partial pair correlation functions, $g_{AA}(r)$, $g_{AB}(r)$, $g_{BB}(r)$, at two different coupling strengths (Assoud et $al.$, 2009a). Except for a fine substructure in $g_{BB}(r)$, there is very good overall agreement. Higher-order structural correlations, e.g., particles with

Complex Plasmas and Colloidal Dispersions

square-like and triangular-like surroundings have also been studied (König, 2005; Assoud *et al.*, 2009a). These building blocks may be important for understanding the onset of glass formation (Assoud *et al.*, 2009b), as we discuss in the next chapter.

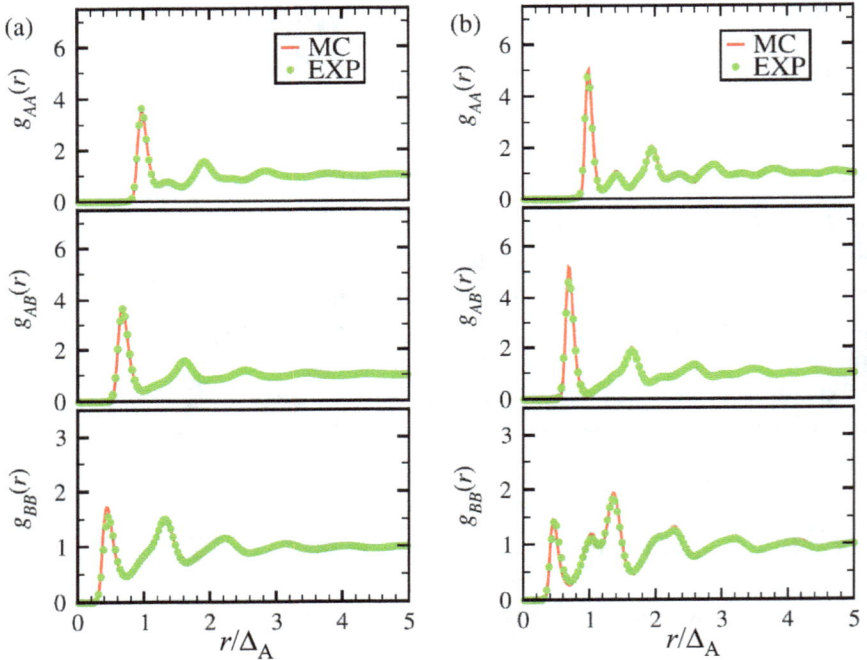

Fig. 8.1   Correlations in binary systems. Partial radial distribution functions for a 2D binary AB mixture of parallel dipoles, $g_{AA}(r)$, $g_{BB}(r)$, and $g_{AB}(r)$, where Δ_A is the mean distance between particles A. Experimental data (EXP) are compared with results of Monte Carlo simulations (MC) for the coupling parameter (of species A) $\Gamma = 22.6$ (a) and $\Gamma = 49.5$ (b) [see Eq. (8.1)]. For both cases, the ratio of magnetic moments is $m_B/m_A = 0.1$ and the composition is fixed to $x_B = 0.44$. From Assoud *et al.* (2009a).

8.2   Crystallization

The number of stable crystalline lattices in a binary mixture is significantly higher than in a one-component system. Finding the stable solid structure is a nontrivial problem even for zero temperature, because minimization of the lattice energy sums needs to be carried out for many candidate structures.

Moreover, the relative stability of a binary crystal at a fixed composition does not guarantee global stability, i.e., a mixed crystal can spontaneously split into two pure crystals. Thanks to new numerical minimization techniques which make it possible to consider huge sets of candidate structures, substantial progress has been achieved here in recent years. At the same time, significant advances in synthesizing binary colloidal mixtures and in analyzing their crystalline structures in real space have led to a flourishing understanding of the equilibrium phase diagrams for crystalline binary systems.

The structures which can be realized in 3D binary mixtures (for a given interaction class) are usually so diverse that they cannot be presented in a form of a (reasonably simple) phase diagram (Chaikin and Lubensky, 1986). In this section, therefore, we use 3D experimental results merely to illustrate enormous variety of crystalline binary mixtures. The theoretical results are focused on 2D mixtures, discussing hard sphere systems, induced dipolar interactions, as well as Yukawa and unscreened Coulomb interactions.

8.2.1   *3D systems*

8.2.1.1   *Hard spheres*

Interactions in a binary hard-sphere mixture are described by the following pair potential:

$$V_{\alpha\beta}(r) = \begin{cases} 0 & \text{if } r \geq \sigma_{\alpha\beta}, \\ \infty & \text{if } r < \sigma_{\alpha\beta}, \end{cases}$$

where $\alpha, \beta = \text{A}, \text{B}$ and $\sigma_{\alpha\beta} = \frac{1}{2}(\sigma_\alpha + \sigma_\beta)$. The equilibrium phase diagram depends on three parameters: The size asymmetry $\tilde{S} = \sigma_\text{B}/\sigma_\text{A}$, the relative composition x_B (of species B), and the (osmotic) pressure p. In the limit $p \to \infty$, a close-packed structure is reached. Parthe (1961) explained many atomic crystal structures using packing arguments of hard spheres. In colloid experiments, Murray and Sanders (1980) and Sanders (1980) observed binary crystal structures in naturally occurring gem opals and interpreted their results in terms of close-packed structures. But finding the close-packed regime with numerical methods is a tricky task even in the $(\tilde{S}, x_\text{B})$-plane, due to its delicate dependence on the size asymmetry (Filion, 2011). It is known that at a given $\tilde{S}$ a multitude of different binary lattices is stable. Among them are AB_n phases with $n = 1, 2, 6, 13, \ldots$ (Trizac *et al.*, 1997; Eldridge *et al.*, 1995, 1993). Possible examples are the structures analogous to the atomic binary crystals of NaCl, CsCl, AlB_2, CrB,

γCuTi, αIrV, HgBr$_2$, AuTe$_2$, Ag$_2$Se, NaZn$_{13}$, and there are many more (Filion and Dijkstra, 2009). Some of these structures, in particular AB$_2$ and AB$_{13}$ lattices, have also been discovered in experiments with binary mixtures of sterically-stabilized colloidal suspensions (Bartlett and Pusey, 1993; Bartlett *et al.*, 1992; Schofield *et al.*, 2005). A detailed comparison to theoretical predictions, however, also reveals some differences which are attributed to non-equilibrium effects (Eldridge *et al.*, 1995).

At finite pressures, when additional subtle effects become important, the variety of stable structures is even larger. For $0.033 < \tilde{S} \leq 0.2$, fcc structures of big spheres with disordered small spheres are stable (Dijkstra *et al.*, 1999). The critical value $\tilde{S} = 2/\sqrt{3}-1 = 0.154...$ marks the geometric threshold for which a small sphere can pass through a triangle formed by three (touching) big spheres and, thus, ergodicity can be restored for any configuration of big spheres. For $0.3 \leq \tilde{S} \leq 0.42$, an interstitial solid solution was found (Filion, 2011). Then, for $0.74 \leq \tilde{S} \leq 0.84$ the so-called *Laves phases* known from atomic metallic mixtures MgCu$_2$, MgNi$_2$, and MgZn$_2$ are stable (Hynninen *et al.*, 2007, 2009). These structures are promising candidates for optical band-gap materials. Finally, for a small size asymmetry in the range of $0.85 \leq \tilde{S} < 1$, eutectic solid solutions, azeotropic, and spindle phase diagrams were obtained (Kranendonk and Frenkel, 1991).

The dynamics of crystallization and nucleation in binary mixtures, which is much less explored, was also found to depend critically on the size ratio. Using MD simulations, Williams *et al.* (2008) discovered that crystallization in mixtures exhibits qualitative differences from classical nucleation theory. Rather, some crystallites form almost immediately and then an interplay between compositional fluctuations and crystal growth is able to dramatically extend the timescale on which further crystallization occurs. Finally, in colloid experiments polydispersity also strongly affects the crystallization kinetics (Martin *et al.*, 2005).

8.2.1.2   *Like-charged colloids*

Charged binary colloidal mixtures are characterized not only by the size asymmetry $\tilde{S}$, but also by the charge asymmetry $\tilde{Q} = Q_B/Q_A$. Moreover, the interaction range in such systems is tunable by adding salt. The resulting dimensionality of the parameter space is therefore higher than that for binary hard-sphere mixtures. This fact naturally complicates comprehensive theoretical calculations: Unlike the 2D case (Assoud *et al.*, 2008), there are no theoretical or simulation data for the 3D freezing phase diagram (even for additive Yukawa mixtures at zero temperature). On the other hand, a wealth of experimental knowledge has been accumulated and the phase boundaries of various charged mixtures have been traced out (Meller and Stavans, 1992; Okubo and Fujita, 1996; Lorenz *et al.*, 2009; Lorenz and Palberg, 2010).

For large asymmetries in size or charge, clustering and phase separation were observed (Wette *et al.*, 2009). In some cases, either the matrix or the clusters had a crystalline structure. For smaller size ratios, $\tilde{S} \leq 0.5$, the system solidifies completely and then details of the phase diagram are determined by the charge ratio $\tilde{Q}$.

Non-equilibrium aspects of crystallization were also explored in experiments with binary mixtures of like-charged colloids. The measured crystal nucleation rates (Wette *et al.*, 2005) and crystal growth velocities (Stipp and Palberg, 2007) strongly deviate from respective results obtained for one-component systems.

8.2.1.3 *Oppositely charged colloids*

Ionic crystals composed of oppositely charged particles are probably the best-known mixed crystals. Based on Madelung sums at zero temperature (Evans, 1966), one finds three basic equimolar 3D structures whose stability depends on the ratio of the ion radii. For increasing asymmetry in the radii, the corresponding stability sequence involves the CsCl, NaCl, and ZnS structures, which are realized in nature for the (ionic) salt crystals. A more detailed study shows that there are more crystalline structures at finite temperature. For example, in the framework of the restricted primitive model (for equal radii) it was shown that "metallic" CuAu crystal structures, disordered fcc crystals, and tetragonal phases become stable at high densities and finite temperatures (Hynninen *et al.*, 2006b).

Oppositely charged colloids provide much more freedom in tuning the valencies and radii. Moreover, unlike molecular and atomic ionic crystals, global charge neutrality is assured by additional counterions in solution. Thus, the charge ratio of the colloids is an independent parameter (as it is for like-charged colloids). A plethora of stable structures emerges in the resulting multiple parameter space. While some of these structures have atomic and molecular analogues, many do not (Leunissen *et al.*, 2005; Hynninen *et al.*, 2006a; Bartlett and Campbell, 2005; Hynninen *et al.*, 2009). Most of the structures found experimentally can be confirmed in Madelung-lattice calculations for an oppositely charged Yukawa model (without any nonadditivity). For a special size asymmetry of $\tilde{S} = 0.31$, a comparison between confocal microscopy images and theoretical predictions is shown in Fig. 8.2. In fact, three different structures which were found to be stable (from ground-state minimization for Yukawa interactions) were confirmed experimentally. All of these structures exhibit strongly asymmetric stoichiometries and have a complex unit cell. The first two structures possess stoichiometry 8 and can be described as LS_8^{hcp} and LS_8^{fcc}, where L and S denote large and small spheres, respectively. The third structure is LS_6 with stoichiometry 6. This corresponds to $S_6 C_{60}^{bcc}$ crystal, where C_{60} denotes a buckyball (fullerene) structure.

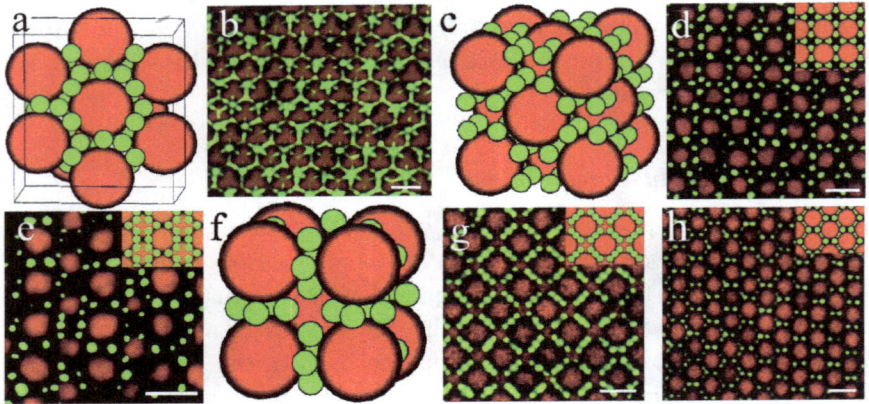

Fig. 8.2 Experimental observations and the corresponding theoretical predictions of crystalline structures in oppositely charged binary colloids. Confocal images of small positive (S, green) and large negative (L, red) PMMA particles: (b) superposition of several layers of LS_8^{hcp}; (d) (100) plane of LS_8^{fcc}; (e) (110) plane of LS_8^{fcc}; (g) (100) plane of $S_6 C_{60}^{bcc}$; (h) (110) plane of $S_6 C_{60}^{bcc}$. All scale bars are 3 μm. The insets in (d),(e), (g), and (h) show the corresponding plane from the theoretical predictions. Also, unit cells of (a) LS_8^{hcp}, (c) LS_8^{fcc}, and (f) $S_6 C_{60}^{bcc}$ structures are depicted. From Hynninen *et al.* (2006a).

Due to the multitude of different metastable crystalline structures in a binary system, crystal nucleation out of a melt can be a very complex process which is not captured by a simple classical nucleation theory. Computer simulations of oppositely charged colloids (Sanz *et al.*, 2007) showed that the nucleating crystalline phase can be metastable and, more surprisingly, the corresponding nucleation free-energy barrier is not necessarily the lowest one in comparison with other competing structures. These results are in direct contradiction with the common assumption inherent in classical nucleation theory.

8.2.2 *2D systems*

We now focus on 2D binary mixtures, to discuss zero-temperature phase diagrams for hard disks and charged particles interacting via screened and unscreened Coulomb interactions. The 2D binary mixture of parallel dipoles is also included as a special case allowing direct comparison between experiments and theory.

8.2.2.1 *Hard disks*

In principle, an additive binary mixture of hard disks could be realized for sterically-stabilized colloids pending at an interface or using gravitational confinement. The interaction potential in this case is similar to that in three dimensions. The central question here concerns the close-packing configuration and the corresponding stable crystalline structures. The close-packing phase diagram was first calculated by Likos and Henley (1993), and revisited recently using novel minimization techniques by Filion and Dijkstra (2009). For small size asymmetries (i.e., for $\tilde{S} \simeq 1$) there is complete phase separation into two pure A and B triangular crystals, while more complicated mixed crystals become stable at higher asymmetry. The stability of most of the phases was recently confirmed at finite pressures (Franzrahe and Nielaba, 2009).

8.2.2.2 *Parallel dipoles*

A binary mixture of super-paramagnetic colloidal particles pending at air-water interface is an excellent realization of a 2D classical many-body system (Zahn *et al.*, 1997, 1999; Zahn and Maret, 2000; Köppl *et al.*, 2007; Mangold *et al.*, 2004) [which can also be prepared by using particles with two different (permanent) dipole moments, Kollmann *et al.* (2002); Hoffmann *et al.* (2006a,b); König *et al.* (2005)]. An external magnetic field B_0 perpendicular to the interface induces parallel dipole moments m_A and m_B at particles A and B, respectively, resulting in an effective repulsive interaction which scales as the inverse cube of the distance r within the monolayer. By defining the magnetic susceptibilities per particle A and B as $\chi_{A,B} = m_{A,B}/B_0$, we obtain the pair potential,

$$V_{\alpha\beta}(r) = \chi_\alpha \chi_\beta \frac{B_0^2}{r^3}.$$

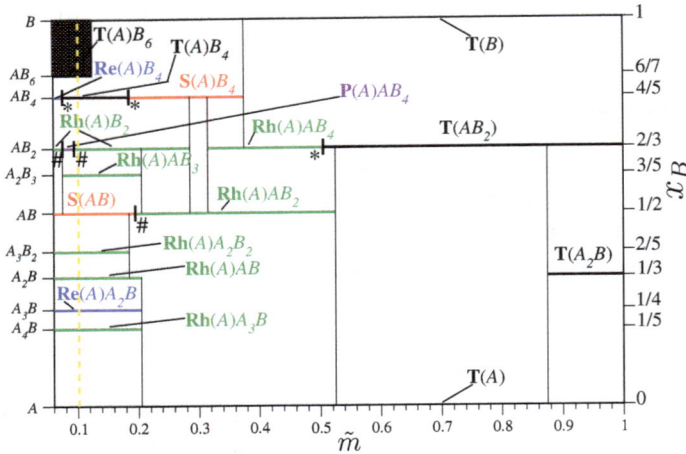

Fig. 8.3   Zero-temperature phase diagram for a 2D binary system of parallel dipoles, plotted in the plane of dipolar asymmetry $\tilde{m}$ and composition x_B. The symbols # and * denote continuous and discontinuous transitions, respectively, the dark box in the upper left corner shows unexplored region, the vertical dashed line indicates the dipolar asymmetry studied in experiments. From Assoud et al. (2007).

Note that for low B_0 the induced dipole moment is linearly proportional to the external field, and then χ_α is field-independent. In this case, for a fixed composition x_B and susceptibility ratio $\tilde{m} \equiv m_B/m_A = \chi_B/\chi_A$, all static quantities depend solely on the coupling parameter (Hansen and MacDonald, 2006),

$$\Gamma = \frac{\chi_A^2 B_0^2}{k_B T \Delta_A^3}, \tag{8.1}$$

where $\Delta_A = (n_{2D,A})^{-1/2}$ is the mean distance between particles A.

At zero temperature (i.e., for $\Gamma \to \infty$), the state of the binary system is completely determined by the ratio $\tilde{m}$ (varying in the range $0 \leq \tilde{m} \leq 1$) and the relative composition x_B of species B (with smaller dipole moment). The resulting 2D phase diagram (Assoud et al., 2007) is shown in Fig. 8.3. A wealth of different stable phases is getting more complex with increasing asymmetry (Fornleitner et al., 2008). The corresponding crystalline structures are illustrated in Fig. 8.4 and summarized in Table 8.1.

For small asymmetries $\tilde{m}$ and intermediate compositions x_B, the system splits into triangular A_2B and AB_2 phases, in marked contrast to the hard-disk mixture (Likos and Henley, 1993) which simply demixes into two pure triangular (A and B) crystals.

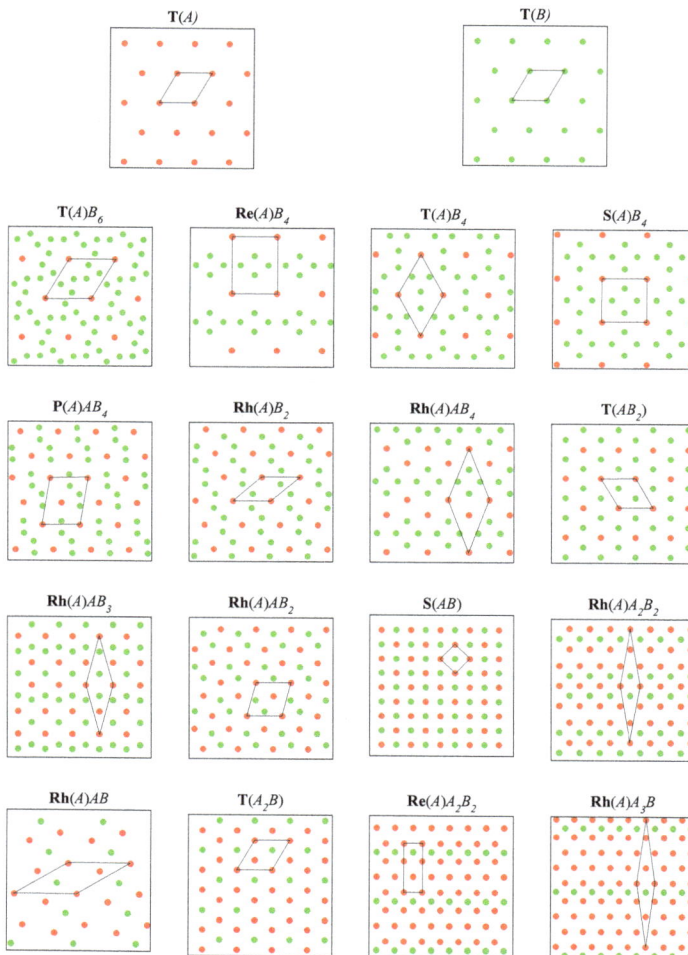

Fig. 8.4   Stable lattice structures and the corresponding primitive cells for a 2D binary system of parallel dipoles. The red and green points represent particles A and B, respectively. From Assoud *et al.* (2007).

Experiments with colloidal dipolar mixtures, which were performed for a strong asymmetry of $\tilde{m} \simeq 0.1$, confirmed the predicted crystalline structures (Ebert *et al.*, 2008) – however, only in a form of small crystallite patches. This provides motivation to study non-equilibrium kinetics which accompanies the quenching from a fluid into the "deep" crystalline state

Table 8.1   Stable phases for a 2D binary system of parallel dipoles.

Phase	Bravais lattice [basis]
$\mathbf{T}(A)$	Triangular for A [one A particle]
$\mathbf{T}(B)$	Triangular for B [one B particle]
$\mathbf{S}(AB)$	Square for A and B together [one A and one B particles]
$\mathbf{S}(A)B_n$	Square for A [one A and n B particles]
$\mathbf{Re}(A)A_mB_n$	Rectangular for A [$(m+1)$ A and n B particles]
$\mathbf{Rh}(A)A_mB_n$	Rhombic for A [$(m+1)$ A and n B particles]
$\mathbf{P}(A)AB_4$	Parallelogram for A [two A and four B particles]
$\mathbf{T}(AB_2)$	Triangular for A and B together [one A and two B particles]
$\mathbf{T}(A_2B)$	Triangular for A and B together [two A and one B particles]
$\mathbf{T}(A)B_n$	Triangular for A [one A and n B particles]

(Assoud *et al.*, 2009b). Within the experimental time scale, the binary mixture does not reach its true ground state but is quenched into a "glass" with patches showing similarities with stable bulk crystals (see Fig. 9.8). In fact, the dynamics in the patchy crystalline regions is significantly slower than in the disordered parts. As discussed in the next chapter, these findings might shed light on certain mechanisms underlying the dynamic heterogeneity in glass-forming systems (Bayer *et al.*, 2007; Hamanaka and Onuki, 2007; Widmer-Cooper and Harrowell, 2006; Kawasaki *et al.*, 2007).

8.2.2.3   *Yukawa particles*

The additive Yukawa model for a 2D binary mixture is characterized by the following pairwise potentials:

$$V_{\mathrm{AA}}(r) = V_0\varphi(r), \quad V_{\mathrm{AB}}(r) = \tilde{Q}V_{\mathrm{AA}}(r), \quad V_{\mathrm{BB}}(r) = \tilde{Q}^2 V_{\mathrm{AA}}(r).$$

The dimensionless function $\varphi(r)$ is given by

$$\varphi(r) = r^{-1}e^{-\kappa r},$$

while the amplitude V_0 sets the only energy scale at zero-temperature. The crystalline ground states were predicted by Assoud *et al.* (2008) in the plane

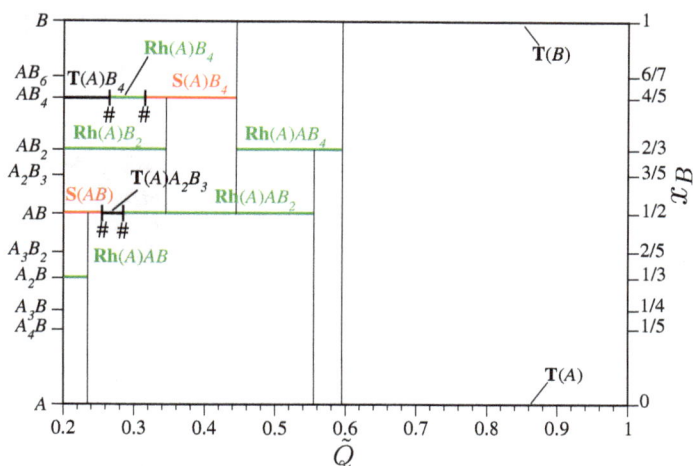

Fig. 8.5   Zero-temperature phase diagram for a 2D binary system of Yukawa particles, plotted in the plane of charge asymmetry $\tilde{Q}$ and composition x_B for a fixed pressure. The symbol # denotes a discontinuous transition. Two equivalent scales are presented along the vertical axis: Possible crystalline structures are shown on the left side, the corresponding compositions are indicated on the right. From Assoud *et al.* (2008).

spanned by the charge asymmetry $\tilde{Q}$ and the composition x_B. The phase diagram (for a fixed pressure) is presented in Fig. 8.5. Similar to the hard-disk interactions, there is a complete phase separation into pure A and B triangular crystals at $\tilde{Q} \simeq 1$. This trend persists over a broad range of asymmetries down to $\tilde{Q} \simeq 0.5 - 0.6$. For smaller $\tilde{Q}$, the phase behavior becomes more complex (Assoud *et al.*, 2008).

8.2.2.4   *Oppositely charged particles*

As a final example, we consider a binary (overall charge-neutral) mixture of oppositely charged particles. Although Coulomb interactions are, strictly speaking, screened for both complex plasmas and colloidal dispersions, at the very small interparticle separations normally found in ionic crystals the screening can be reasonably neglected. The corresponding pair potential is then

$$V_{\alpha\beta}(r) = \begin{cases} \dfrac{Q_\alpha Q_\beta}{r} & \text{if } r \geq \sigma_{\alpha\beta}, \\ \infty & \text{if } r < \sigma_{\alpha\beta}, \end{cases}$$

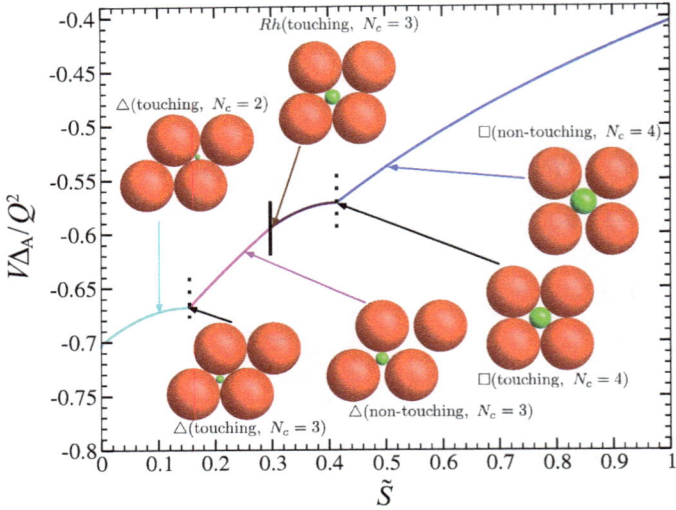

Fig. 8.6   Stable structures of oppositely charged spheres, shown for the (scaled) energy V per ion versus the size asymmetry $\tilde{S}$. The solid and broken bars denote the discontinuous and continuous transitions, respectively. Unit cells of the corresponding stable phases are indicated, N_c refers to the number of touching spheres in the unit cell. From Assoud *et al.* (2010).

where the charges are $Q_A = -Q_B = Q$. A typical setup is to consider zero imposed pressure (Hynninen *et al.*, 2006a,b).

Figure 8.6 demonstrates stable crystalline structures as a function of the size asymmetry $\tilde{S}$ (Assoud *et al.*, 2010). These structures minimize the potential energy per particle pair, and are also classified according to their "connectivity" and their coordination number. In particular, a "touching" configuration implies connected big spheres and N_c measures the contacts between big and small spheres per big particle. By increasing the size asymmetry, a cascade with six different structures is obtained.

8.2.2.5   *Quasi-2D systems*

We conclude this section with a characteristic example of a binary complex plasma crystal which illustrates a distinction between strictly-2D and quasi-2D crystalline systems. Such a distinction can be made even if only very weak perturbations of the third (vertical) coordinate are allowed (i.e., we are not talking here about the crossover from 2D to 3D crystallization discussed in Sec. 7.4.1).

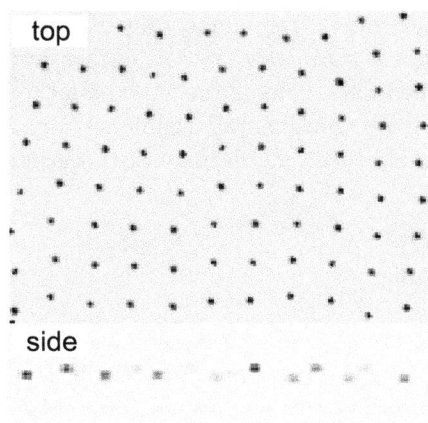

Fig. 8.7   Top and side view on a small region of a binary plasma crystal. Mixture of particles ($x_B \simeq 0.5$, diameters 9.19 μm and 11.36 μm) form a square lattice with the structure $\mathbf{S}(AB)$. Courtesy of C. Du.

Figure 8.7 shows the side and top views of a binary crystal formed by Yukawa-interacting particles of rather small size disparity. The vertical harmonic confinement in this experiment (determined by the balance of gravity and inhomogeneous vertical electric field) was not particularly stiff and therefore smaller and larger particles levitated at slightly different height (the height difference $\simeq 150$ μm was about one quarter of Δ). While for a strictly 2D case the ground state of such system would undoubtedly be a triangular lattice (see Fig. 8.5), the dominant structure observed in the experiment is a square (slightly distorted, since the crystal consists of patchy regions of different orientation and shape). This is because these two configurations are characterized by very similar energy levels, and therefore a slight increase in the coupling energy associated with the transition from triangular to square lattice (in a 2D plane) is well compensated by the energy decrease due to buckling (and hence increased interparticle distance). A similar transition also occurs for one-component crystals in a harmonic confinement:[1] Figure 7.21 shows that the width of the triangle lattice zone of the buckled monolayer is very narrow – by decreasing the confinement by less than 10 % below the buckling line one encounters the transition to a square lattice.

---

[1] As we pointed out in the end of Sec. 7.4.2.3, a buckled configuration in this case is equivalent to a 2D binary mixture of particles with negative nonadditivity.

This example illustrates a very important fact that the theoretical results on the structure of 2D binary crystals (which were presented above) can be directly applied only to experiments where the confinement is sufficiently strong. While experimental techniques for colloidal dispersions are relatively well established, 2D plasma crystals still require very careful preparation.

8.3   Fluid Demixing

Fluid phase separation is a ubiquitous phenomenon (Onuki, 2004) observed in various systems ranging from molecular fluids (Rowlinson and Swinton, 1982) to colloidal suspensions (Hansen and Löwen, 2000) (see also discussion in Sec. 6.3). This phenomenon, despite its long research history, remains of fundamental importance. In particular, the behavior of binary mixtures in the vicinity of the critical point belongs to the same universality class as that of, e.g., conventional liquid-vapor phase transition (provided interparticle interactions are sufficiently short-range), the ferromagnetic transition in uniaxial magnets or the 3D Ising model (Fisher, 1974).

8.3.1   *Phase equilibrium*

Generally, the phase equilibrium in a binary fluid consisting of species A and B is commonly characterized by the mean-field order parameter $\psi = n_A - n_B$, where $n_{A,B}$ are the *local* densities of the species (Onuki, 2004; Puri and Wadhawan, 2009). For a given value of the control parameter, which can be the temperature T or/and total density n, there are two possibilities: The system can stay in a one-phase (mixed) state with the order parameter ψ_0, or undergo separation (demixing) into two distinct (A-rich and B-rich) phases characterized by the order parameters ψ_1 and ψ_2, respectively. The demixing is obviously the process with *conserved kinetics* (in contrast to, e.g., the spontaneous magnetization), when the order parameter is conserved, i.e., $\psi_0 = x_A\psi_1 + (1 - x_A)\psi_2$, where x_A is the fraction of A-rich phase (this determines the so-called "lever rule"). In the framework of the simplest mean-field approximation neglecting surface and inhomogeneity contributions, the free energy of the demixed state (for a given order parameter) is then $F_{demix} = x_A F(\psi_1) + (1 - x_A)F(\psi_2)$. The conditions of the phase coexistence are determined from the minimization of the function $\tilde{F}_{demix} = F_{demix} + \lambda[x_A\psi_1 + (1 - x_A)\psi_2 - \psi_0]$, with the

Lagrange multiplier λ representing the chemical potential of the system. From the conditions $\partial \tilde{F}_{\text{demix}}/\partial x_A$ and $\partial \tilde{F}_{\text{demix}}/\partial \psi_{1,2}$ we readily derive the well-known Maxwell's double-tangent construction (Landau and Lifshitz, 1978),

$$\frac{F(\psi_2) - F(\psi_1)}{\psi_2 - \psi_1} = F'|_{\psi=\psi_{1,2}} \ (=-\lambda). \tag{8.2}$$

(In a homogeneous case, $\psi_2 - \psi_1 \to 0$, this is reduced to identity.) The locus of points $\psi_{1,2}$ derived from Eq. (8.2) for different values of the control parameter determine the coexistence curve (binodal) for two phases. The condition $\partial^2 F/\partial \psi^2 < 0$ identifies the mechanically unstable region, and the line which bounds this (absolutely unstable) region is called the spinodal. Of course, the same formalism also applies to one-component systems undergoing phase separation (e.g., liquid-vapor phase transition).

To illustrate the basic thermodynamic features of fluid demixing let us consider the Ising model, which is very instructive for demonstrating the role of correlations and hence for comparing the simplest mean-field picture with the behavior of an interacting many-body system (Onuki, 2004). The model is represented by a set of N spins whose positions are fixed sites of a lattice (of arbitrary dimensionality); the spins can have two orientations ("up" or "down") and interact with the nearest neighbors only. Being originally proposed for magnetic systems, this setup can be easily mapped onto the lattice gas, to study demixing of binary AB mixtures: By introducing the occupation number $n_{\alpha i}$ for each site of the lattice ($\alpha = A$ or B, so that $n_{Ai} + n_{Bi} = 1$ for $i = 1, \dots N$) one can relate the local "magnetization" (ψ_i) to the species density via $\psi_i = 2n_{Ai} - 1 = 1 - 2n_{Bi}$, i.e., the up and down spins are represented by the species A and B, respectively. We assume that the pair interaction energy for the species α and β is $\epsilon_{\alpha\beta}$. Then the corresponding (exact) Hamiltonian of the system is

$$H = \sum_{(i,j)_{\text{nn}}} [\epsilon_{AA} n_{Ai} n_{Aj} + \epsilon_{BB} n_{Bi} n_{Bj} + \epsilon_{AB} (n_{Ai} n_{Bj} + n_{Bi} n_{Aj})]$$

$$= -J \sum_{(i,j)_{\text{nn}}} \psi_i \psi_j + \text{const}, \tag{8.3}$$

where $J = \frac{1}{4}(2\epsilon_{AB} - \epsilon_{AA} - \epsilon_{BB})$, the summation is for the nearest neighbors only (nn), and const denotes configuration-independent terms (note that H in this case is symmetric with respect to the A$\leftrightarrow$B permutation). The mean-field consideration of the problem is obtained by discarding the correlations between states of the neighboring sites. Then the local order parameter is replaced by the mean value, $\langle \psi_i \rangle = \psi$, so that the mean densities are $n_{A,B} = \frac{1}{2}(1 \pm \psi)$ and T is the control parameter. The resulting (mean-field) Helmholtz free energy is $F(T, \psi) = E(\psi) - TS(\psi)$, where $E \simeq -\frac{1}{2}NqJ\psi^2$ is the interaction energy (here q is the coordination number of the lattice site) and $S \simeq -Nk_B \ln(n_A^{n_A} n_B^{n_B})$ is the entropy of the lattice, i.e.,

$$F(T, \psi)/N = -\frac{1}{2}qJ\psi^2 + \frac{1}{2}k_B T \left[(1+\psi)\ln(1+\psi) + (1-\psi)\ln(1-\psi) \right].$$

For the Ising model $F(\psi)$ is an even function and, hence, from Eq. (8.2) we obtain that the binodal line $\psi_{bin}(T)$ is simply determined by the condition $\partial F/\partial \psi = 0$. This yields $\psi_{bin} = \tanh[(T_{cr}/T)\psi_{bin}]$, whereas the spinodal line (locus of inflection points $\partial^2 F/\partial \psi^2 = 0$) is given by $\psi_{sp}(T) = \sqrt{1 - T/T_{cr}}$. Here $T_{cr} = qJ$ is the critical temperature: For $T > T_{cr}$ there is a single extremum (minimum) at $\psi = 0$, whereas for $T < T_{cr}$ there is a double minimum at $\psi_1 = -\psi_2 \equiv \psi_{bin}(T)$ (and a local maximum at $\psi = 0$). Hence, T_{cr} identifies the bifurcation point for the free

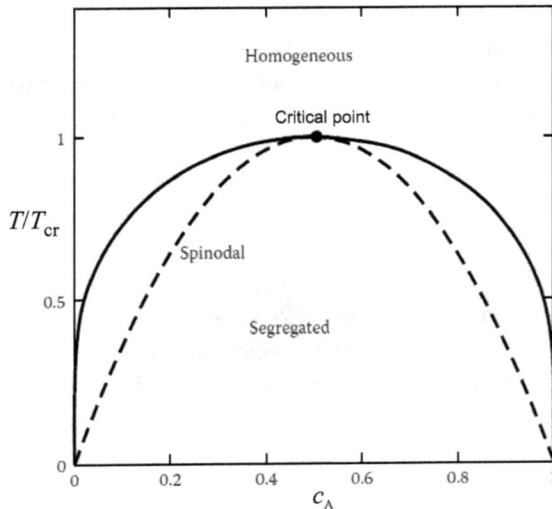

Fig. 8.8   Sketch of a mean-field phase diagram for (symmetric) AB binary mixtures. The system parameters are the concentration c_A of species A and the temperature T (normalized by the critical temperature T_{cr}, the critical concentration is at $c_A = 0.5$). Above the coexistence curve (solid line) the system is in a homogeneous (mixed) state, below the curve it is in a phase-separated state (characterized by A-rich and B-rich regions). The dashed line denotes the spinodal curve. The homogeneous system is metastable between the coexistence and spinodal curves and unstable below the spinodal lines. From Puri and Wadhawan (2009).

energy, where $F(\psi)$ changes from V- to W-shape form. Note that the phase separation is only possible for $J > 0$, i.e., in the case of *positive nonadditivity* of pair interactions (in Sec. 8.3.4 we present a more detailed discussion of this condition).

Figure 8.8 illustrates the phase diagram of a symmetric binary mixture (note the similarity with the diagram of liquid-vapor phase transition, Fig. 6.5a). If the composition is symmetric in the mixed state, i.e., $\psi_0 = 0$ (critical composition), then after quenching into the unstable regime $T < T_{cr}$ the order parameter varies continuously with temperature, following the classical mean-field scaling $\psi_{bin}(T) \simeq \sqrt{3}\ \psi_{sp}(T)$ in the vicinity of T_{cr}. For asymmetric ("off-critical") quenching the demixing is a discontinuous process – in accordance with the lever rule we have $\psi_0 = (2x_A - 1)\psi_{bin}(T)$ and hence the phase separation starts when $\psi_{bin} \geq \psi_0$.

8.3.2   *Demixing kinetics*

The common "mean-field view" on the demixing kinetics of binary fluids (similar to that of the liquid-vapor phase transition) is that there are two different pathways for the phase separation to evolve. This can either be the *nucleation and growth* process – if the system is quenched into the metastable region between the binodal and spinodal curves, or the process of *spinodal decomposition* – for the quenching below the spinodal. Droplets of minority phase growing in the process of nucleation (which can be realized when the quenching is off-critical) must overcome a free-energy barrier. The latter is associated with the competition of the bulk and surface contributions to the free energy, i.e., emerging nuclei eventually form a new macroscopic phase if they grow beyond the critical size, or dissolve otherwise. In the regime of spinodal decomposition the phase separation is initiated by the spontaneous growth of long-wavelength fluctuations.[2] The dynamics of domain evolution in both regimes is accompanied by a sequence of growth stages which are believed to be self-similar in time, i.e., the domain morphology (at each stage) is preserved (Onuki, 2004; Puri and Wadhawan, 2009; Bray, 2002). This implies a single time-dependent characteristic length which obeys a power-law growth, $L(t) \propto t^\alpha$, with different growth exponents α peculiar to each stage. Let us briefly summarize the growing stages in both regimes.

---

[2]The distinction between different regimes of the phase separation can be poorly defined outside the range of applicability of the mean-field theory – see next section.

The nucleation and growth regime consists of (at least) two distinct consecutive stages: The first stage is due to the so-called evaporation-condensation mechanism proposed by Lifshitz and Slyozov (1961), when nuclei of the minority phase grow out of the supersaturated solution. Larger droplets, with the size above the critical, grow on expense of smaller ones which eventually dissolve (and, therefore, sometimes this stage is also called coalescence). The degree of supersaturation gradually decreases and, correspondingly, the critical size increases, which gives rise to the self-similar growth law, where the mean droplet size increases as $L(t) \simeq (\gamma D\phi_A t/nk_B T)^{1/3}$, where γ is the surface tension of the two-phase interface, D is the diffusion coefficient, ϕ_A is the volume fraction of the minority (A) phase at the coexistence line, and n is the total number density. Interestingly, when the droplets become sufficiently large the thermal fluctuations facilitate the second growth mechanism – the coalescence driven by Brownian motion of the droplets (Bray, 2002). The growth law at this stage can be easily estimated using the mass conservation for droplets, $NL^3 = $ const, and assuming that the droplets merge in pair coalescence events, so that the number density of droplets is governed by $\dot{N} \sim -DLN^2$. Taking into account the Stokes-Einstein relation $DL \sim k_B T/\eta$ (here η is the dynamic viscosity) we obtain $L(t) \sim (k_B Tt/\eta)^{1/3}$, i.e., the growth exponent remains unchanged and equal to 1/3.

One of the most broadly used mean-field theories to describe the dynamics of spinodal decompositions is the Cahn-Hilliard model (Cahn and Hilliard, 1958, 1959) for the order parameter $\psi(\mathbf{r}, t)$ which is now also a function of position. The model employs the free energy in the Ginzburg-Landau form (Onuki, 2004),

$$F[\psi] = \int d\mathbf{r} \left[F(\psi) + \frac{1}{2}K(\nabla\psi)^2 \right], \qquad (8.4)$$

so that the chemical potential is expressed via the functional derivative, $\mu(\mathbf{r}, t) = \delta F[\psi]/\delta\psi = F'(\psi) - K\nabla^2\psi$. The gradient term in Eq. (8.4) takes into account large-scale correlations in Eq. (8.3) (which are completely neglected in the mean-field consideration described in the previous section), and the coefficient $K > 0$ is a measure of the surface tension.[3] Strictly speaking, such gradient expansion is only accurate when the correlation

---

[3]By multiplying μ with $\partial\psi/\partial n$ (where $\mathbf{n}$ is the normal coordinate at the interface) and integrating, one readily obtains the Gibbs-Thomson boundary condition $\mu = -\frac{1}{2}\kappa_{int}\gamma$, where $\gamma = K \int dn(\partial\psi/\partial n)^2$ is the surface tension and $\kappa_{int} = \nabla \cdot (\mathbf{n}/n)$ is the interface curvature (Bray, 2002).

length ξ is sufficiently large, i.e., Eq. (8.4) is only exact in a vicinity of the critical point where ξ diverges.

The evolution of the order parameter in the Cahn-Hilliard model is governed by the continuity equation, where the diffusive flux $\mathbf{J}$ is given by the first Fick's law,[4] $\mathbf{J} = -(D/k_B T)\nabla \mu$. Using Eq. (8.4) we readily obtain the Cahn-Hilliard equation (Onuki, 2004; Puri and Wadhawan, 2009; Bray, 2002),

$$\frac{\partial \psi}{\partial t} + \mathbf{v} \cdot \nabla \psi = (D/k_B T)\nabla^2 \mu, \qquad (8.5)$$

where the advective derivative is added for the case when the hydrodynamic flow is important.

The initial (linear) stage of spinodal decomposition is obtained from Eq. (8.5) by linearizing the chemical potential, $\mu = (F''|_{\psi_0} - K\nabla^2)\delta\psi$, where $\delta\psi = \psi - \psi_0$. Assuming $\delta\psi \propto e^{\lambda t + i\mathbf{k}\cdot\mathbf{r}}$ we obtain $\lambda \propto -k^2(F''|_{\psi_0} + Kk^2)$. In the spinodal regime the second derivative of the free energy is negative and therefore the long-wavelength perturbations are unstable at $k < k_{cr}$, where $k_{cr} = \sqrt{-K^{-1}F''|_{\psi_0}}$, and the maximum growth rate is attained at $\frac{1}{\sqrt{2}}k_{cr}$. Thus, the fastest growth at this stage is observed for domains of size $L \sim k_{cr}^{-1}$ which is time-independent.

When the domain growth becomes highly nonlinear and sufficiently sharp interfaces between the phases are formed (so that one can talk about surface tension), the sequence of nonlinear phase separation regimes sets in. At the first nonlinear stage the hydrodynamics plays no role and the advective term in Eq. (8.5) can be neglected. This stage is termed *diffusive* and is characterized by the universal growth exponent $\alpha = 1/3$, both for 2D and 3D systems (Bray, 2002). In fact, Huse (1986) showed that the dependence $L(t)$ derived by Lifshitz and Slyozov for the nucleation and growth also operates at the diffusive stage of spinodal decomposition: The Gibbs-Thomson boundary condition yields $\mu \sim \gamma/L$, whereas the l.h.s. of Eq.(8.5) scales as $\partial\psi/\partial t \sim nL^{-1}dL/dt$, which yield $L(t) \sim (\gamma Dt/nk_B T)^{1/3}$. Although this mechanism appears to be fairly universal, in some cases it can be preempted, e.g., due to long-range interactions (see below). Figure 8.9 demonstrates typical morphology of evolving domains after critical (left) and off-critical (right) quench.

The next nonlinear stage of the demixing is termed *viscous hydrodynamic*, because hydrodynamics plays the major role now. The Cahn-Hilliard equations must be coupled to the Navier-Stokes equation, with

---

[4]To avoid confusion with the chemical potential, here we do not use the mobility coefficient explicitly.

Fig. 8.9   Typical example of spinodal decomposition in a critical ($n_A = 0.5$, left) and off-critical ($n_A = 0.3$, right) symmetric binary mixture which is quenched below the spinodal curve. The pictures are obtained from a numerical solution of the Cahn-Hilliard equation [Eq. (8.5) without advection] for the Ising model [Hamiltonian from Eq. (8.3)]. A-rich and B-rich regions are marked by black and white, respectively. From Puri and Wadhawan (2009).

the additional term $-\psi\nabla\mu$ describing the force due to the presence of interfaces, and to the continuity equation for the mass density $\rho = mn$ (Puri and Wadhawan, 2009; Bray, 2002),

$$\rho\left[\frac{\partial\mathbf{v}}{\partial t} + (\mathbf{v}\cdot\nabla)\mathbf{v}\right] = -\nabla p + \eta\nabla^2\mathbf{v} - \psi\nabla\mu,$$

$$\frac{\partial\rho}{\partial t} + \nabla\cdot(\rho\mathbf{v}) = 0.$$

When the quenching composition is not too off-critical the growing minority phase can percolate the system, i.e., there is a connectivity between the neighboring domains. In this case the surface tension drives the merging of droplets which is balanced by the viscous friction (Siggia, 1979). By comparing the surface tension term on the r.h.s. of the Navier-Stokes equation, $\sim \gamma/L^2$, with the viscous term, $\sim \eta L^{-2}dL/dt$, we obtain the linear growth $L(t) \sim (\gamma/\eta)t$. The crossover from the diffusive to viscous stage occurs at $L \sim (D\eta/nk_BT)^{1/2}$. Note that for highly off-critical mixtures (when the volume fraction of the minority-phase droplets is sufficiently small) the droplets are no longer connected and hence the viscous growth mechanism is disabled. Then we have the same sequence as discussed above for the nucleation and growth regime – the diffusive stage crosses over into the coalescence driven by Brownian motion of droplets (Siggia, 1979), with the growth law $L(t) \sim (k_BTt/\eta)^{1/3}$.

For completeness we just mention that at much later demixing stages, when the domains become so large that characteristic Reynolds numbers are comparable to unity and hence the inertia is no longer negligible, the surface tension force is balanced by the inertia term at the l.h.s. of the Navier-Stokes equation, $\sim \rho L^{-1}(dL/dt)^2$. This yields $L(t) \sim (\gamma t^2/\rho)^{1/3}$, and the crossover to the inertial stage occurs at $L \sim \eta^2/\rho\gamma$.

In experiments and numerical simulations, the evolving domain size $L(t)$ is usually deduced from the time-dependent average structure factor $S(k,t)$ calculated for the order parameter $\psi(\mathbf{r},t)$ (Aarts *et al.*, 2005; Furukawa, 1984), so that the position of maximum of $S(k)$ is identified as $2\pi/L$. For $k \to 0$ the universal scaling is $S(k) \propto k^2$, whereas at large k it depends on the demixing regime (Bray, 2002): The initial (linear) regime of the spinodal decomposition (when $L = $ const) yields $S(k) \propto k^{-2}$, the subsequent nonlinear stages are characterized by the Porod tail $S(k) \propto k^{-(D+1)}$ at $k \to \infty$. The measured $S(k)$ in the latter case is fitted by an interpolation formula (Furukawa, 1984); for 3D systems this gives $S(k) \propto (kL/2\pi)^2/[2 + (kL/2\pi)^6]$.

8.3.3   *Limitations of mean-field approach*

Simple mean-field considerations presented above provide a rather general view on the phase equilibria and the demixing kinetics. Such considerations, however, leave aside many important effects which can dramatically change the phase behavior.

The behavior near the spinodal line [which can be considered as a line of critical points, see review by Binder (1987) and references therein] is very complicated. The spinodal itself is only well defined for interactions which are sufficiently long-range, otherwise the fluctuations in its vicinity smear out the border between the spinodal decomposition and nucleation, i.e., there must be a gradual transition between the two regimes. There have been several attempts made to develop theories describing the gradual crossover [either based on the nonlinear Cahn-Hilliard equation with a random fluctuation source, e.g., by Langer *et al.* (1975), or employing a nucleation theory extended to a conservative cluster dynamics, e.g., by Binder (1977)], but unfortunately all revealed significant drawbacks (Binder, 1987). Furthermore, the position of the spinodal (even for long-range interactions) depends on the size of the considered system, $\mathcal{L}$, with respect to the correlation length ξ: The free energy of such a "coarse-grained" system is close to the mean-field form only when $\mathcal{L}/\xi \ll 1$ (and, hence, the phase separation is inhibited within the system), whereas for $\mathcal{L}/\xi \gtrsim 1$ it depends strongly on $\mathcal{L}$, so that in the limit $\mathcal{L} \to \infty$ the spinodal completely disappears (Lebowitz and Penrose, 1966; Penrose and Lebowitz, 1971; Langer, 1974).

The influence of the interaction range on the demixing kinetics is particularly striking on the initial stage, when the magnitudes of both the order parameter and its gradient are small and the linearized Cahn-Hilliard model is supposed to be applicable. Nevertheless, it was shown that for systems with short-range interactions the basic predictions of the linearized theory of spinodal decomposition [e.g., time-independent position of maximum of $S(k)$ and the exponential growth] are not confirmed (Binder, 1987; Marro *et al.*, 1975).

These examples illustrate the importance of systematic particle-resolved studies carried out in model systems where the interaction range can be tuned from a short-range to a very-long-range form. In the following section we present some very recent results in this direction.

8.3.4   *Particle-resolved studies*

The tendency for particles of different types to mix or demix is basically determined by the relative strengths of their interactions. It is noteworthy that the phase separation in such multicomponent systems does not require an attraction in the interparticle interactions – which is a necessary condition for the fluid phase transition in single-species systems.

The phase equilibrium can be conveniently specified by relations that are often referred to as the Lorentz-Berthelot mixing rules (Henderson and Leonard, 1971). These rules can have different formulations for different classes of interaction potentials [in some cases, the rules can also be obtained for interactions with an attractive long-range component, e.g., for the Lennard-Jones potential (Boda and Henderson, 2008)], but they have the most simple representation for repulsive interactions. For a binary composition of particles with long-range interactions $V_{\alpha\beta}(r)$ $(\alpha, \beta = \text{A,B})$, when the particle size is sufficiently small (viz., much smaller than both the screening length and the interparticle distance, which is typical for complex plasmas) and they can be considered as point-like, the Berthelot mixing rule has the following form:

$$V_{\text{AB}} = (1 + \delta_V)\sqrt{V_{\text{AA}}V_{\text{BB}}}. \tag{8.6}$$

In the opposite limit of short-range interactions (hard-sphere-like, typical to colloidal dispersions) characterized by the effective diameter $\sigma_{\alpha\beta}$ the

Lorentz rule is

$$\sigma_{AB} = (1 + \delta_\sigma) \frac{\sigma_{AA} + \sigma_{BB}}{2}. \tag{8.7}$$

Here $\delta_{V,\sigma}$ are the so-called nonadditivity parameters, and the associated asymmetry in the mutual interactions is usually referred to as "interaction nonadditivity". The Berthelot rule represents a natural condition that mixing/demixing is preferred when the AB interaction is less/more repulsive than the (geometric) mean of AA and BB interactions (i.e., when δ_V are negative/positive). Qualitatively, Eq. (8.6) follows from a combination (of Fourier-transformed interactions) which enters eigenvalues of the Ornstein-Zernike equation complemented with the random-phase approximation closure [the eigenvalues determine asymptotic stability of binary mixtures at small k (Hansen and MacDonald, 2006)]. The reasoning for the Lorentz rule is fairly analogous, with the only difference that now the mechanism behind demixing is entropy-driven. The origin of Eq. (8.7) can be understood from the Ising model (where the relative particle positions are fixed, which provides a certain analogy to hard-sphere interaction): In a system described by the Hamiltonian of Eq. (8.3), the phase equilibrium is determined by the sign of the interaction constant J, which naturally results in Eq. (8.7).

One can easily explain the generic thermodynamic mechanism that should drive the phase separation in binary complex plasmas (Ivlev *et al.*, 2009). Let us consider a pair of charges Q_i and Q_j: Their interaction energy is $V_{ij} = \frac{1}{2}(Q_i \varphi_{ji} + Q_j \varphi_{ij})$, where φ_{ij} is the potential produced by charge Q_i at the position of charge Q_j. The potential can be written as $\varphi_{ij} = Q_i^* Y(r_{ij})$, where $Y(r)$ is the far-field radial profile of the potential (independent of the actual particle charge Q_i) and $Q_i^*(Q_i)$ is the corresponding far-field charge (which is generally a function of Q_i). By introducing the renormalizing charge ratio $\nu_i = Q_i^*/Q_i$ we get $V_{ij} = \frac{1}{2}Q_i Q_j(\nu_i + \nu_j)Y(r)$. By substituting in Eq. (8.6) for $i,j =$ A,B we immediately obtain

$$\delta = \frac{(\sqrt{\nu_B/\nu_A} - 1)^2}{2\sqrt{\nu_B/\nu_A}} \geq 0, \tag{8.8}$$

(here and below, the subscript for the nonadditivity parameter is omitted). Thus, we see that for any nonlinear relation $Q_i^*(Q_i)$ interparticle interactions in binary complex plasmas exhibit a positive nonadditivity which always stimulates phase separation in isotropic (bulk) conditions. Equation (8.8) shows that the tendency to phase separate does not depend on a particular form of the interaction potential (which can be affected by

numerous processes operating in complex plasmas, e.g., collisions of ions
with neutrals and/or the ion absorption on the microparticle surfaces).
The interaction nonadditivity in complex plasmas is solely determined by a
nonlinear relation between the actual charge carried by a particle and an effective charge that characterizes the interaction potential at large distances.
One can derive the spinodal line for isotropic binary complex plasmas and
show that for typical experimental conditions the regime of the spinodal
decomposition is easily achievable (Ivlev *et al.*, 2009).

In order to illustrate phase separation in binary complex plasmas, we refer to the recent experiments by Wysocki *et al.* (2010) performed under microgravity conditions in the PK-3 Plus rf discharge chamber (see Fig. 5.5).
The initial stage of this experiment is shown in Fig. 10.13, where the formation of interpenetrating small- and big-particle lanes was observed. The
later stage is illustrated in Fig. 8.10: When the small particles approached
the center of the chamber and thus the driving field practically vanished,
an apparent phase separation was observed accompanied by the formation
of a small-particle droplet with a well-defined ellipsoidal shape (note that
the droplet eventually "coalesces" with the particle-free void and becomes
smeared out over its surface).

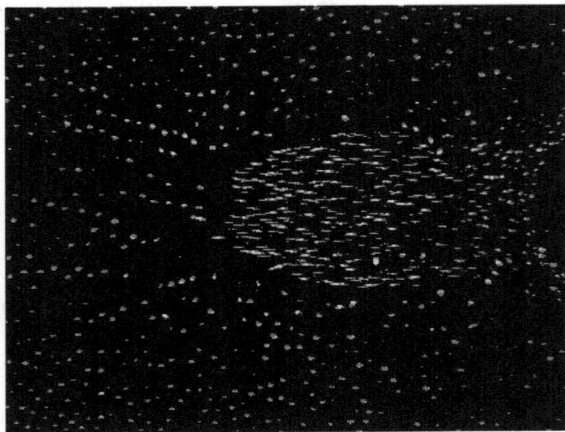

Fig. 8.10   Phase separation in binary complex plasmas. Particles of 3.4 μm diameter
form an ellipsoidal droplet in a stationary cloud of 9.2 μm particles (later stage of the experiment shown in Fig. 10.13). Particle positions during 0.1 s are depicted, superimposed
and color coded from green to red. From Ivlev *et al.* (2009).

We should point out that competing interactions can play a very important role in the morphology of separating phases, resulting in rich variety of domain patterns (Seul and Andelman, 1995; Stradner *et al.*, 2004; Royall *et al.*, 2005; Klix *et al.*, 2010). The asymptotic evolution in systems with such interactions is governed by the competition between the long-range repulsion stimulating subdivision of domains and the short-range (effective) attraction resulting in the growth of interfacial energy. There has been a great deal of analytical/numerical research of this process [see, e.g., Sagui and Desai (1994); Groenewold and Kegel (2001); Archer and Wilding (2007)] as well as a number of particle-resolved experiments (Stradner *et al.*, 2004; Royall *et al.*, 2005; Campbell *et al.*, 2005; Dibble *et al.*, 2006; Klix *et al.*, 2010) in particular addressing the role of competing interactions.

In complex plasmas, the interaction can also be characterized by two dominating asymptotes – both having the repulsive Yukawa form [Eq. (2.5)]. The corresponding effective screening lengths are normally very different and therefore referred to as "short-range" (SR, λ_{SR} is usually smaller than the mean interparticle distance Δ) and "long-range" (LR, λ_{LR} is much larger than Δ). This "double-Yukawa" potential yields the following pair interaction energy ($\beta = \mathrm{SR}, \mathrm{LR}$): $V_{ij}(r) = \frac{1}{2r} \sum_{\beta} (Q_i Q_{\beta,j}^* + Q_j Q_{\beta,i}^*) \mathrm{e}^{-r/\lambda_\beta}$. Note that the (effective) nonadditivity parameter δ now depends on r.

Figure 8.11 illustrates the phase diagram obtained by Wysocki *et al.* (2010) for binary fluids with double-Yukawa interactions from the mean-field theory.[5] Let us analyze the relative contribution of the SR and LR interaction parts to the phase equilibrium. The two basic similarity parameters of the interaction are the effective charge ratio $\epsilon_Q = Q_{\mathrm{LR}}^*/Q_{\mathrm{SR}}^*$ (which is typically small) and the screening length ratio $\Lambda = \lambda_{\mathrm{LR}}/\lambda_{\mathrm{SR}}$ (> 1). For $\epsilon_Q \Lambda^2 \ll 1$ the SR interactions dominate and the spinodal coincides with that derived for binary fluids with (single) Yukawa interactions (Ivlev *et al.*, 2009). The LR interactions dominate in the opposite limit $\epsilon_Q \Lambda^2 \gg 1$, where the spinodal has the same (rescaled) form. For the chosen parameter set the orange curves illustrate the "SR-dominated" case and the black (and green) curves are for the "LR-dominated" case. This plot demonstrates that the LR interactions – although relatively weak – can significantly enhance demixing. One can also see that for $\epsilon_Q \Lambda^2 \gg 1$ the SR nonadditivity has practically no effect on the equilibrium phase diagram (cf. black and green curves).

---

[5]Note the asymmetry of the phase diagram in comparison to, e.g., Fig. 8.8, caused by the charge asymmetry $\tilde{Q} \neq 1$.

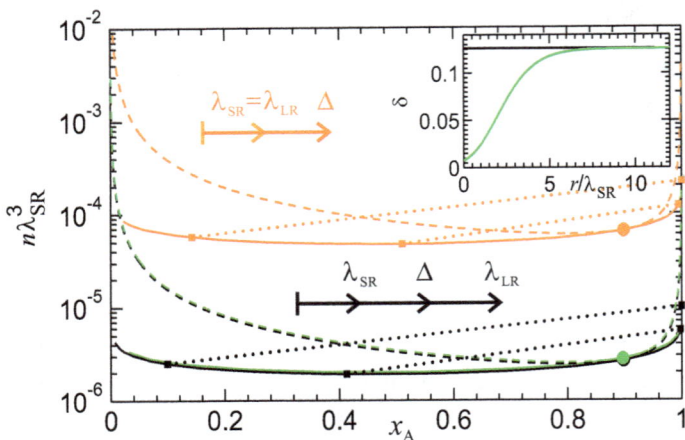

Fig. 8.11   Phase diagram for binary fluids with the double-Yukawa repulsion [Eq. (2.5)]. The phase variables are the total particle density, $n\lambda_{SR}^3$ (in units of the cubed SR screening length), and the relative composition x_A of species A. Thick solid lines are the binodals, dashed lines are spinodals, the critical points are marked by bullets, and dotted tie lines connect coexisting fluid states marked by squares (all obtained from the Ornstein-Zernike equation with the random-phase approximation closure). The calculations are for the coupling parameter (of species A) $\Gamma = 6.4 \times 10^3$, the effective charge ratio $\epsilon_Q = Q_{LR}^*/Q_{SR}^* = 0.2$ (for both species), and the charge asymmetry $\tilde{Q} = Q_B/Q_A = 2.7$. Three cases are illustrated, corresponding to different values of $\Lambda = \lambda_{LR}/\lambda_{SR}$ and $\tilde{\nu}_{SR}$: Orange lines are for $\Lambda = 1$, black are for $\Lambda = 12$, in both cases $\tilde{\nu}_{LR} = \tilde{\nu}_{SR} = 2.7$ (both LR and SR interactions are nonadditive). Green lines represent the case $\Lambda = 12$, but with $\tilde{\nu}_{LR} = 2.7$ and $\tilde{\nu}_{SR} = 1$ (only LR interactions are nonadditive). For the orange and black lines, arrows show schematically the relations between λ_{SR}, λ_{LR}, and the mean interparticle distance $\Delta = n^{-1/3}$. The inset is the effective nonadditivity parameter δ vs. distance between particles (for black and green lines). From Wysocki *et al.* (2010).

To demonstrate the role of competing interactions on the demixing kinetics, experiments with binary complex plasmas were combined with Langevin dynamics simulations (Wysocki *et al.*, 2010). Figure 8.12 summarizes the results. It turned out that for binary fluids with competing short-range and long-range interactions the onset of the spinodal decomposition is different from that in fluids with short-range interactions. Instead of the regular growth sequence discussed in the previous section, the LR-dominated interactions ($\Lambda = 12$) can shortcut the initial evolution: The diffusive growth with $\alpha = 1/3$ is replaced with the emergence of domains of (almost) time-independent length "preset" by the interaction range. During this "plateau" regime L is much larger than the short-range correlation length (which usually characterizes the initial scale at the diffusive growth),

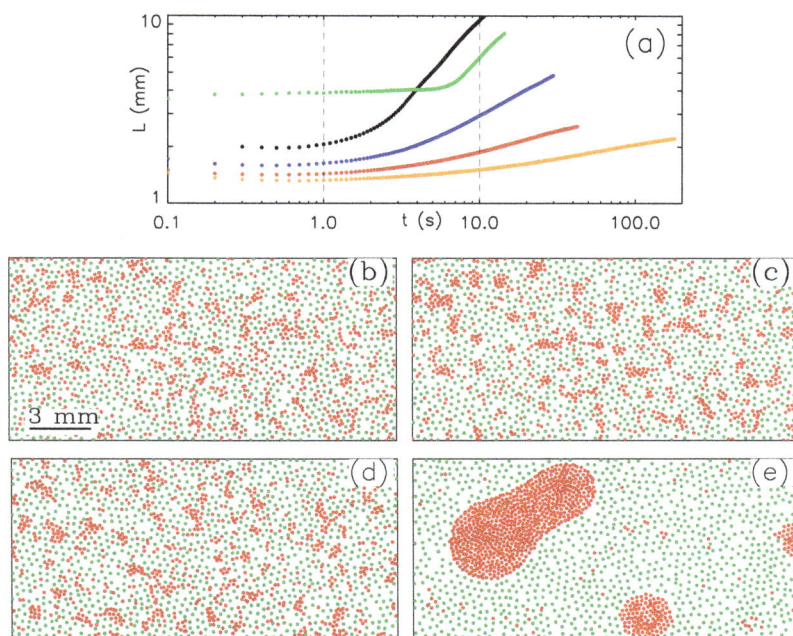

Fig. 8.12   Kinetics of phase separation in fluids with the double-Yukawa repulsion [Eq. (2.5)].   The results are from MD simulations for the off-critical composition $x_A = 1/2$, with $Q_A = 4000$ e, $T = 0.024$ eV, $\lambda_{SR} = 150$ μm, and $\Delta = 0.3$ mm, the charge ratios are $\epsilon_Q = 0.2$ and $\tilde{Q} = 2.7$ (as in Fig. 8.11).   The upper panel (a) shows the characteristic length of growing domains (of the minority phase A) versus time, $L(t)$, obtained for four different ratios of the screening lengths: $\Lambda = 1$ (orange), 2 (red), 4 (blue), and 12 (black).   Green dots are for $\Lambda = 12$ with $\tilde{\nu}_{SR} = 1$ (only LR interactions are nonadditive).   The lower panel shows the domain morphology for $\Lambda = 1$ (b,c) and $\Lambda = 12$ (d,e), particles A are color coded in red, particles B are green (slices of 0.3 mm are depicted).   Left column represents the early stage of the phase separation, $t = 1$ s, right column is for the developed stage, $t = 10$ s.   From Wysocki *et al.* (2010).

and interfaces between the domains sharpen until the surface tension drives the transition in a coarsening viscous growth. Note that the growth in the SR-dominated case shown here ($\Lambda = 1$ and 2) is characterized by rather small exponents, $\alpha \lesssim 0.2$. The reason for that is, apparently, a strong coupling between particles (Jiang *et al.*, 2011) – the interaction nonadditivity is too weak to drive the phase separation in this case, so that dynamic arrest inhibits the demixing and domains remain fuzzy. By decreasing the coupling between particles by 1–2 orders of magnitude one restores the classical $\alpha = 1/3$ diffusive exponent at the initial stage.

The described competition between the short-range and long-range interactions can be a generic process playing an important role in fluid phase separation occurring in different systems. The onset of this process – where the discreteness effects play essential role – requires separate particle-resolved studies focused on the comparison with the results of the coarse-grained approach.

8.4   Summary

What is attractive about particle-resolved studies of binary colloids and complex plasmas (in particular, of binary hard spheres and Yukawa systems) is that their remarkably diverse behavior can be described theoretically and in simulations in a relatively straightforward manner (as compared to molecular systems). Among the key insights is that even binary hard spheres reveal a bewildering complexity of different crystal structures. Furthermore, beyond capturing much of the richness of ionic chemistry, binary charged colloids can exhibit novel phases, since the charge is balanced by small ions and thus charge neutrality at the colloid level is not required.

Compared to the one-component case, studies of non-equilibrium phenomena are largely in their infancy. One important exception is studies of slow dynamics (where binary systems are a frequently used route to suppress crystallization), as we shall explore in the next chapter. Leaving the glass transition aside, there is, e.g., only one particle-resolved study of fluid-fluid phase separation dynamics – and that is in complex plasmas, and likewise one particle-resolved example of nucleation – in oppositely charged colloids. We can be clear that, going forward, there is a wealth of opportunity in exploiting binary systems. Among the key areas of interest – and one with significant practical application in the form of photonic crystals – is why binary hard spheres can fail to crystallize in experiments despite predictions from simulations and theory.

The results discussed above demonstrate that binary complex plasmas can be a suitable model system to study atomistic dynamics of fluid phase transitions and the associated phenomena, such as surface tension, critical behavior, etc. Similar processes can be also investigated with colloid-polymer mixtures (see Sec. 6.3). Thus, using the two model systems one can observe the onset of the spinodal decomposition, the coursing and crossover between different growth regimes, etc. in greatest details at the fully resolved kinetic level.

Chapter 9

Slow Dynamics

The slow dynamics associated with the cooling of a liquid (or increasing the density) is undoubtedly one of the major outstanding problems in condensed matter physics. The nature of how the viscosity of a liquid increases with remarkable rapidity upon cooling, and whether this has any thermodynamic origin or is a purely kinetic phenomenon – these basic fundamental issues remain unresolved (Cavagna, 2009).

Among the main challenges of the glass transition is its inaccessibility. Molecular systems are termed glasses when the viscosity and hence the relaxation time are increased by about 14 orders of magnitude relative to the normal liquid, as illustrated in Fig. 9.1. This regime is clearly inaccessible to present-day numerical simulations (which are currently limited to a dynamic slowdown of around 6 decades), yet any "true" glass transition would exhibit dynamics which is slower still. In other words, *the glass transition cannot be reached, and so cannot be probed.*

In this respect, particle-resolved studies can play a truly crucial role, since system sizes accessible in experiments (especially with colloidal dispersions) are typically much larger than those in numerical simulations. Furthermore, the role of weak neutral damping operating in complex plasmas can become constructive indeed – by varying the gas pressure, one can control the quenching rate in very broad range and hence bring the system to a desirable degree of supercooling.

Concerning the classification of slow dynamics adopted here, we note that soft matter can exhibit two distinct regimes: vitrification and gelation. In this chapter we first discuss vitrification, as gelation may be interpreted as the interplay between vitrification and spinodal phase separation. Furthermore, for the purposes of our book we assume the following definition which allows us to distinguish between gels and glasses: The approach to

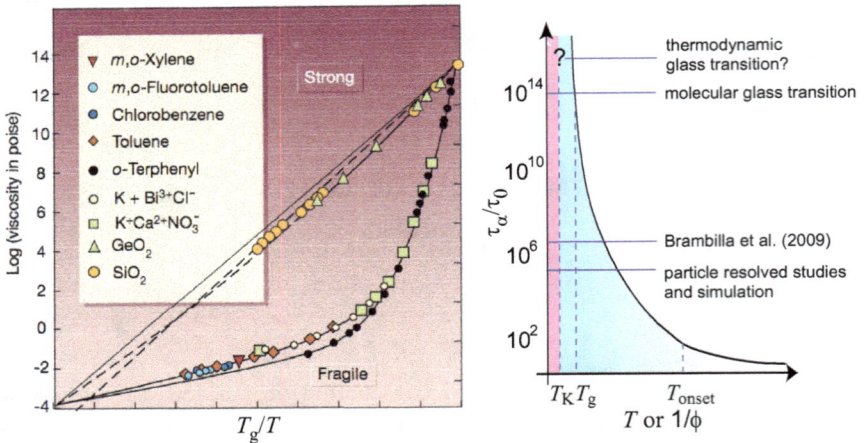

Fig. 9.1   Supercooled liquids: molecular versus soft matter systems. Left panel shows the so-called Angell plot – Arrhenius representation of liquid viscosities, with temperature scaled by T_g. *Strong liquids* exhibit Arrhenius behavior characterized by an approximately straight line, indicative of a temperature-independent activation energy. *Fragile liquids*, on contrary, reveal super-Arrhenius behavior, when activation energy grows as temperature decreases (Angell, 1988; Debenedetti and Stillinger, 2001). On the right, the corresponding increase of the relaxation time τ_α (relative to the normal liquid value τ_0) is sketched in the supercooled regime $T \leq T_{onset}$ (where hypothetic "thermodynamic" glass transition is marked by the Kauzmann temperature T_K). Relaxation times reached for particle-resolved and light-scattering (Brambilla *et al.*, 2009) experiments with colloidal "glasses" are also indicated.

the glass transition involves dynamical arrest where the system stays in equilibrium on the experimental timescale. At some arbitrary point (for example, 100 s for molecular glasses – see discussion in Sec. 9.1.4), the structural relaxation time is said to exceed to experimental timescale, and the system is termed a glass. Gelation, on the other hand, corresponds to quenching the system through a liquid-gas spinodal, and the resulting phase separation leads to a jump in density which in turn causes dynamical arrest, as illustrated in Fig. 6.5b. The main issue here lies in that gels are intrinsically metastable systems – they form percolating structures, which is required for dynamic arrest and ensures that the phase separation does not develop significantly on experimental timescales. This requirement clearly distinguishes gelation from conventional gas-liquid phase separation (Lu *et al.*, 2008). Note that our definition does not necessarily associate gels with networks: Not all loose network structures are gels, since by reducing the number of bonds that a particle can make one may produce network-like structures which, thermodynamically, are equilibrium liquids (Bianchi *et al.*, 2006).

Finally, we should stress that it would go far beyond the remit of this book to attempt any in-depth discussion of the glass transition in general, nor even the contribution that soft matter studies have made to it. There is an ever-increasing number of excellent reviews on this topic: For the glass transition in general, of particular note is that of Cavagna (2009) and Berthier *et al.* (2011), while for soft matter we suggest Cipelletti and Ramos (2005).

9.1   Principal Characteristics and Concepts

Many features characterizing supercooled fluids (especially, dynamic heterogeneity) are quite general. The particular form of the interaction potential plays only a minor role here – corresponding properties have been reported for many model atomic systems, e.g., colloids, viscous silica, network-forming liquids, and granular matter (Reichman *et al.*, 2005; Berthier *et al.*, 2011). This gives us grounds to expect the behavior of supercooled fluids to be fairly universal. In this section we briefly summarize some of the most important generic characteristics and concepts which are commonly used to describe supercooled liquids and the transition to a glassy state.

9.1.1   *Supercooled liquids*

When a liquid is cooled (or its density increased), the characteristic relaxation time τ grows. The regime in which the dynamics is said to be "slow" – while it remains possible to fully equilibrate the system in the experimental time window – is referred to as a "supercooled liquid". In the supercooled regime, decoupling of "rapid" (or "molecular") relaxation from "slow" (the so-called β- and α-) relaxation occurs, with the former being related to motion on length scales less than the particle size or interparticle distance, and the latter to full structural relaxation (Stillinger, 1995; Götze and Sjogren, 1992; Götze, 1999). One usually quantifies the dynamic slowdown by using the intermediate scattering function (ISF) $F(t)$ defined in Eq. (6.2).

Different relaxation regimes have essentially different scaling laws (Götze and Sjogren, 1992; Götze, 1999): The β-regime, which is manifested by a pronounced "plateau" in $F(t)$ as the glass transition is approached,

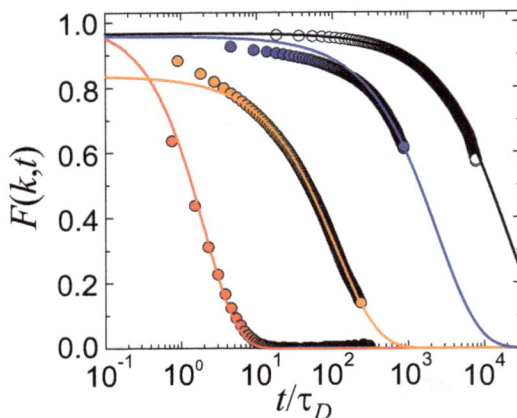

Fig. 9.2   Intermediate scattering function (ISF) calculated from real-space analysis of a colloidal "hard-sphere" supercooled liquid. Results are for $k \simeq 2\pi/\sigma$ and different values of the packing fraction (increasing from left to right). Lines are fits to the tail of the ISF following Kohlrausch-Williams-Watts law, τ_D is the Brownian diffusive timescale. From Royall *et al.* (2011).

usually obeys a power-law scaling, whereas the ultimate α-relaxation, which determines overall transport properties, typically exhibits a stretched-exponential decay $[F(t) \propto e^{-(t/\tau_\alpha)^b}$, the so-called Kohlrausch-Williams-Watts law with $0 < b < 1]$. Eventually, beyond the glass transition, the long-time α-relaxation is frozen. This is accompanied by complete arrest of structural relaxations and only local in-cage motions remain. Different relaxation regimes are illustrated in Fig. 9.2 showing typical ISFs from particle-resolved studies of colloidal dispersions (Royall *et al.*, 2011).

Angell (1988) made the most obvious reasonable assumption that the α-relaxation timescale follows an Arrhenius-like behavior. That is to say $\tau_\alpha \propto e^{E_a/k_B T}$, where E_a may be viewed as an activation energy. Then in the plot shown in Fig. 9.1, the viscosity (which is directly proportional to τ_α) should be represented by a straight line. For some materials, notably silica, this is indeed close to a straight line. However, for very many systems, a *fragile* or super-Arrhenius behavior is found. The increase in relaxation time is often well described by the empirical Vogel-Fulcher-Tamman (VFT) law, $\tau_\alpha \propto e^{\tilde{E}_a/k_B(T-T_0)}$, where T_0 ($\simeq T_K$) is somewhat lower than the glass transition temperature T_g and $\tilde{E}_a$ is a measure of the *fragility* – the degree to which the relaxation time increases.

9.1.2   *Dynamic heterogeneity and dynamic length scales*

Approaching the glass transition, *dynamic heterogeneity* becomes significant: Particles move in an increasingly cooperative manner creating dynamically correlated mesoscopic domains (Fischer, 1993; Sillescu, 1999; Cavagna, 2009). Understanding the properties of supercooled fluids in this regime is one of the most controversial issues in contemporary physics of fluids, with a number of mutually exclusive interpretations of various aspects of the complex supercooled fluids behavior being discussed (Richert, 2002; Cavagna, 2009; Cipelletti and Ramos, 2005; Chandler and Garrahan, 2010).

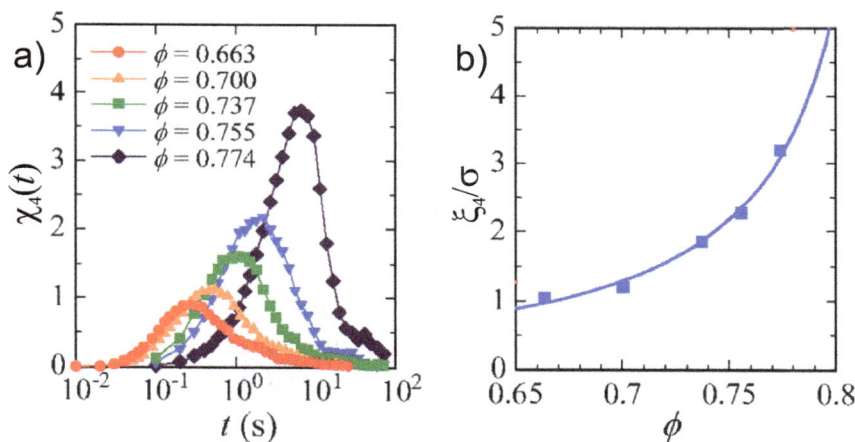

Fig. 9.3   Dynamic susceptibility $\chi_4(t)$ (a) and dynamic correlation length ξ_4 (b) of four-point correlations, for different packing fractions ϕ. Results are from particle-resolved studies of quasi-2D granular media, by Watanabe and Tanaka (2008).

It is possible to associate a *dynamic length scale* with the increasing dynamic heterogeneity – a measure which characterizes the size of growing cooperative motion. This cooperative motion can be quantified by using so-called four-point correlation functions. A time-dependent order parameter which measures the overlap between two configurations separated by time t is represented by the following binary correlation function:

$$R(t) = \int d\mathbf{r}_1 d\mathbf{r}_2 \rho(\mathbf{r}_1, t)\rho(\mathbf{r}_2, 0)w(|\mathbf{r}_1 - \mathbf{r}_2|) = \sum_{i,j}^{N} w\left(|\mathbf{r}_i(t) - \mathbf{r}_j(0)|\right). \quad (9.1)$$

Here, $w(r)$ is an "overlap" function defined as unity for $r \leq \sigma_*$ and zero otherwise (where $\sigma_* \sim 0.3\sigma$ is the length scale associated with the in-cage motion). The fluctuations are then characterized by the *dynamic susceptibility*,

$$\chi_4(t) = \frac{1}{Nnk_BT} \left[\langle R^2(t) \rangle - \langle R(t) \rangle^2 \right], \tag{9.2}$$

which is obtained by integrating a four-point time-dependent density correlation function over volume (Lacevic *et al.*, 2003). For a given packing fraction ϕ (or temperature T), the fluctuations (i.e., the susceptibility) attain a maximum at certain $t = t_{max}$, and then die away, as illustrated in Fig. 9.3a.

The exact nature of divergence of the dynamic length scale remains unclear, and depends upon the system under consideration. For simulations using the Lennard-Jones and related models (Tanaka *et al.*, 2010), this divergence has been likened to critical phenomena in liquids, which are characterized by divergence of the static density-density correlation length (see Sec. 6.3). In this spirit, one can define a four-point structure factor $S_4(k,t)$, so that the dynamic heterogeneities are maximized at $t = t_{max}$. One may then fit $S_4(k, t_{max})$ with the Ornstein-Zernike relation near the critical point [Hansen and MacDonald (2006)] to obtain a dynamic correlation length ξ_4,

$$S_4(k, t_{max}) = \frac{S_4^0}{1 + (k\xi_4)^2}, \tag{9.3}$$

where S_4^0 is a fitting parameter. Figure 9.3b shows that ξ_4 increases with ϕ approaching the glass transition, and can be fitting to critical-like exponents for some systems (Tanaka *et al.*, 2010).

9.1.3   *Boson peaks*

An alternative way to tackle glasses lies in considering them as solids (which they are). Crystalline solids are characterized by a density of states $D(\omega)$ for which the Debye model predicts $D(\omega) \propto \omega^2$. Indirect measurements on conventional materials suggest an anomalous behavior for glasses – the so-called "boson peaks" which appear as an excess contribution to the Debye density of states and which affect short-time (molecular) relaxation in glasses [see, e.g., papers by Schirmacher *et al.* (1998) and Zorn (2003)].

As noted in Sec. 4.2.2, due to their overdamped dynamics colloidal dispersions do not strictly have a density of states. Nonetheless, the dynamic matrix can still be measured, and from this one can deduce $D(\omega)$ which the system would have if the dynamics were Newtonian. This has been done in experiments with 3D systems (Ghosh *et al.*, 2010) and 2D systems (Chen *et al.*, 2010b; Kaya *et al.*, 2010) where boson peaks were indeed

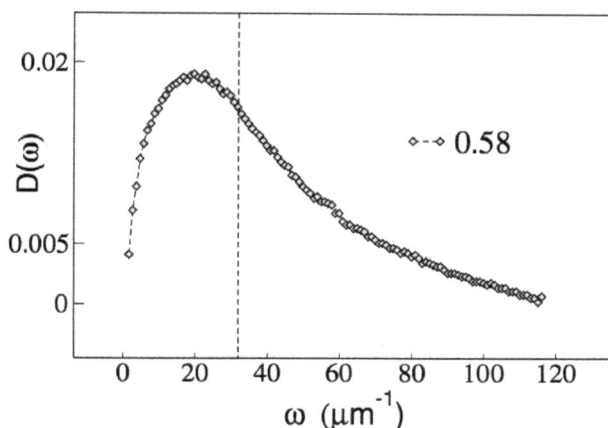

Fig. 9.4   Density of states $D(\omega)$ for a colloidal "hard sphere" glass at packing fraction $\phi = 0.58$. The dashed line shows the upper limit of frequency until which the spectrum is reliable. From Ghosh *et al.* (2010).

found, as shown in Fig. 9.4. It would be highly desirable to perform similar experiments also in complex plasmas, where the short-time dynamics is practically undamped and therefore the "molecular" relaxation can be studied directly.

9.1.4   *Glass transition and soft matter*

Among the key misconceptions is that the glass transition is somehow absolute. This is connected to a drastic difference in timescales between soft matter and molecular fluids, and is illustrated in Fig. 9.1. Molecules are said to vitrify when the relaxation time exceeds 100 s, which is an entirely anthropomorphic quantity. This is around 14 orders of magnitude longer than relaxation timescale in high-temperature liquids. On the other hand, for instance colloids can have relaxation times of ~ 100 s without exhibiting any signs of slow dynamics. This obviously presents a major problem: If we employ for colloids or complex plasmas the same criteria as those applied to molecular fluids, we find that the "molecular" 100 s corresponds to $\sim 10^8$ *years*. In other words, in order to obtain the equivalent *dynamic range* of 14 orders of magnitude in particle-resolved techniques would have had to be invented in the Jurassic period, and be imaged continuously ever since! This, clearly, is an example where slow dynamics of micron-size particles works against us.

This dynamic disparity between soft matter and molecules has led to states being termed "glasses", which in other fields would only be called slow liquids (with the viscosity of honey, for example). Consequently, hard-sphere systems (which slow down as a function of packing fraction, rather than temperature) were held to vitrify at a packing fraction around $\phi \simeq 0.58$, where the mode coupling theory predicts an absolute glass transition (see Sec. 9.2.3). However, recent experiments running to longer times than had previously been accessed (as illustrated in Fig. 9.1) showed that such systems could relax up to $\phi \simeq 0.61$ (Brambilla *et al.*, 2009). The implications of this work are profound: previously, measuring relaxation times over just a few orders of magnitude, agreement was found between experiment and the mode-coupling theory. The work of Brambilla *et al.* (2009) showed that, qualitatively, this is not the case. The power-law divergence of the mode coupling theory did not fit the data, while a generalized VFT form did. The importance of this work is that it delivered colloid data with sufficient dynamic range which showed that, at the very least, some refinement to the mode coupling theory was needed. This brought the colloid results into line with what had been known for some time in the molecular glass community where dynamic ranges of 10^{14} are routine (Berthier *et al.*, 2011).

9.2   Basic Theories of Liquid-to-Glass Transition

It is said that "...there are more theories of the glass transition than there are theorists who propose them" (Chang, 2008). Here we will briefly mention three of them. Note that these consider supercooled liquids, i.e., amorphous materials with slow dynamics, that are nonetheless able to relax completely on experimental timescales.

9.2.1   *Adam-Gibbs theory*

This approach is among the first modern theories of the glass transition (Adam and Gibbs, 1965). The basic idea is the assumption that liquids in the supercooled state organize themselves into cooperative regions whose size $\xi(T)$ increases as the temperature goes down. Next, it is supposed that each cooperative region has a few preferred states M independent of ξ. Then the configurational entropy per unit volume is given by $S_{\text{conf}}(T) \sim k_{\text{B}} \xi^{-3} \ln M$. The final assumption is that an energy barrier E_{a}

for rearranging a cooperative region – which enters the Arrhenius scaling of the α-relaxation time – is proportional to region's volume, $E_a = C_0\xi^3$. This yields a super-Arrhenius (VFT-type) form for the relaxation time,

$$\tau_\alpha = \tau_0 \exp\left[\frac{C_0 \ln M}{T S_{\text{conf}}(T)}\right],$$

(since $S_{\text{conf}} \propto T - T_0$), leading to reasonable agreement with experiments (Bouchaud and Biroli, 2004).

Due to several strong assumptions made in the Adam-Gibbs theory, it cannot be considered as completely quantitative approach: Indeed, it is unlikely that M does not depend on ξ (in fact, it can have an exponential scaling on ξ), and that the energy barrier is proportional to ξ^3 (if the cooperative region is too large, the rearrangement probably does not involve a finite fraction of all particles). Nevertheless, the Adam-Gibbs theory follows a simple and intuitive framework and introduces several key concepts, such as cooperative relaxation and growing dynamic length scale.

9.2.2 *Random first-order/mosaic theory*

A complementary view, proposed by Kirkpatrick, Thirumalai and Wolynes makes a connection with spin glasses (Kirkpatrick *et al.*, 1989). Here, we consider the temperature regime between T_0, a presumed thermodynamic phase transition to a true glass (which is a solid and whose dynamic correlation functions do not relax on any timescale) and T_{MCT}, a higher temperature at which the mode coupling theory (discussed below) predicts a glass transition. It is argued that in this intermediate regime $T_0 \leq T \leq T_{\text{MCT}}$ relaxation occurs in a nucleation-like way. The system is pictured as a mosaic of regions, each of which is pinned by immobile neighboring regions providing an activation barrier to relaxation. However, unlike (classical) nucleation theory, relaxation events are entropically driven: For a domain of size ξ the entropy gain is $T S_{\text{conf}}(T)\xi^3$ while the energy loss, which due to mismatch between the nucleating and ambient states, is proportional to effective surface tension γ_{eff} and scales as $\gamma_{\text{eff}}\xi^\theta$ (where $\theta \leq 2$). The characteristic critical length of the mosaic state is determined from the balance of the two processes, which leads to the energy barrier $E_a \sim \gamma_{\text{eff}}(\gamma_{\text{eff}}/T S_{\text{conf}})^{\theta/(3-\theta)}$. The resulting α-relaxation time is then given by the generalized VFT law which coincides with the Adam-Gibbs results for $\theta = 3/2$. Both Adam-Gibbs and mosaic theories lead to the concept of dynamic heterogeneity which is thought to be the dominant mode of relaxation for deeply supercooled materials.

9.2.3   *Mode coupling theory*

In the colloid community, undoubtedly the most well-known theory of the glass transition is the mode coupling theory (MCT) (Götze and Sjogren, 1992; Götze, 1999, 1991, 2009) which can be directly tested against a given system, provided its interaction potential is amenable to liquid state theory. The simple interactions of colloidal systems are thus ideal for MCT, which has been used on numerous occasions to account for, e.g., dynamic arrest in hard-sphere glasses (van Megen *et al.*, 1998) and the reentrant glass transition in sticky spheres [Puertas *et al.* (2002) and Pham *et al.* (2002), where hard-sphere "glasses" and "glasses" in colloid-polymer mixtures bound a mobile liquid state]. In hard spheres, MCT predicts a transition at a critical packing fraction of $\simeq 0.52$, which can be rescaled to fit with experimental data at $\simeq 0.58$. However, the deviations from the power-law divergence in relaxation time at the critical packing fraction (or temperature), have recently been found in colloidal experiments and simulations (Brambilla *et al.*, 2009). We thus regard MCT as a useful tool to predict the onset of slow dynamics rather than a complete theory.

The idealized MCT deals with the structural dynamics of simple liquids (Götze, 2009). A self-consistent treatment of the cage effect, which is thought to be the reason for the onset of slow dynamics, leads to a closed set of integro-differential equations for the density correlator $\Phi_k(t) = F(k,t)/S(k)$, i.e., for the ISF [Eq. (6.2)] normalized by the structure factor [Eq. (6.3)]. The equation governing the evolution of $\Phi_k(t)$ is obtained by using the Zwanzig-Mori projection-operator technique (Hansen and MacDonald, 2006),

$$\frac{\partial^2 \Phi_k(t)}{\partial t^2} + \nu \frac{\partial \Phi_k(t)}{\partial t} + \Omega_k^2 \left[\Phi_k(t) + \int_0^t dt_1 m_k(t-t_1) \frac{\partial \Phi_k(t_1)}{\partial t_1} \right] = 0, \quad (9.4)$$

where $\Omega_k^2 = k^2 v_T^2/S(k)$ is the characteristic frequency and ν is the damping rate. The memory function $m_k(t)$, which plays the role of a generalized friction coefficient in this "equation of motion", describes the coupling of mode k (viz., density fluctuations at k) with other modes. In the MCT, one assumes that the major contribution is provided by pair density fluctuations, which leads to the uncontrolled approximation that the memory function is described by the following functional:

$$m_k(t) = \mathcal{F}_k \left[\Phi_k(t) \right] = \int \frac{d\mathbf{q} d\mathbf{p}}{(2\pi)^3} V(\mathbf{k}, \mathbf{q}, \mathbf{p}) \Phi_q(t) \Phi_p(t). \quad (9.5)$$

The vertex V is uniquely determined by the equilibrium structure factor and thus contains information about interparticle interactions,

$$V(\mathbf{k}, \mathbf{q}, \mathbf{p}) = \frac{n}{2} \frac{S_k S_q S_p}{k^4} (\mathbf{k} \cdot \mathbf{q} c_q + \mathbf{k} \cdot \mathbf{p} c_p)^2 \delta(\mathbf{k} - \mathbf{p} - \mathbf{q}), \qquad (9.6)$$

where $nc_k = 1 - S_k^{-1}$ is the equilibrium direct correlation function and $S_k \equiv S(k)$. At a critical value of the control parameter (temperature or density) Eqs (9.4)-(9.6) exhibit a bifurcation in the long-time behavior of the density correlator. This bifurcation singularity is identified in MCT with a liquid-glass transition: At the critical point, the Debye-Waller factor $f_k = \lim_{t \to \infty} \Phi_k(t)$ (also called the nonergodicity parameter) jumps from zero to the critical value $0 < f_k < 1$ which is the solution of the following nonlinear equation (Hansen and MacDonald, 2006):

$$\frac{f_k}{1 - f_k} = \mathcal{F}_k[f_k].$$

The characteristic behavior of $\Phi_k(t)$ is illustrated in Fig. 9.2. Note that beyond the critical point f_k changes continuously with the control parameter.

Thus, MCT yields ISFs using the static structure factor as the only input. Since for colloidal systems S_k may readily be calculated from liquid state theory (Hansen and MacDonald, 2006), and $\Phi_k(t)$ can be measured by dynamic light scattering and confocal microscopy, the agreement found with MCT has provided powerful support. It is nonetheless important to note again that the central approximation within MCT lies in the memory function, and that this approximation is uncontrolled. Thus the agreement with experiments and simulations even over three or so decades in relaxation time may be seen as truly remarkable. While it is reasonable to say that MCT is a powerful tool for predicting the onset of slow dynamics for a given system, it yields little insight into the relaxation dynamics deep in the glass. This is where particle-resolved studies are among the best (and in some cases the only) experimental techniques which can provide crucial new insights. For further information, the reader is referred to the excellent review by Charbonneau and Reichman (2005).

In conclusion we note that the MCT can also be used for the tagged-particle correlator, $\Phi_k^s(t) = \langle n_\mathbf{k}^s(t) n_{-\mathbf{k}}^s(0) \rangle$, where $n_\mathbf{k}^s(t) = e^{-i\mathbf{k} \cdot \mathbf{r}(t)}$ is the Fourier-transformed single-particle density (Fuchs *et al.*, 1998; Voigtmann *et al.*, 2004). The evolution of $\Phi_k^s(t)$ is governed by Eq. (9.4) with the tagged memory function $m_k^s(t)$: Now, one has to replace in Eq. (9.5) the product of the density correlators with $\Phi_q^s(t) \Phi_p(t)$, while the vertex, $V(\mathbf{k}, \mathbf{q}, \mathbf{p}) = n S_q (\mathbf{k} \cdot \mathbf{q} c_q^s / k^2)^2 \delta(\mathbf{k} - \mathbf{p} - \mathbf{q})$, is determined by the single-particle direct correlation function $c_k^s = \langle n_\mathbf{k}^s(0) n_{-\mathbf{k}}^s(0) \rangle / (n S_k)$. The mean squared displacement can be obtained from the modified Eq. (9.4) by exploiting the small-k behavior of the tagged-particle density correlator, $\Phi_k^s(t) = 1 - \frac{1}{6}k^2\delta r^2(t) + O(k^4)$. By neglecting the short-time dynamics for simplicity, one derives the integral equation,

$$\delta r^2(t) + D_0 \int_0^t dt_1 m_0(t - t_1)\delta r^2(t_1) = 6D_0 t,$$

where $m_0(t) = \lim_{k\to 0} k^2 m_k^s(t)$ (the tagged kernel scales as $\propto k^{-2}$ at small k). This expression allows us to calculate the transient particle dynamics (see Sec. 6.2), e.g., to deduce the diffusion exponent $\alpha(t)$ and obtain the asymptotic diffusion coefficient $D = \frac{1}{6}\lim_{t\to\infty} \delta r^2(t)/t$,

$$D = \frac{D_0}{1 + D_0 \int_0^\infty dt\, m_0(t)}.$$

This ratio is less than unity, which is an obvious manifestation of the cage effect.

9.3   Real-Space Analysis of Supercooled Liquids

Few real-space analysis experiments have had more impact than those carried out by Weeks *et al.* (2001), who measured dynamic heterogeneity in colloidal supercooled liquids. The key results are reproduced in Fig. 9.5a. First, these results represent a major technical achievement, being the first 3D time-resolved data. Second, they established the key insight available with real-space analysis that is accessible to no other experimental technique – *the ability to probe local behavior directly, in both space and time.* Simultaneously, Kegel and van Blaaderen (2001) reported on a 2D analysis (in a 3D system) showing the same effect, and were quickly followed by Cui *et al.* (2001) who demonstrated dynamic heterogeneity in 2D.

Dynamic heterogeneity had been found in computer simulations by Butler and Harrowell (1991), and indirect measurements in molecular systems had been also made (Ediger, 2000). However, the colloid experiments firmly established the ability of real-space analysis to provide direct, clear experimental evidence. The nature of data that real-space analysis delivers enabled Weeks and Weitz (2002) to probe questions, e.g., was there any local difference in the dynamically heterogeneous regions, were there some local fluctuations in density (and, indeed, "fast" regions turned out to have slightly lower density).

Complex plasmas also exhibit dynamic heterogeneity in 2D and quasi-2D strongly coupled systems (Lai and I, 2002; Nunomura *et al.*, 2006; Juan

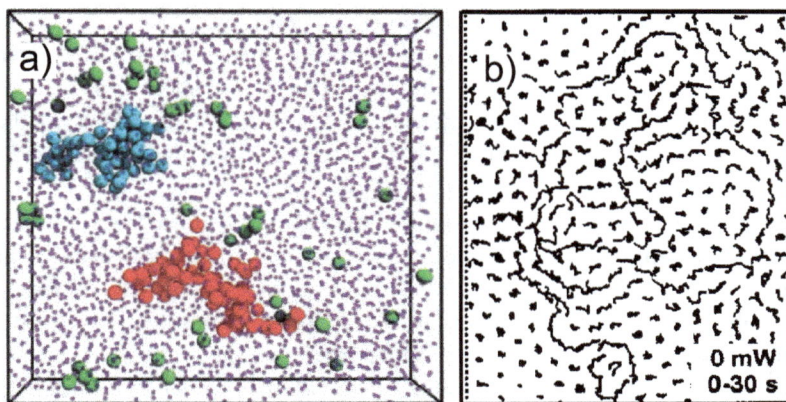

Fig. 9.5   Dynamic heterogeneity in experiments with colloidal dispersions and complex plasmas. (a) Colloidal hard spheres of 2.36 μm diameter imaged in 3D with confocal microscopy. Colored spheres represent fastest particles (displacement > 0.67 μm during 720 s exposure), "slow" particles are shown smaller than actual size (the red cluster consists of 69 particles, the light blue cluster contains 50 particles). From Weeks *et al.* (2001). (b) Quasi-2D liquid complex plasma. Shown are trajectories of $\simeq 7$ μm diameter particles at 30 s exposure time. From Juan *et al.* (2001).

et al., 2001; Woon and I, 2004; Huang and Lin, 2007). Figure 9.5b shows a snapshot from experiment by Juan *et al.* (2001) where most particles are mutually confined by (quasi-ordered) neighbors, and exhibit caged motion with small amplitude oscillations. However, there is a certain fraction of particles that undergo rearrangement. Particle-resolved studies enable direct observation of cage-escape events – strings or vortices surrounding crystallites (ordered domains) with the size of a few Δ. Usually, a local rearrangement ceases after the particles involved move a distance of $\simeq \Delta$ (equivalent to $\sim \sigma$ in hard-sphere colloids) and then reenter the new caged state. Particles may start coherent rearrangement only after accumulating sufficient stress, and then transferring the excess energy to the neighbors through mutual interactions. The cooperative motion is rapidly smeared out unless further constructive perturbations occur at a timescale smaller than the momentum relaxation time. Thus, the dynamic heterogeneities die out at long times (cf. Fig. 9.3a).

The effect of confinement on the glass transition temperature (or, in the case of hard spheres, packing fraction) is poorly understood, and littered with inconsistent measurements in the case of, e.g., polymer films. Nugent *et al.* (2007) showed that the glass transition occurs at a lower packing fraction under confinement. More recently, Jenkins *et al.* (2010) were able to use real-space data to determine load-bearing bridges in colloidal glasses, by identifying putative contacts between colloidal particles. Contacts leading to jamming were directly identified with larger (granular) emulsion droplets which are large enough ($\gtrsim 10 \ \mu$m) to be resolved with confocal microscopy (Brujic *et al.*, 2007).

Around the turn of the millenium, experimental data from scattering experiments supported the prediction of a reentrant glass transition (Pham *et al.*, 2002). This occurs when short-ranged attractions are added to a hard-sphere glass, leading to enhanced mobility ("melting") and subsequent dynamic arrest upon further addition of polymer. The reentrant glass transition was also observed in real space, where the coordinate tracking revealed no measurable structural changes in the mobile intermediate or either arrested states (Simeonova *et al.*, 2006).

9.4   Real-Space Analysis of Gels

Gels are perhaps the most poorly defined everyday material. Certainly, colloidal gels exist, and are generally accepted to be a state which exhibits slow dynamics. In colloid-polymer mixtures, the most frequently used gel formers for particle-resolved studies, the role of the range of attractive interactions is all-important. As discussed in Sec. 6.3, long interaction ranges, corresponding to polymers of relatively large gyration radius (Ilett *et al.*, 1995; Lekkerkerker *et al.*, 1992) provide stable gas-liquid behavior for colloids. In the concept developed by Tanaka (2000), which is termed "viscoelastic phase separation", the lifetime of the phase separation process is governed by the viscosity of the slowest phase. Since the colloid-rich phase typically has higher viscosity, it normally dominates the dynamics. As the interaction range is reduced (by reducing the polymer size relative to that of the colloid, or by increasing the polymer concentration) the packing fraction of the colloidal liquid increases (Elliot and Hu, 1999). When this approaches the packing of $\simeq 0.58$ the phase separation becomes arrested and the system undergoes gelation.

To our knowledge, the first real-space studies of gelation were carried out by Dinsmore and Weitz (2002). The local structure was determined in terms of the number of neighbors, local fractal dimension, and filamentary nature of the chains from which the gel network was comprised. Dinsmore *et al.* (2006) went on to consider elasticity of the gel network, by measuring the fluctuations of particles in the "arms" of the gel.

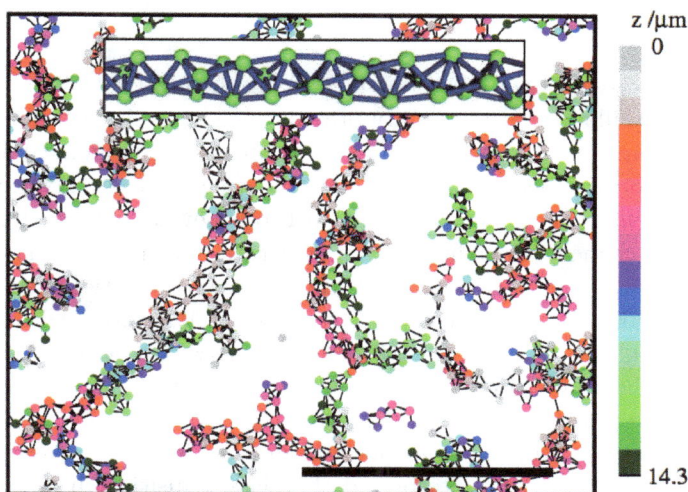

Fig. 9.6   2D projection of the particle centers within a slab of gel in a colloidal system with competing interactions. Particles are colored as a function of their depth within the sample and drawn 40 % of their actual size for clarity. The inset shows a spiral chain formed from tetrahedra of particles sharing faces. From Campbell *et al.* (2005).

The density-matching solvents necessary for confocal microscopy can result in some remarkable behavior during colloidal gelation. As mentioned in Sec. 3.2, these solvents lead to significant electrostatic interactions. The combination of short-ranged depletion-driven attraction and long-ranged electrostatic repulsion creates a system with competing interactions, a class of materials hitherto unexplored in the colloidal domain, but which encompasses spin glasses and exotic matter [such as the pasta phase in neutron stars (Horowitz *et al.*, 2004) and strongly-correlated electron systems (Dagotto, 2005)].

These competing interactions have a profound influence on the arrested phase-separation behavior (see also Sec. 8.3.4), leading to elongated structures (which minimize the long-ranged electrostatic repulsions). Campbell *et al.* (2005) were able to show that the elongated structures are Bernal spirals depicted in Fig. 9.6. Stronger repulsions can also cause dilute colloidal fluids to undergo arrest, forming a "Wigner glass" and exhibiting a first-order-like non-equilibrium transition between the Wigner glass and gel (Klix *et al.*, 2010).

9.5   Local Structure and Dynamic Arrest

The work by Weeks *et al.* (2001) on identifying the local nature of dynamic heterogeneity illustrated the ability of real-space analysis to tackle key issues of dynamic arrest. One open question concerning dynamical arrest is the role of local structure. Indeed, the divergence of the relaxation time seen in Fig. 9.1 is enormous, and it is hard to believe that there are no accompanying changes occurring in local structure. On the other hand, only negligible changes had been detected via averaged quantities such as $g(r)$ and $S(k)$. More than half a century ago, Frank (1952) proposed that local structure in glasses might exhibit fivefold symmetry, predicting the formation of icosahedra of 13 particles forming upon cooling. The fivefold symmetry was held to suppress crystallization and hence to stabilize the glass. Fivefold symmetry (or, at least, local structure) was also considered by Tarjus *et al.* (2005), who interpreted the glass transition in terms of an avoided transition to an ideal glass with icosahedral symmetry. Furthermore, Tanaka (1999) emphasized the importance of local structure through his two-order-parameter model, which uses the concept of a second non-conserved order parameter in the form of energetically favored local structures (for example, icosahedra).

Extracting such local structural measures in conventional experiments is challenging, to say the least. However, indirect evidence of fivefold symmetry has been found using novel X-ray absorption techniques (Di Cicco *et al.*, 2003). Coordinate-level data from computer simulations (Steinhardt *et al.*, 1983; Jonsson and Andersen, 1988) using bond-orientational order parameters based on the W_6 invariant provided direct evidence for fivefold symmetry in glasses (negative values of W_6 are indicators of fivefold symmetry, icosahedra have $W_6 = -0.18$). Steinhardt *et al.* (1983) proposed that W_6 and related bond-order parameters would enable a kind of spectroscopy for local structure in fluids.

In their pioneering work on real-space analysis of colloidal glasses, van Blaaderen and Wiltzius (1995) took advantage of the coordinate data – they used the W_6 bond-order parameter for the first time in an experiment and found a negative W_6 in the colloidal glass. Campbell *et al.* (2005) also found a negative W_6 in gels of Bernal spirals. Particle-resolved studies are thus the only experimental technique which can directly reveal this sort of local structural information.

The role of local structure is particularly prevalent in 2D glass forming systems, since the local hexagonal order is the same as that of the crystal.

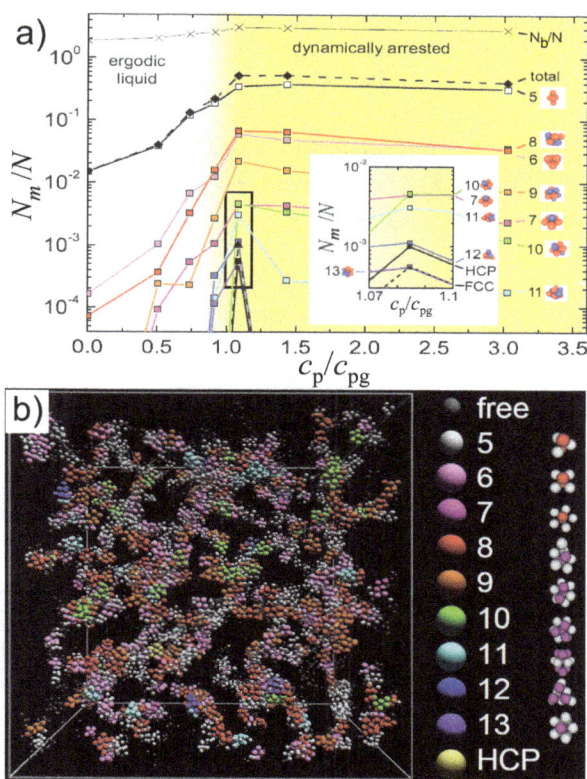

Fig. 9.7   A colloidal gel formed of energetically favored local structures (clusters). (a) Fraction of particles in clusters of size m, N_m/N, is shown as a function of polymer concentration c_p/c_{pg} (normalized by the concentration required for gelation). The dependencies are represented by squares, color-coded in accordance with the cluster size, the corresponding nearest-neighbor structures are depicted on the right. Black diamonds represent the total fraction of particles identified as being a member of a local structure, total bonds per particle N_b/N are shown by black crosses. The inset demonstrates traces of 13-particle fivefold-symmetric clusters as well as of fcc and hcp lattices detected in the vicinity of $c_p/c_{pg} \simeq 1$. (b) Example of a dilute gel with a percolating structure. Free particles (gray) are shown 0.4 actual size, crystalline (yellow) have 0.8 actual size, members of the structures of size m are color-coded as in (a) and shown 0.8 actual size (except of $m = 5$, shown 0.6 actual size). From Royall *et al.* (2008).

By employing their 2D tuneable dipolar setup, König *et al.* (2005) began using binary systems which prevent crystallization, and noticed local structural motifs in glasses.

The first systematic studies of local structure and its relation to dynamic arrest provided strong evidence that, in the case of gelation, the locally favored structures actually unpin dynamic arrest. Royall *et al.* (2008) extended Frank's concept of icosahedra to consider a zoo of clusters of various sizes which provide energetic ground states for small numbers of colloidal particles, as summarized in Fig. 9.7. They developed a novel topological method to directly identify these locally favored structures from coordinate data, and established clear links between structure and dynamic arrest as well as between structure and local dynamics.

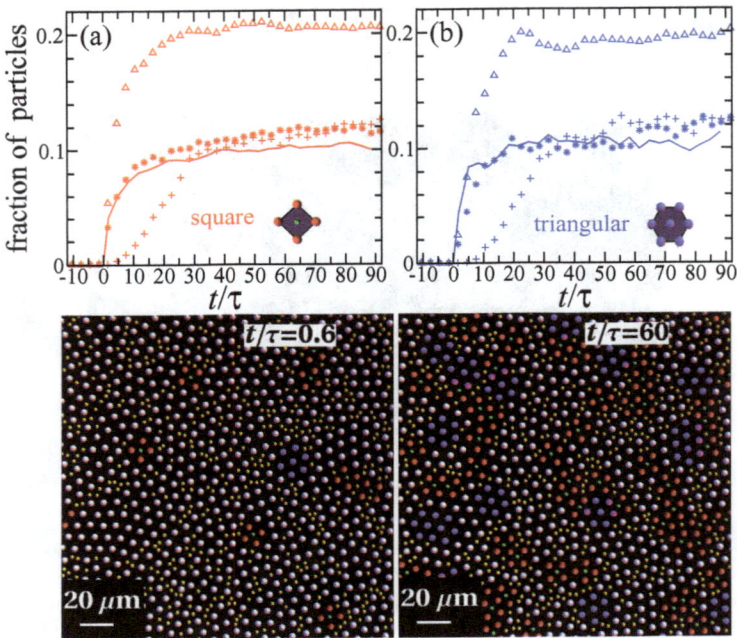

Fig. 9.8   Local structure of a supercooled binary colloidal mixture. Upper panel presents fraction of particles which belong to a square (a) and hexagonal (b) crystalline neighborhood, shown versus the reduced time t/τ_D for an "instantaneous" quench from $\Gamma \simeq 1$ to $\Gamma \simeq 71$ (τ_D is the Brownian diffusive timescale). The solid lines are experimental data while the symbols (∗) are data from the corresponding Brownian dynamics simulations. Lower panel shows two experimental snapshots, short after the quench ($t/\tau_D = 0.6$, left configuration) and a later stage ($t/\tau_D = 60$, right configuration). Large particles are shown in blue (red) if they have a hexagonal (square) surrounding. Few big particles belonging to both hexagonal and square surroundings are shown in pink, all other big particles are depicted in white color. Small particles are shown in green if they belong to a square center of big particles, otherwise they appear in yellow. Also included are simulation data for an instantaneous quench from $\Gamma = 1$ to $\Gamma = \infty$ ($\triangle$) and for a linear increase from $\Gamma = 1$ to $\Gamma = 71$ during time $30\tau_D$ (+). From Assoud *et al.* (2009b).

The longer timescales of colloidal dispersions and complex plasmas enable us to observe phenomena which cannot be practically realized in molecular systems. A good example here is the "ultra-fast quenching" work carried out by Assoud *et al.* (2009b), where the external field which controls
the particle effective temperature was switched on at a timescale far shorter
than the colloidal relaxation, allowing practically instantaneous quenching.
Figure 9.8 shows experimental 2D snapshots of a colloidal mixture with
$\tilde{m} = 0.1$ and $x_B = 0.4$ (see Sec. 8.2.2 for parameter definition). Before
the quench, the system is exposed to a low magnetic field which provides
an effective coupling parameter $\Gamma \simeq 1$. Then the field is changed "instantaneously" to a higher value, resulting in $\Gamma \simeq 71$. The left snapshot
in Fig. 9.8 illustrates the structure observed short after the quench, the
right one is for a later time. We see that the fraction of blue and red
colored particles representing the hexagonal and square neighborhood, respectively, gradually increases with time and reaches saturation. These
experiments reveal a significant number of crystalline patches in the supercooled "glassy" state which are characterized by structural order. Clearly,
these patches contribute to the dynamical heterogeneity thus pointing to
a possible connection between structural and dynamical heterogeneity in
glasses.

9.6 Aging of Glasses

Supercooled fluids are regarded as glasses when they no longer relax on
the experimental timescale. Nonetheless, properties of glasses change over
time, and this process is termed aging. During aging, (incomplete) relaxation events often reduce in frequency, as the system "settles" in its energy
landscape. We recall that soft matter systems are termed glassy under
weaker conditions than atomic and molecular systems, due to the reduced
dynamic range, and this is particularly true for larger particle sizes (required for particle-resolved studies).

In this context, three studies of colloidal glasses are relevant. The first
one is by Cianci *et al.* (2006), where 3D confocal studies were employed
to consider both structural and dynamic behavior of aging systems. While
the dynamics slowed, only small changes were revealed in the structure (by
decomposing into tetrahedra, although a weak correlation between tetrahedra and mobility was found). A second study was performed in 2D by
Assoud *et al.* (2009b), focused on the relevant local structures. They noted

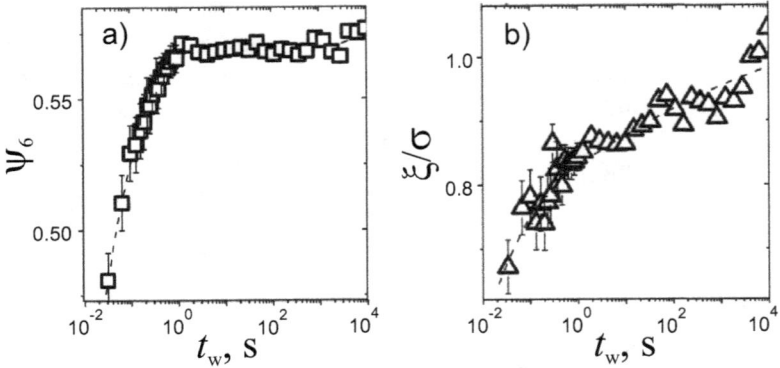

Fig. 9.9   Local bond-orientational order parameter ψ_6 and in corresponding structural correlation length ξ in an aging 2D colloidal glass as functions of the waiting time t_w. From Yunker *et al.* (2009).

systematic structural changes, both in their specific local order and even in quantities such as peaks of the pair correlation function $g(r)$. The latter is interesting, as these could be obtained (in principle) from reciprocal-space measurements of atomic and molecular systems. Finally, long-time experiments with a 2D bi-disperse hard-sphere-like microgel system carried out by Yunker *et al.* (2009) identified irreversible structural relaxations, which could be attributed to a growing structural length scale, as shown in Fig. 9.9.

We note that aging mechanisms in gels are profoundly different. Since gels are metastable with respect to gas-liquid phase separation, aging can typically take the form of ultra-slow phase separation. The "arms" of the gel coarsen with time, which was observed by Lu *et al.* (2008). The authors found a power-law dependence of the "arm" thickness on time.

9.7   Summary

Dynamical arrest is among the deepest unsolved problems in condensed matter physics – and one where real-space analysis has an important role to play. In particular, the ability to observe local dynamics and structure, where system sizes do not suffer from the limitations of computer simulation, cannot be overstated. From the first direct observations of dynamical heterogeneity to identifying long-sought structural mechanisms for dynamical arrest, the impact of particle-resolved studies has been truly profound.

Among the key challenges going forward must be to try to approach the glass transition itself. As we have observed, even in conventional materials ~ 14 decades increase of relaxation time does not reveal the thermodynamic glass transition – if such a transition exists at all. However, exciting new developments suggest that the use of a pinning field (where a set of particles is prevented from moving) may actually bring the glass transition – or aspects of it – within the timescales accessible to particle-resolved studies. The use of, e.g., optical tweezers to produce such a pinning field provides a new way to tackle dynamical arrest at the individual-particle level.

Chapter 10

Driven Systems

Whilst the macroscopic (hydrodynamic) behavior of fluids is well understood, a microscopic theory at the single-particle level is a major outstanding problem, which makes particle-resolved studies tremendously important. Of particular interest is the onset and nonlinear development of
hydrodynamic instabilities at the individual particle level. Furthermore,
one can address the "discreteness issue" of continuous media which is formulated as follows: "What is the smallest scale at which the conventional
hydrodynamic description breaks down?" Apparently, the answer depends
on the particular problem under consideration: It is determined by the
similarity variables (and hence the related physical parameters) that play
the major role in the description of the macroscopic problem – e.g., for a
planar shear flow these are, primarily, the Reynolds and Mach numbers.[1]
Here we present some recent advances in the particle-resolved studies of
driven systems, and focus on examples which enable us to identify generic
"atomistic" mechanisms governing numerous non-equilibrium processes in
liquids and solids in the presence of external fields (which are essentially
different from the equilibrium behavior).

10.1   Stable Shear Flows

Shear flows appear as an inevitable ingredient of more complicated flows.
Even in the simplest case of laminar shear flows, many fundamental questions immediately arise: What is the kinetic structure of the flow (e.g., how
does the transverse momentum relaxation occur)? What is the kinetics

---

[1]Of course, one should remember that the basic parameters entering hydrodynamics
such as viscosity or surface tension are quantities which are well defined only for sufficiently large systems.

of non-Newtonian fluids (e.g., what determines the relevant timescales in the viscoelastic fluids)? What happens at shear fluid boundaries (e.g., how good is the Navier ansatz for the slip velocity and what is the corresponding slip length)? In the case of multiphase flows, many more fundamental problems turn up, especially those related to the shear boundaries. Probably, the most "obvious" one is the contact-line singularity problem: a movable intersection of the fluid-fluid interface with the solid wall is incompatible with the no-slip boundary condition (Qian *et al.*, 2006).

Let us take a closer look at the individual particle trajectories in a fluid that exhibits a shear flow. In complex plasmas, one can easily induce various types of flows with controllable characteristics by applying laser beams or creating flows in the neutral gas (Fortov *et al.*, 2005). A clear advantage of such methods of particle manipulation is that the background plasma, and hence parameters of the interparticle interaction, remain unchanged, yet the characteristics of the particle flow – especially the flow shear rate – can be varied over an exceptionally broad range.

Figure 10.1 shows an example of a shear flow experiment performed in a 2D complex plasma by Nosenko and Goree (2004). Initially, particles formed an (almost perfect) hexagonal crystalline monolayer, and the shear flow was created by applying two counter-propagating laser sheets. By increasing the laser power and, hence, the shear stress, the authors observed that the particle suspension passed through four distinct stages: elastic deformation, defect generation while in a solid state, onset of plastic flow, and fully developed shear flow. The figure presents data for the latter two stages. At the onset of plastic flow, Fig. 10.1a, the particles hopped between equilibrium lattice sites. Domain walls developed, and moved continuously. The crystalline order of the lattice in the shearing region deteriorated, broadening the peaks in the static structure factor (not shown here). At still higher levels of shear stress, Fig. 10.1b, the lattice fully melted everywhere, and a shear flow developed. Note that particles in Fig. 10.1 are globally confined, so that after flowing out of the field of view on one side they circulate around the suspension's perimeter and reenter on the opposite side.

In terms of the applied laser power (and hence the resulting stress), the onset of the plastic flow is a rather distinct phenomenon with well-defined yield stress (Nosenko and Goree, 2004). It is noteworthy that the "percolating" trajectories of individual particles that identify the onset of the plastic flow are quite peculiar: These have a zigzag-like shape, jumping along the local principal vector of the hexagonal lattice, i.e., apparently in

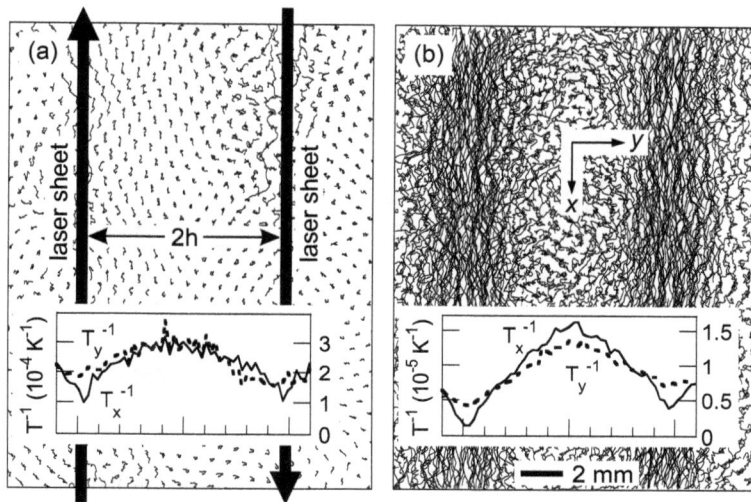

Fig. 10.1   Planar Couette flow in a 2D complex plasma. Initially a crystalline configuration of particles of 8.09 μm diameter is sheared by two counter-propagating laser sheets. At the onset of plastic flow (a), the particles hop between equilibrium lattice sites. In a fully developed shear flow (b), the particle motion is highly irregular on smaller scales comparable with the interparticle spacing, but on larger scales it is like a laminar flow in a fluid. Trajectories over $\simeq 1.7$ s are shown. From Nosenko and Goree (2004).

the direction where the macroscopic lattice has the least yield stress. At the stage of fully developed shear flow, the particle motion is still highly irregular on small spatial scales (comparable to the interparticle distance), but on larger scales it is like a laminar flow in a fluid. In this case, the liquid-like order of the particle suspension can be clearly identified from the diffusiveness of the structure factor.

For all values of the laser power used in the experiment by Nosenko and Goree (2004) the local velocity distribution of particles is (with very high accuracy) Maxwellian, although at the highest shear rates the mismatch between the longitudinal and transverse temperatures is as high as $\sim$ 30 %. This means that the internal momentum and energy equilibration in the particle ensemble is fast enough to balance the heat released due to the shear flow and, hence, the concept of equilibrium viscosity (as a function of local temperature) is well justified. On the other hand, this again underlines the principal distinction between complex plasmas and colloidal dispersions in terms of their individual-particle dynamics: Complex plasmas under shear allow us to observe specific kinetic effects occurring in conventional fluids (e.g., shear-induced heating), whereas colloids, where the kinetic temperature is fixed, are an ideal model system for comparison with rheological models based on statistical mechanics principles.

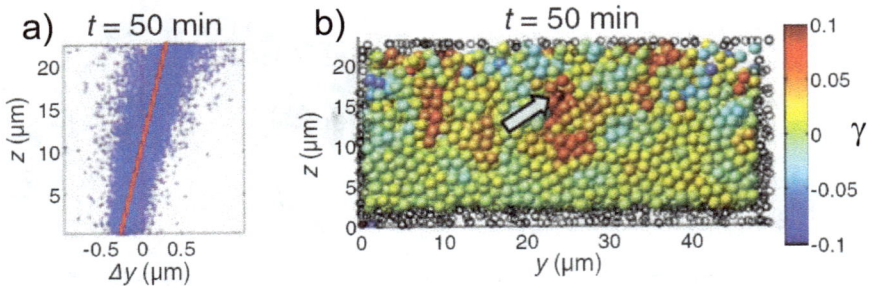

Fig. 10.2   Response of a 3D colloidal glass to shear strain. (a) Shear-induced displacement of particles Δy at height between $z = 0$ and $z = 23$ mm, after 50 min of shear. (b) Distribution of cumulative strain γ after shear. Scale denotes degree of strain. Arrow points to a shear transformation zone, which subsequently relaxed. From Schall *et al.* (2007).

Colloidal dispersions can be likewise brought to flow in a controlled way using shear cells (see Sec. 3.5.6). Such studies make it possible to access local phenomena, in a similar spirit to dynamical heterogeneities in glassy systems (see discussion in Chap. 9). We note that 3D imaging of sheared systems, which is perhaps the most technically challenging of all particle-resolved studies, has required the development of special techniques. The first 3D particle-resolved work of sheared glassy systems was carried out by Besseling *et al.* (2008) who found that structural relaxation was isotropic – a non-trivial result in a shear field. Moreover, while the overall rheology of their sheared hard-sphere-like colloidal suspension followed a Herschel-Bulkley behavior with a smooth flow profile, locally strong shear-thinning effects were observed. This kind of local information underscores the need to use real space analysis in very many situations, even when the bulk behavior appears reasonably straightforward. The same group went on to demonstrate shear banding, where a spatial breakdown in shear rate occurs (Besseling *et al.*, 2010).

Particle-resolved studies also allow us to understand principal mechanisms of relaxation in sheared glassy systems (Schall *et al.*, 2007). Since the strain can be determined at the level of individual particles, and local relaxation events can be directly observed, these can be correlated. This is illustrated in Fig. 10.2, where particles are colored according to local strain. One can clearly see that the strain is highly inhomogeneous, so that a bulk-like mean-field description is expected to break down in such a situation.

10.2 Rheology

The flow of supercooled fluids under external stress can be described by employing the concept of viscoelasticity. The essential feature of a viscoelastic flow is that it displays elastic deformation on short temporal and spatial scales but looks more like a viscous flow on larger scales. In the framework of the simplest linear Maxwell model (Landau and Lifshitz, 1986), the strain γ is a superposition of two components: The elastic contribution responds to the stress through Hooke's law, $\Sigma = G\gamma$, and the viscous contribution through Newton's relation, $\Sigma = \eta\dot{\gamma}$, where G and η are the high-frequency Young's modulus and static shear viscosity, respectively. From these limiting relations one obtains the differential equation $G^{-1}\dot{\Sigma} + \eta^{-1}\Sigma = \dot{\gamma}$. The general solution expresses the stress as a linear response on the time history of the strain rate, with an exponentially decaying response function characterized by the Maxwell timescale $\tau_\mathrm{M} = \eta/G$. This classical model implies a separation between the elastic and hydrodynamic responses controlled by the Weissenberg number $\dot{\gamma}\tau_\mathrm{M}$: The response is viscous at timescales $t \gg \tau_\mathrm{M}$ and elastic for $t \ll \tau_\mathrm{M}$, at intermediate timescales we have a complex Young's modulus (or viscosity). By measuring the response to external stresses at different frequencies one can obtain the complex Young's modulus, derive τ_M, and compare it with the results retrieved from the diffusion measurements.

One should note that such a simple rheological model can usually be employed only to describe the qualitative behavior, and is applicable for fluids which are sufficiently far from the glass transition. The model normally breaks down at the crossover from the Arrhenius to super-Arrhenius scaling (Fischer, 1993), and then more sophisticated nonlinear rheological models should be implemented.

Below we consider two more advanced theoretical approaches which can be used to calculate the rheology of strongly coupled fluids. One approach provides the calculation of the stress from the non-equilibrium pair correlation function $g(\mathbf{r}, t)$, another one uses the *equilibrium* structure factor $S(k)$ as the input characteristics. We will see below that the former – relatively simple – approach does not provide a self-consistent calculation of $g(\mathbf{r})$, whereas the latter approach is self-consistent but certain simplifications are required in order to make it tractable. Note that, in their current formulation, both approaches entirely neglect nonreciprocal (hydrodynamic) interactions (see Sec. 4.3).

10.2.1 *Theoretical approaches*

The first approach, which expresses the stress tensor via the pair correlation function, can be employed for arbitrary pairwise interactions. This yields an exact expression (Kirkwood *et al.*, 1949) where the stress is built up as a sum of partial contributions due to forces exerted by neighbors,[2]

$$\mathbf{\Sigma}(t) = -nk_{\mathrm{B}}T\mathsf{l} - \frac{1}{2}n^2 \int d\mathbf{r}\frac{\mathbf{rr}}{r}\frac{dV}{dr}g(\mathbf{r}, t), \qquad (10.1)$$

where $g(\mathbf{r}, t)$ is the *momentary non-equilibrium* pair correlation function. Let us consider the case of a homogeneous *steady-state* shear flow with the velocity $\mathbf{v}(\mathbf{r}) = \mathbf{\Upsilon} \cdot \mathbf{r}$, where $\mathbf{\Upsilon} = \dot{\gamma}_{\alpha\beta}$ is the velocity gradient tensor. In the limit of a weak shear, when the corresponding Peclet number Pe is small, the non-equilibrium correlation function is given by (Russel *et al.*, 1989; Batchelor, 1977)

$$g(\mathbf{r}) = g_{\mathrm{eq}}(r)\left[1 + Pe\frac{\mathbf{r}\cdot\bar{\mathbf{\Upsilon}}\cdot\mathbf{r}}{r^2}f(r) + O(Pe^2)\right],$$

where $\bar{\mathbf{\Upsilon}} = \mathbf{\Upsilon}/|\mathbf{\Upsilon}|$ and $|\mathbf{\Upsilon}| = \sqrt{\frac{1}{2}\mathrm{Tr}\{\mathbf{\Upsilon}\cdot\mathbf{\Upsilon}\}}$ is the tensor trace norm. For a planar shear flow (in the xy-plane, so that $\mathbf{\Upsilon} = \dot{\gamma}\delta_{\alpha x}\delta_{\beta y}$) we have $g(\mathbf{r}) = g_{\mathrm{eq}}(r)[1 + Pe(xy/r^2)f(r) + O(Pe^2)]$. The function $f(r)$ is determined by the particular form of the (spherically-symmetric) pair interaction potential $V(r)$. This approach can be also employed for a non-stationary shear, then f is a function of $\mathbf{r}$ and t. A self-consistent calculation of $g(\mathbf{r}, t)$ is, however, problematic – it depends on a particular closure relation and neglects nonlocal time correlations (Brader, 2010).

The time correlations are taken into account in the *alternative* second approach which allows us to express the stress via $S(k)$. This approach is based on the recently developed "integration through transients" (ITT) theory (Fuchs and Cates, 2002, 2009) which employs the formal solution of the Smoluchowski equation $\partial\Psi/\partial t = \Omega\Psi$ [see Eqs. (4.4) and (4.5)], in the following form:[3]

---

[2]Here and below, the symbol of dyadic product is omitted for brevity.

[3]The time-ordered exponent e_- is used in Eq. (10.2) because the time-dependent operator $\Omega(t)$ generally does not commute with itself at different t.

$$\Psi(t) = \Psi_{\text{eq}} + \Psi_{\text{eq}} \int_{-\infty}^{t} dt_1 \text{Tr} \left\{ \Upsilon(t_1) \cdot \hat{\Sigma} \right\} e_{-}^{\int_{t_1}^{t} dt' \Omega^{\dagger}(t')} \tag{10.2}$$

where Ψ_{eq} is the equilibrium distribution function in the absence of shear, $\hat{\Sigma} = -\sum_i F_{i\alpha} r_{i\beta}$ is the stress tensor (here $r_\alpha = x, y, z$), and $\Omega^{\dagger}$ is the adjoint Smoluchowski operator,

$$\Omega^{\dagger}(t) = \sum_i \left[\mu_0 (k_B T \nabla_i + \mathbf{F}_i) + \mathbf{r} \cdot \Upsilon^T(t) \right] \cdot \nabla_i,$$

which is readily obtained from Eq. (4.4) by adding to $\mathbf{J}_i$ the advective term $\mathbf{v}_i \Psi$ due to the shear flow. The stress is then obtained by averaging $\hat{\Sigma}$ over the transient distribution function $\Psi(t)$.

To demonstrate the essential elements of ITT theory and its distinction from the first approach, we consider the case of a planar shear flow which is switched from zero to a constant value (Zausch *et al.*, 2008), i.e., $\dot{\gamma}(t) = \dot{\gamma}\Theta(t)$ (then Ω is time independent at $t > 0$ and hence is commutative). Such a simple setup, which can easily be implemented in experiments, allows us to study the transition from the equilibrium to steady-state sheared system and observe the whole range of relevant transient phenomena. For the shear stress we obtain the following expression:

$$\Sigma_{xy}(t) = \dot{\gamma} \int_0^t dt_1 G(t_1), \tag{10.3}$$

where $G(t) = V^{-1} \langle \hat{\Sigma}_{xy} e^{t\Omega^{\dagger}} \hat{\Sigma}_{xy} \rangle$ is the generalized dynamic shear modulus (stress autocorrelation function) and $\langle \ldots \rangle$ denotes the canonical average over Ψ_{eq} (Fuchs and Cates, 2002; Brader *et al.*, 2007, 2008). Thus, the ITT theory provides a nonlinear generalization of the Green-Kubo relation for the stress in non-equilibrium regime.

The shear modulus should be related to the evolving microstructure, which is obtained by implementing the mode coupling theory (MCT, see Sec. 9.2.3). Within the MCT-like closure, $G(t)$ is approximated by projecting the dynamics onto density-pair modes corresponding to all possible wave vector pairs and directions. The numerical complexity of the resulting equation can be significantly reduced by employing the so-called *isotropic approximation* for the modulus (Fuchs and Cates, 2002, 2009; Brader *et al.*, 2009),

$$G(t) = \frac{k_B T}{60\pi^2} \int dk \frac{k^5}{k(t)} \frac{S'_k S'_{k(t)}}{S_k^2} \Phi_{k(t)}^2(t), \tag{10.4}$$

(the prime denotes the derivative with respect to k), where the transient density correlator is

$$\Phi_{k(t)}(t) = \frac{1}{N S_k} \langle n_k^* e^{t\Omega^\dagger} n_{k(-t)} \rangle.$$

The idea of the isotropic approximation is to replace the so-called advected wave vector, $\mathbf{k}(t) = \mathbf{k} \cdot e^{t\Upsilon} = (k_x, k_y + \dot{\gamma} t k_x, k_z)$ [which removes the trivial decorrelation of density fluctuations due to pure affine flow, i.e., $\Phi_{k(t)}(t) = 1$ in the absence of interactions and random motion] by the scalar wave vector $k(t) \approx k \sqrt{1 + (\dot{\gamma} t / \gamma_c)^2}$, where $\gamma_c \sim 1$ is the fitting (crossover) strain parameter (Zausch *et al.*, 2008). The evolution of the transient correlator is governed by the Zwanzig-Mori-type equation, Eq. (9.4), with neglected short-time dynamics,

$$\frac{\partial \Phi_k(t)}{\partial t} + \frac{\Omega_{k(t)}^2}{\nu} \left[\Phi_k(t) + \int_0^t dt_1 m_{k(t_1)}(t - t_1) \frac{\partial \Phi_k(t_1)}{\partial t_1} \right] = 0, \qquad (10.5)$$

where $\Omega_k^2 / \nu = k^2(t) D_0 / S_{k(t)}$ is the "free" decay rate for correlations at given k. Similar to the quiescent state described by Eqs (9.5) and (9.6), in the presence of shear MCT approximations provide the following explicit form for the memory function (Zausch *et al.*, 2008; Fuchs and Cates, 2009; Brader *et al.*, 2009):

$$m_k(t) = \frac{n}{16\pi^3} \int d\mathbf{q} d\mathbf{p} \delta(\mathbf{k} - \mathbf{p} - \mathbf{q}) \frac{S_k S_q S_p}{k^4}$$
$$\times [\mathbf{k} \cdot \mathbf{q} c_q + \mathbf{k} \cdot \mathbf{p} c_p] [\mathbf{k} \cdot \mathbf{q} c_{q(t)} + \mathbf{k} \cdot \mathbf{p} c_{p(t)}] \Phi_q(t) \Phi_p(t). \qquad (10.6)$$

Equations (10.4)-(10.6) form a closed ITT-MCT theory for the calculation of $G(t)$ and hence of the stress.

Formally, the two approaches discussed above are equivalent. The difference between them is revealed by considering transient phenomena, when the competition between structural relaxation and flow becomes essential. The approach based on the pair correlation function cannot be considered as a closed theory since it does not include the self-consistent calculation of $g(\mathbf{r}, t)$. In particular, it does not explicitly take into account nonlocal time correlations and therefore cannot provide a consistent description of sufficiently strongly coupled systems, e.g., close to the glass transition (Brader, 2010). This problem is resolved in the ITT-MCT approach which takes into account the competition between the intrinsic long-time relaxation occurring in quiescent dense liquids due to the cage effect (α-relaxation at the timescale τ_α, see Sec. 9.1.1) and the shear-induced decorrelation of density fluctuations characterized by the timescale $\sim \dot{\gamma}^{-1}$ [this allows us to neglect in Eq. (10.5) the short-time dynamics which is limited by much shorter damping timescales ν^{-1}].

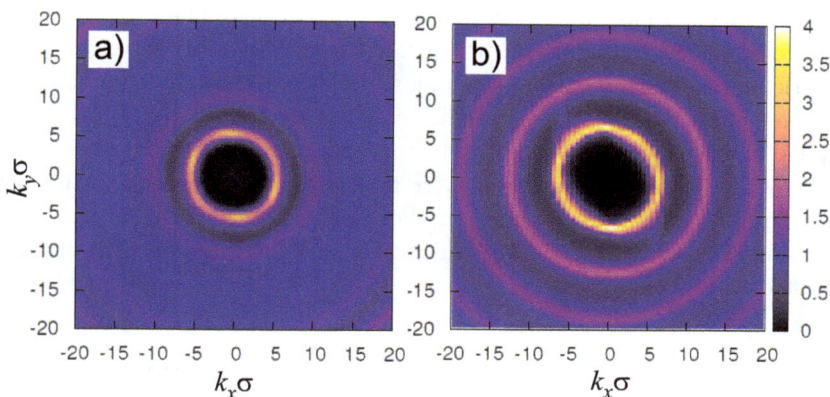

Fig. 10.3   Contour plot of steady-state distorted structure factor $S(\mathbf{k})$ from simulation (a) and theory (b). Shown are the results for a 2D system of hard discs, in a glass at high shear rates: $\phi = 0.79$ and $Pe_0 = 2$ for simulations and $\phi - \phi_g \simeq 10^{-3}$ and $Pe_0 = 10^{-2}$ for MCT (where ϕ_g is the glass transition packing fraction). From Henrich *et al.* (2009).

It is noteworthy that the ITT approach also allows us to calculate the shear-distorted structure factor $S(\mathbf{k}, t)$. This enables direct comparison with the microstructure obtained from the Smoluchowski equation for $g(\mathbf{r}, t)$, Eq. (4.6). The MCT projection operator formalism yields the following expression for the disturbance $\delta S = S(\mathbf{k}, t) - S_k$ (Brader, 2010; Fuchs and Cates, 2009):

$$\delta S(\mathbf{k}, t) = \int_0^t dt_1 \frac{\partial S_{k(t_1)}}{\partial t_1} \Phi_{k(t_1)}^2(t_1), \qquad (10.7)$$

where the transient density correlator is given by the solution of Eqs. (10.5) and (10.6). Equation (10.7) shows that flow-induced structural changes are accumulated by integrating the (shear-distorted) equilibrium structure over the entire flow history. The temporal nonlocality of Eq. (10.7) is clearly contrasting with the local (Markovian) evolution of the pair correlation function. Otherwise, similar to Eq. (10.1) relating the stress to the pair correlation function, the ITT-MCT approach provides the connection between the shear-distorted structure factor and the resulting stress disturbance (Brader, 2010; Brader *et al.*, 2008),

$$\delta \mathbf{\Sigma}(t) = -\frac{n k_B T}{16\pi^3} \int d\mathbf{k} \frac{\mathbf{k}\mathbf{k}}{k} c_k' \delta S(\mathbf{k}, t).$$

This formula has important practical implications. The evolving structure factor is a characteristic which is directly accessible in experiments (Clark and Ackerson, 1980; Vermant and Solomon, 2005), and therefore the flow curve $\Sigma_{xy}(\dot{\gamma})$ can be readily calculated. Figure 10.3 illustrates a typical form of distorted structure factor for a planar shear (Henrich *et al.*, 2009). The isolines are ellipses at $\simeq \pi/4$ angle with respect to the flow direction [the deviation towards smaller angles is a measure of the shear thinning (Hanley *et al.*, 1983)], which provides the major contribution to the stress.

10.2.2   *Comparison with experiments and simulations*

The "original" ITT-MCT theory formulated for anisotropic wave vectors is nearly intractable, and even after the substantial isotropic approximation, Eq. (10.4), the practical implementation remains rather complicated (Brader, 2010; Brader *et al.*, 2009). This is the reason why so-called "schematic model" has been recently proposed (Brader *et al.*, 2009; Fuchs and Cates, 2003). The essential simplification step was to discard the k-dependence in (10.4)-(10.6), which is equivalent to the replacement of S_k by a delta-function at some "effective" $k \sim \sigma^{-1}$. The shear modulus becomes $G(t) \propto \Phi^2(t)$ and the density correlator is described by Eq. (10.5) with the omitted k-dependence. Furthermore, the coupling in the memory function $m(t)$ is usually replaced by a polynomial expansion over $\Phi(t)$ whereas the explicit dependence on the shear rate is represented by a common scaling factor $\propto [1 + (\dot{\gamma}t/\gamma_c)^2]^{-1}$.

Figure 10.4 shows the comparison of the steady-state shear stress Σ_{xy} and viscosity η derived from the schematic model with the results of an experiment performed with colloidal dispersions near the glass transition (Fuchs and Ballauff, 2005). The effective volume fraction of thermosensitive microgel particles, ϕ, was controlled by changing the temperature, so that ϕ decreased with T (see Sec. 3.2). One can see that the (relatively simple) schematic model provides reasonable agreement with the measured data and reveals all principal rheological features of dense liquids – the pronounced shear shinning (occurring when the Weissenberg number $\dot{\gamma}\tau_\alpha$ reaches the critical value of the order of a few tenths) and onset of the yield stress at the glass transition (associated with the divergence of τ_α and η, which is removed at infinitesimal $\dot{\gamma}$).

Now let us consider the transient dynamics of dense colloidal dispersions under the shear. We illustrate this with recent results (Zausch *et al.*, 2008)

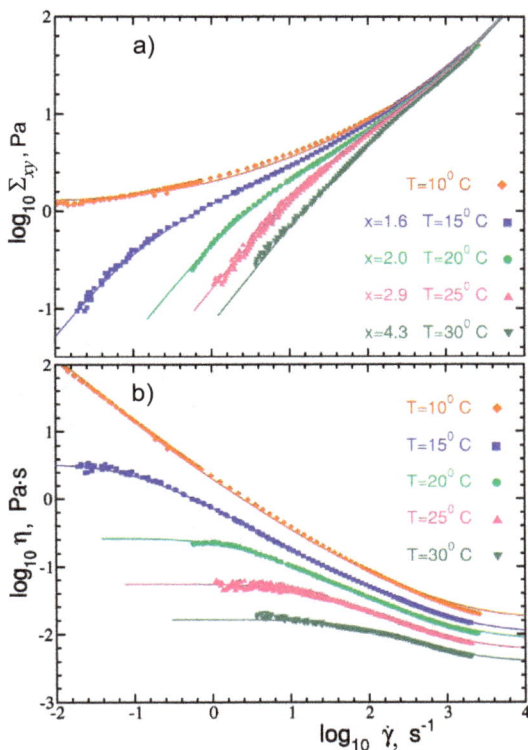

Fig. 10.4   Steady state flow curves and viscosity of dense colloidal dispersions. Shown are the stationary transverse stress Σ_{xy} (a) and the viscosity η (b) as functions of the shear rate $\dot{\gamma}$, for the thermosensitive colloids close to glassy arrest, at various temperatures T from top to bottom as denoted. Solid lines represent fits with the schematic model to the flow curves at small $\dot{\gamma}$ where the structural dynamics dominates. The glass transition temperature lies close to $T_g = 11°$ C. Note the rescaling by the factor x made in (a) at large $\dot{\gamma}$ in order to eliminate effects of hydrodynamic interactions (not included in the model, so that rescaled curves converge). From Fuchs and Ballauff (2005).

obtained for a 3D hard-sphere system (MCT) and 3D hard-core Yukawa system (simulations). A discontinuous switch from the quiescent state to a constant shear rate was investigated, to demonstrate how the transition to a steady-state shear flow evolves near the glass transition. Figure 10.5a shows the shear modulus $G(t)$ in equilibrium and for three sheared states. The equilibrium curve has the usual shape peculiar to other correlation functions in the vicinity of the glass transition. The nonlinear shear thinning accelerates the ultimate relaxation occurring at the characteristic timescale

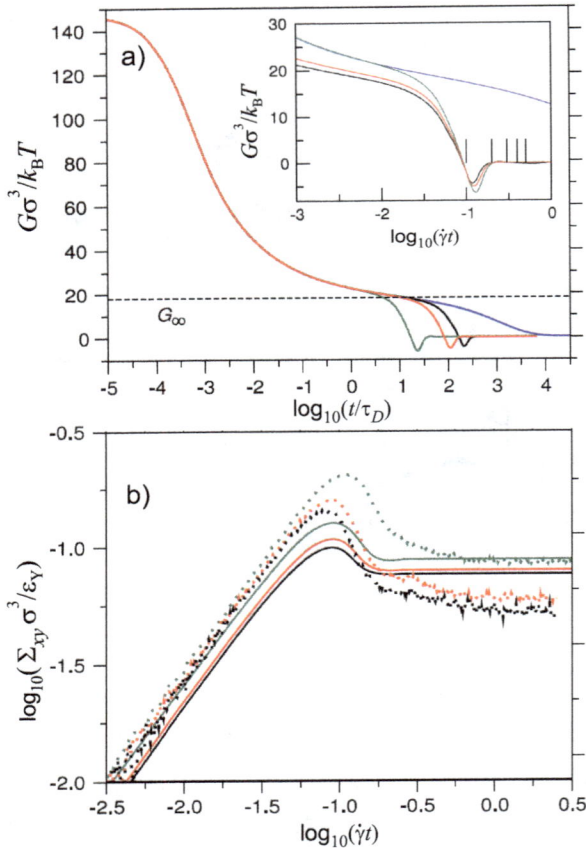

Fig. 10.5   Rheology of hard spheres upon step-like shear. (a) Generalized shear modulus $G(t)$ from the MCT, calculated for supercooled state at the volume fraction $\phi - \phi_g = -1.16 \times 10^{-3}$, for $\dot{\gamma} = 0$ (rightmost curve) and $\dot{\gamma}\tau_D = 5.5 \times 10^{-3}$, 1.1×10^{-3}, and 5.5×10^{-4} (left to right, τ_D is the Brownian diffusive timescale). The plateau value which would develop closer to the glass transition is indicated by the horizontal dashed line. The inset shows the moduli as a function of strain. (b) Shear stress $\Sigma_{xy}(t)$ for the same values of $\dot{\gamma}$ calculated from the MCT (solid lines) and simulations (dots). Note different time normalization in (a) and (b) as well as the stress normalization in (b) with the Yukawa energy scale ϵ_Y. The stress overshoot is related to a negative dip in $G(t)$ at long times. From Zausch *et al.* (2008).

$\dot{\gamma}^{-1}$. The transient shear modulus has a characteristic negative dip for strains at around $\dot{\gamma}t \sim 0.1$. The integral of $G(t)$, Eq. (10.3), yields the transient stress $\Sigma_{xy}(t)$ shown in Fig. 10.5b. At small times corresponding to strains $\dot{\gamma}t \lesssim 0.1$ the curve $\Sigma_{xy}(\dot{\gamma})$ is practically linear. This reveals the regime of elastic response, with the slope determined by the elastic constant G_∞ and therefore practically independent of $\dot\gamma$. On the other hand, at large strains, $\dot\gamma t \gtrsim 1$, the stress approaches a constant asymptote which corresponds to the steady-state flow and hence yields the steady-state viscosity for a given $\dot\gamma$. The prominent feature of the transient dynamics is the stress overshoot occurring between the elastic and steady-state regimes, which is the direct consequence of the negative dip in $G(t)$. The stress overshoot, which can be observed both in supercooled liquids and glasses[4] and whose "microscopic" origin is the shear-induced breakup of particle caging, is the remarkable manifestation of the crossover from the elastic to fluid behavior (Zausch *et al.*, 2008; Pham *et al.*, 2008; Laurati *et al.*, 2011).

In complex plasmas, the fact that the particle dynamics is weakly damped results in the self-consistent variation of the local kinetic temperature associated with the viscous heating – the generic process operating in conventional (molecular) fluids under shear. We demonstrate the importance of this process with the examples below.

Numerical simulations by Saigo and Hamaguchi (2002) and Salin and Caillol (2002) predict that the shear viscosity of complex plasmas stongly depends on the concentration of microparticles, which is one of the essential features of complex fluids. Moreover, recent experiments and simulations (Nosenko and Goree, 2004; Gavrikov *et al.*, 2005; Donko *et al.*, 2006; Ivlev *et al.*, 2007a) verified that the viscosity can exhibit significant shear thinning (and even thickening). This non-Newtonian behavior of complex plasmas occurs because the viscosity η is a strong function of the kinetic particle temperature which, in turn, is determined by the local viscous heat released due to shear flow and is proportional to $\eta\dot\gamma^2$. Based on a simple rheological model by Ivlev *et al.* (2007a), one can identify three distinct regimes for a qualitative dependence of the viscosity and the shear stress $\Sigma = \eta\dot\gamma$ on the shear rate $\dot\gamma$: (i) At sufficiently low $\dot\gamma$ the viscosity remains constant and stress grows linearly with $\dot\gamma$, which corresponds to Newtonian fluids. (ii) Above a certain critical value of $\dot\gamma$ shear thinning is observed, which can be quite significant – the viscosity can decrease by an order of magnitude. (iii) At even higher $\dot\gamma$ the crossover to the shear thickening occurs. A remarkable rheological feature is that the viscosity decrease in the second regime can be so rapid that the $\Sigma(\dot\gamma)$ dependence may have an anomalous N-shaped profile. In this case the part of the curve with $d\Sigma/d\dot\gamma < 0$ becomes unstable and the flow is accompanied by a discontinuity in $\dot\gamma$. This causes

---

[4]In strong attractive glasses several stress overshoots can be seen, indicating multiple yield points (Pham *et al.*, 2008).

the formation of shear bands – a phenomenon often observed in complex fluids (Salmon *et al.*, 2003). Thus, liquid complex plasmas can exhibit the essential rheological features peculiar to "classic" non-Newtonian fluids.

Moreover, by combining different methods to induce shear flows – e.g., inhomogeneous gas flows and laser beams – one can directly measure the shear viscosity in the entire range of shear rates – all the way to the limit where the discreteness enters and a fluid cannot be formally considered as a continuous medium. Probably, the most surprising result of such an investigation was that at "extreme" shear rates (up to $\dot{\gamma} \sim U/\Delta$, where U is the magnitude of the flow velocity and Δ is the interparticle distance) the formal hydrodynamic description with the Navier-Stokes equation still provides fairly good agreement with experiment (Ivlev *et al.*, 2007a).

It is worth mentioning that the transport coefficients of fluid complex plasmas, including the viscosity, could be calculated numerically for an arbitrary rate of the frictional dissipation (Vaulina *et al.*, 2002a; Vaulina and Dranzhevskii, 2007). However, in contrast to steady-state structural properties (see Chap. 6 and 7), the dynamics of individual particles in strongly dissipative systems is completely different from that in conventional single-species fluids. Therefore, systems with strongly damped dynamics (e.g., colloidal dispersions or complex plasmas at high gas pressures) are not appropriate for investigating, e.g., the kinetics of the momentum/energy transfer in shear flows. In the next subsection, where we focus on the kinetics of the energy transport, the importance of weak frictional dissipation becomes particularly clear.

10.3   Heat Transport

Thermal conductivity is an important property that is essential in many engineering applications. At the same time, the behavior of thermal conductivity in various situations is governed by diverse fundamental processes that occur at the atomistic (kinetic) level. Measurements of the thermal conductivity in molecular condensed matter are only possible at a macroscopic scale, and therefore they cannot help in identifying elementary processes which govern the heat transport. The obvious reason for this is the lack of experimental techniques to study the motion of individual atoms. Therefore, here too, liquid or solid complex plasmas occupy the important position of an experimental model system where the motion of individual "atoms" can be observed in real time.

Analysis of heat transport, especially in 2D crystalline systems, is a controversial problem that has a long history: Some authors claim that thermal conductivity in such systems diverges in the thermodynamic limit. Liquid systems are far less studied – one can mention numerical simulations of frictionless hard disks by Shimada *et al.* (2000), where the thermal conductivity slowly diverged as well, and a theoretical study by Ernst *et al.* (1970), where the lack of a valid thermal conductivity was conjectured.

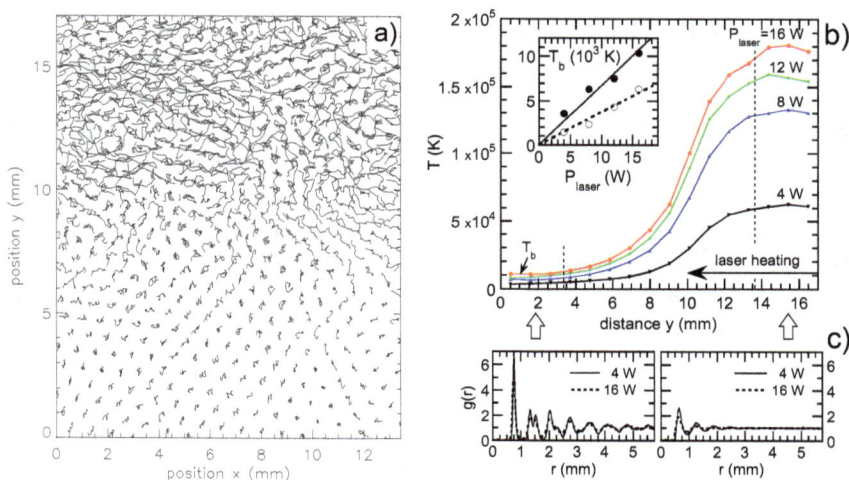

Fig. 10.6   Heat transport in 2D complex plasmas. (a) Example of the particle trajectories (8.09 μm diameter, duration $\simeq$ 1.7 s, courtesy of V. Nosenko) in a 2D plasma crystal heated by a laser at power of P_{laser} = 16 W (heated region y > 13.6 mm). (b) Profiles of the kinetic particle temperature as a function of the transverse coordinate y, for different values of P_{laser}. The inset shows the background particle temperature T_b. (c) Pair correlation function $g(r)$ far from the heated region (left) and inside the heated region (right), suggesting crystalline and liquid states, respectively. From Nosenko *et al.* (2008).

Recently, kinetics of heat transport in liquid and solid complex plasmas was experimentally investigated (Fortov *et al.*, 2007; Nosenko *et al.*, 2008). Below we focus on the experiment performed by Nosenko *et al.* (2008) in a 2D complex plasma that is undergoing a phase transition and therefore constitutes a mixture of crystalline and liquid phases. To melt the lattice locally and to control the temperature of the resulting liquid, the laser-heating method was employed, so that particles were pushed randomly by the radiation pressure force. To produce a quasi-1D temperature gradient, with temperature varying mostly in the y direction, a narrow area which extended fully across the particle suspension in the x direction was heated, as shown in Fig. 10.6a. Under these conditions, heat was mainly transferred by thermal conductivity in the region where the temperature gradient was high.

Figure 10.6b shows the resulting profiles of the kinetic particle temperature, $T(y)$, measured for different values of the laser power. The particle suspension was melted in this temperature range, as can be seen from the analysis of the pair correlation function $g(r)$: Far from the laser-heated area, $g(r)$ has the characteristic appearance of the solid phase with peak splitting and pronounced long-ranged oscillations, whereas inside the laser-heated area, $g(r)$ is typical for a liquid phase with a few peaks.[5] Also, estimates of the melting temperature according to the KTHNY theory (see Sec. 7.1.2) yield the values well within the temperature range achieved in the experiment. Irrespective of the applied heating power P_{laser}, all the measured temperature profiles are very well fitted by the exponential function $T(y) \propto \exp(y/L_{heat})$, where the heat transport length L_{heat} turns out to be practically constant (in the parameter range studied in the experiment). In the framework of the continuous approach to the heat transport, such scaling implies that L_{heat} is identical to the friction length $L_{fr} \equiv \sqrt{\chi/2\nu_{dn}}$ (see Sec. 4.2.2.1), and hence the thermal conductivity χ is practically independent of T.

Thus, the heat transport in a 2D system that undergoes a phase transition turns out to be quite interesting: On the one hand, the experiment yielded the expected result that the thermal conductivity does not exhibit any major discontinuity at the liquid-solid phase boundary – such behavior is well known in conventional condensed matter (March and Tosi, 2002). On the other hand, the values of χ obtained for different particle temperatures are almost the same. This result is not trivial, since individual phases, solid and liquid, are expected to have a temperature-dependent χ. Nosenko *et al.* (2008) suggested that the dominant mechanism of thermal conduction in such systems is phonon scattering on heterogeneous fluctuations that occur in the melting region.

It is important to emphasize that although the measured temperature profile (viz., the value of L_{heat}) is determined by friction, the thermal conductivity itself is solely determined by generic internal properties of the strongly-coupled medium (Yukawa system in our case) and does not depend on the damping: It was shown that the effective length of the phonon

---

[5]Background temperature of the crystal, T_b, was naturally increasing with the applied laser power P_{laser}.

scattering ℓ_{ph} that actually determines the heat conduction ($\chi \simeq \frac{1}{2}C_1\ell_{ph}$) is at least an order of magnitude smaller than the length of the frictional phonon decay ($\simeq C_1/\nu_{dn}$), where C_1 is the measured longitudinal acoustic velocity (see Sec. 4.2.2.2). This allows us to extrapolate the knowledge about the kinetics of heat transport (obtained with weakly damped complex plasmas) to conventional liquids and solids and, thus, to gain deeper understanding of the underlying generic mechanisms operating at the atomistic level. In colloidal suspensions, where particles are fully thermalized by the surrounding solvent and therefore the thermal conductivity (associated with the mutual interactions between colloids) cannot be measured, the temperature gradient imposed to a system leads to the particle drift – a phenomenon well known as the Soret effect (Piazza and Parola, 2008).

10.4   Sedimentation and Basic Hydrodynamic Instabilities

10.4.1   *Sedimentation*

The most straightforward motion of colloidal particles under an external field would be low-density quiescent (Stokes) flow, which can be realized in infinite systems in the regime of low Peclet numbers. At finite densities under confinement, however, the situation is more complex (Royall *et al.*, 2007b). In the framework of the dynamical density functional theory (Archer and Evans, 2004) one can write a continuity equation for the one-body density $n(\mathbf{r}, t)$,

$$\frac{\partial n}{\partial t} = \nabla \cdot \left(\mu n \nabla \frac{\delta F[n]}{\delta n} \right), \tag{10.8}$$

where μ is the particle mobility. In the simplest case, where we assume that the system is weakly driven, one can take $F[n]$ to be an equilibrium Helmholtz free energy [e.g., one can use Rosenfeld's fundamental measure theory for hard spheres (Rosenfeld, 1989)]. A simple treatment of the mobility, which includes the hydrodynamic interactions at the simplest possible mean-field level, is to assume a bulk form. For example, the Hayakawa-Ichiki expression for the mobility reads (Hayakawa and Ichiki, 1995): $\mu(\phi) = \mu_0(1-\phi)^3/[1+2\phi+1.492\phi(1-\phi)^3]$, where ϕ is the packing fraction. It is then possible to predict the evolution of a density profile in a confined system at finite densities.

The simplest way to treat sedimentation would be to *presume* homogeneity in the transverse direction (with respect to the applied field), such

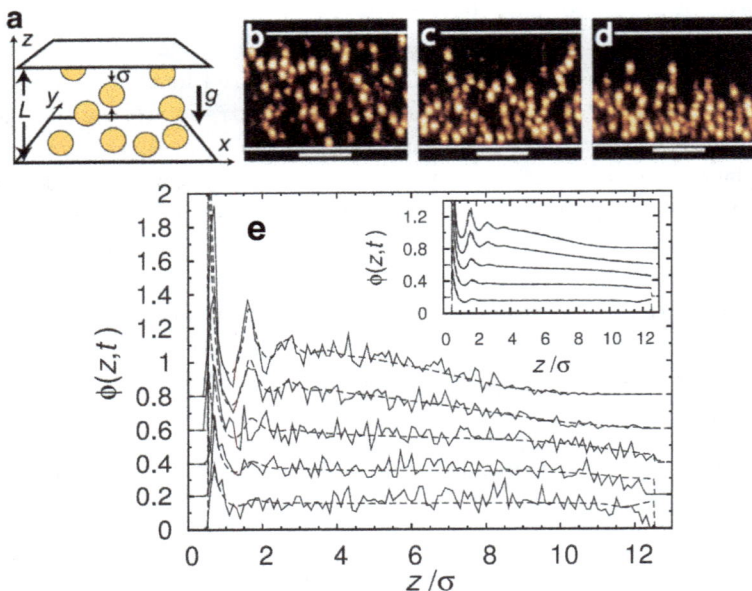

Fig. 10.7   Non-equilibrium sedimentation of colloids.   (a) Sketch showing colloidal spheres under gravity, vertically confined between two walls. (b-d) Time series of confocal micrographs taken in the vertical (xz) plane at times $t/\tau_D = 3, 26, 200$, respectively (normalized by the diffusive timescale τ_D). The scale bars denote 20 μm, the horizontal lines indicate positions of the walls. (e) Time evolution of the density distribution. Shown is the local packing fraction ϕ as a function of the (scaled) height coordinate z/σ for $t/\tau_D = 3, 7, 14, 39, 100$ (from bottom to top). Full lines represent experimental results, dashed lines are from 1D solution of Eq. (10.8) with a density-dependent mobility (the curves are vertically offset by 0.2 for clarity). The inset shows results of Brownian dynamics computer simulations obtained for constant mobility. From Royall *et al.* (2007b).

that the problem collapses to one dimension. In this case, Eq. (10.8) accurately describes sedimentation of colloidal hard spheres across a capillary with *no fitting parameters* (Royall *et al.*, 2007b). Figure 10.7 shows experimental snapshots (for $Pe \simeq 0.625$) and the comparison of measurements with the theory. Considering the simplicity of the approach, the agreement between theory and experiment is truly impressive.

The initial conditions here assume a homogenous density profile, neglecting local packing effects at the wall. Entirely different behavior is seen in precisely the *same sample* for different initial conditions (Wysocki *et al.*, 2009). If the initial distribution is inhomogeneous, a hydrodynamic instability sets in, as pictured in Fig. 10.8. In fact, this behavior is remarkably

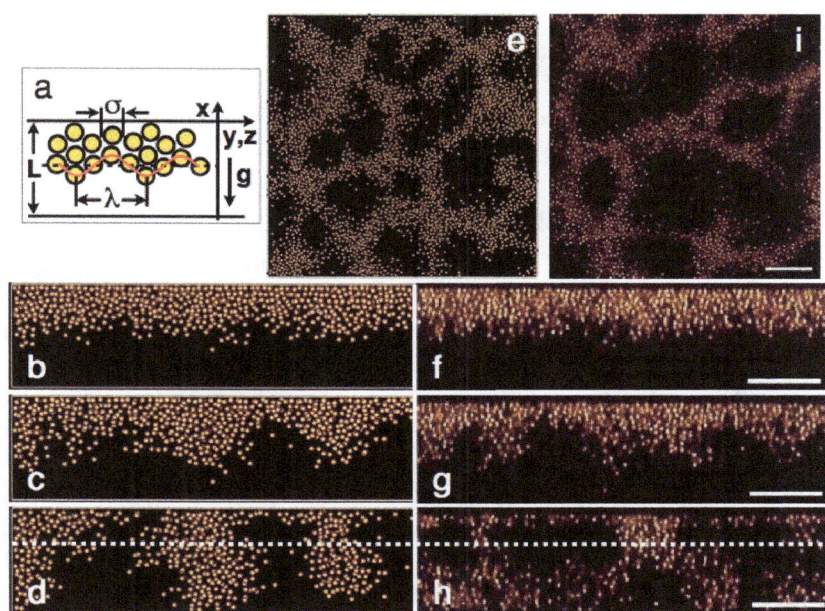

Fig. 10.8   Rayleigh-Taylor-like instability in sedimenting colloids. (a) A sketch illustrating the spatial parameters. (b-e) Simulation snapshots of a system with the Peclet
number $Pe = 1.6$, time t is normalized by the diffusive timescale τ_D: Side view (xy-
plane) at $t/\tau_D = 3.2$ (b), 6.4 (c), 9.6 (d); (e) top view (yz-plane) at $t/\tau_D = 9.6$.
(f-i) Experimental realization of the instability (confocal micrographs) for the parameters $\phi = 0.15$ and $Pe = 1.1$: (f-h) Side view at $t/\tau_D = 1.43$ (f), 5.5 (g), 11.22 (h); (i) top
view at $t/\tau_D = 11.22$. In (f-i) the scale bars denote 40 μm. Cross sections for the top
view shown in (e) and (i) are indicated by the dashed lines in (d) and (h), respectively.
From Wysocki *et al.* (2009).

similar to the Rayleigh-Taylor (RT) instability familiar from immiscible
fluids, which could also be found in colloid-polymer mixtures (see section
Fig. 6.6). However the "hard sphere" system under consideration has no
surface tension. There is no formal interface, merely a density profile which
varies on the particle scale (since the length scale for the density gradient
in equilibrium is determined by the gravitational length $\ell_g = Pe^{-1}\sigma$, see
Sec. 3.5.2, and $Pe \sim 1$). Nevertheless, this density variation leads to a
system that very much resembles a "true" liquid-gas interface shown in
Fig. 6.6.

Characteristics of the observed RT-like instability were estimated using
a linear analysis (Wysocki *et al.*, 2009), and a reasonable agreement with

experiment was found if the term accounting for diffusion is included. This contributes (along with the confining geometry) to stabilizing the large-scale density fluctuations, i.e., in hard-sphere systems the effect of diffusion can, to some extent, "replace" the surface tension.

10.4.2   *Rayleigh-Taylor instability*

Let us discuss the effect of the surface tension on hydrodynamic instabilities in more detail. For this, we focus on the progress achieved recently in particle-resolved computer simulations of large systems forming an interface. In particular, the microscopic origin of the RT instability was explored in molecular dynamics simulations. In a paper by Kadau *et al.* (2004) the initial stage of the RT instability has been simulated with a few million particles interacting via the Lennard-Jones potential and obeying undamped (Newtonian) dynamics. The appearance of the mixing layer as well as major scalings obtained from the simulations were shown to be in excellent agreement with classic hydrodynamic results. Wysocki and Löwen (2004) performed the complementary simulations of the RT instability in the fully damped (Brownian) regime of colloidal suspensions. In these simulations, two different scenarios were observed that occur either for "high" or "low" surface tension, as shown in Fig. 10.9.

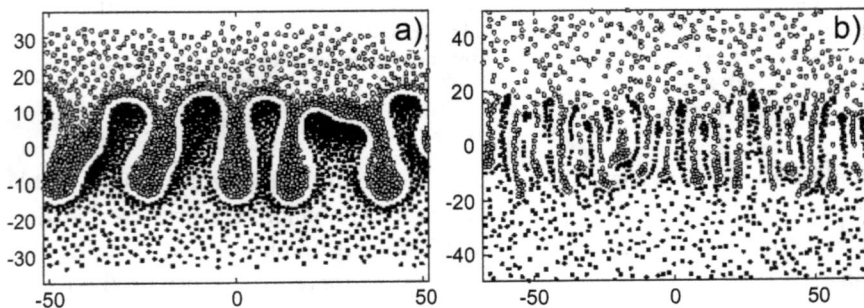

Fig. 10.9   Brownian dynamics simulation snapshots for an interface with externally driven colloids. Distances are measured in units of the particle hard-core radius, the starting configuration was an equilibrated horizontal interface at zero height. Particles in the upper fluid phase are driven downwards, in the lower phase they are driven upwards (all forces are of the same magnitude). Examples of "high" (a) and "low" (b) surface tension are presented, with the lane formation onset seen in the latter case. From Wysocki and Löwen (2004).

The high-surface-tension scenario (Fig. 10.9a) is characterized by interfacial instability which is similar in spirit to the classical RT instability (Chandrasekhar, 1961). As in the undamped case, the classical threshold value for the wavelength of unstable interface perturbations is confirmed. The development of the mixing layer at the initial stage has a similar appearance to that seen in undamped Newtonian liquids (Kadau *et al.*, 2004), although at the later stage the interpenetrating bubbles and spikes do not develop into "mushrooms" but keep growing further aperiodically. Also, the thickness of the mixing layer increases as $\propto t$ with time instead of the classic $\propto t^2$ scaling. Phenomenologically, such a difference is because the governing equation for the Brownian dynamics does not contain the time derivative of the velocity. Therefore, in analogy with the classic case when spatial perturbations driven by a uniform external field exhibit "free" kinematic growth $\propto t^2$, the RT development in overdamped systems should reveal the linear time scaling.

When the interfacial surface tension is sufficiently low (Fig. 10.9b), a completely different behavior is observed: The particles penetrate the interface easily as a result of the driving field and form microscopic lanes (see Sec. 10.5.1). These results are obtained in the regime when the classical RT threshold for the unstable wavelength (calculated for given values of the surface tension and driving force) is smaller than the interparticle distance and hence a breakdown of hydrodynamics is expected. Therefore, the microscopic appearance of the RT instability might be completely different as the discreteness enters, and this conclusion is quite intuitive: The surface tension is the major stabilizing mechanism of the instability, and once this mechanism becomes weak enough to allow the growth at hydrodynamic scales smaller than the discreteness limit, the hydrodynamics itself becomes meaningless. On the other hand, the instability should develop in some form anyway, and the only imaginable picture for that are the interpenetrating strings, as observed.

10.4.3 *Shear-induced instabilities*

In order to investigate hydrodynamic instabilities in further detail, let us consider examples of particle-resolved shear flows observed in complex plasmas (Morfill *et al.*, 2004) and shown in Fig. 10.10. Different flow topologies were observed, with the (average) flow lines being either straight (a) or curved with a radius of curvature of about $80 - 100\ \Delta$ (b). The lower part of the microparticle cloud is at rest. The observations suggest that the

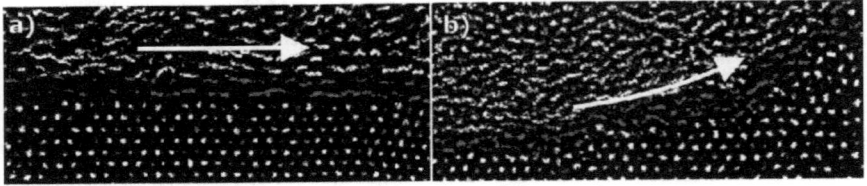

Fig. 10.10   Two examples of highly resolved complex plasma flows. The figures show a shear flow over a flat-surface (a) and curved-surface (b) plasma crystal. Note the small-angle perturbations of the particle trajectories seen inside the transition (mixing) layer in (a), and the considerably larger scattering for the curved flow in (b). Particles are of 6.8 μm diameter, the flow velocity is ~ 1 mm/s. From Morfill *et al.* (2004).

width and the structure of the transition (mixing) layer strongly depends on the geometry. For the planar flow the interface is quite smooth, with the flow along a particular monolayer. The trajectories of individual flowing particles experience only weak deflections and the overall flow appears to be stable and laminar. In contrast, for the curved flow the interface has a curious rough structure, the "microscopic" flow is not laminar. Apparently, the mixing layer becomes unstable at the individual-particle level. The microscopic behavior may be interpreted as the centrifugally driven RT instability. By analyzing longer sequences of such images, one can quantify "elementary" (discrete) perturbations in two ways – the fraction of inter-penetrating (say, $\gtrsim \Delta$) particles, and the fraction of particles undergoing "large angle" (say, $\gtrsim 30°$) collisions in the transition layer. For instance, for the straight flow the quantities are (almost) 0 % and ~ 3 %, respectively, for the curved flow these are ~ 3 % and ~ 30 %. The latter can be understood kinetically in terms of the higher collision frequency with smaller impact parameter, due to particle inertia at a curved surface. This has also been confirmed by numerical simulations (Morfill *et al.*, 2004) conducted for similar geometry and flow conditions as in the experiments. The topology of the mixing layer found in the simulations corresponds closely to the measurements, which supports the kinetic interpretation.

Following these considerations, one can argue that the Kelvin-Helmholtz (KH) instability at the discreteness limit also has a different appearance. In order to illustrate this point, we consider another example of the hydro-dynamic behavior of liquid complex plasmas (Morfill *et al.*, 2004) shown in Fig. 10.11. Particles were flowing around an "obstacle" – the void of size ~ 100 Δ. One can see stable laminar shear flow around the obstacle, the development of a downstream "wake" exhibiting stable vortex flows, and

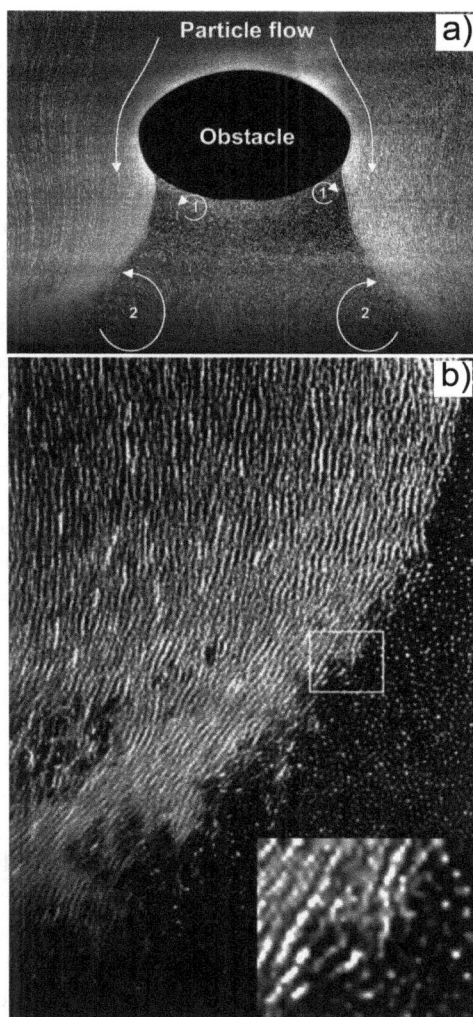

Fig. 10.11 Flow past an obstacle in fluid complex plasmas. (a) Overall topology of the 3.7 μm particle flow, the system is approximately symmetric around the vertical axis (exposure time 1 s). The flow leads to a compressed laminar layer, which becomes detached at the outer perimeter of the wake. The steady vortex flow patterns in the wake are illustrated. The boundary between the laminar flow and wake becomes unstable; a mixing layer is formed, which grows in width with distance downstream. (b) An example of the mixing layer (an enlargement of the left side, exposure time 0.05 s). The points (lines) represent traces of slow (fast) moving microparticles. The inset shows trajectories of individual particles in the mixing layer (in the region marked by the rectangular frame). From Morfill *et al.* (2004).

a mixing layer between the flow and the wake. The enlargement of the mixing layer (Fig. 10.11b) shows that the flow is quite unstable at the kinetic level, with instabilities rapidly becoming nonlinear. The width of the mixing layer grows monotonically with distance from the border where the laminar flow becomes detached from the obstacle. The length scale of the growth is of the order of a few Δ, i.e., much smaller than the hydrodynamic scales, $n(dn/dx)^{-1}$ or $u(du/dx)^{-1}$, which would be expected macroscopically in fluids and which refer to the RT or KH instability, respectively. This rapid onset of surface instabilities followed by mixing and momentum exchange at scales $\sim \Delta$, i.e., the smallest interaction length scale available, is not consistent, therefore, with conventional macroscopic instability theories. While this is not expected at the kinetic level, it clearly points to new physics and, possibly, a hierarchy of processes governing fluid flows: First, "strong" binary collisions provide effective particle and momentum exchange on discrete scales (a few Δ), then collective effects (due to the correlations defining fluid flows) take over and propel this "discrete" instability to macroscopic scales, creating cascades of growing clumps characterized by enhanced vorticity.

Although the onset of the instability shown in Fig. 10.11 occurs at scales $\sim \Delta$, its further development is in amazing agreement with the simplest conceptual picture of continuous jet turbulence: It is well known that the mixing between a jet and its surroundings occurs in two stages (Tennekes and Lumley, 1972). During the first stage (which is a distinct peculiarity of jets), a shear layer is formed immediately downstream of the jet source, between jet stream and surroundings. As one moves downstream, there is an early linear-instability regime, involving exponential growth of small perturbations introduced at the jet source. Beyond this development stage, in the nonlinear KH instability regime, the dynamics of large-scale vortex formation and merging become the defining feature of the transitional shear flow. Apparently, the observed clump cascading fully mimics this scenario, which suggests – again – that the similarity of the coarse-grained hydrodynamics is preserved down to the physical discreteness limit.

Unfortunately, in experiments with complex plasmas so far it has been impossible to observe the second stage typical to any developed turbulence – when vortices (clumps) break down leading to a more disorganized flow regime characterized by smaller-scale vortices. The spectral energy content at this stage should be consistent with the Kolmogorov inverse cascade theory of turbulence. These processes develop at much longer timescales, when the neutral friction starts playing an important role and simply "freezes out" free hydrodynamic motion. In order to observe this turbulent stage in experiments, one needs to decrease the neutral gas pressure substantially and to obtain large-scale complex plasma flows.

These examples suggest a naive microscopic picture of the hydrodynamic instabilities: It is not unreasonable to conclude that many instabilities have a kinetic (discrete) trigger, and that the most effective trigger mechanism is provided by "strong" (large-angle) scattering in localized structures and/or inhomogeneities of scales comparable to the particle correlation length. However, the mathematical techniques required to quantify the kinetic behavior and to transfer this to macroscopic scales still need to be developed.

10.5 Non-Equilibrium Phase Transitions

So-called "open" systems are those which may exchange energy and matter. These can be systems in contact with reservoirs, driven and constrained systems, etc. (Hoover, 1985; Evans and Morriss, 1990; Tuckerman *et al.*, 2001). A remarkable property of nonlinear open systems is *self-organization* (Prigogine, 1980) – a spontaneous emergence of stable spatial (or temporal) structures, which are often referred to as "dissipative structures", since dissipation plays a constructive role in their formation. Dissipative structures are the manifestation of non-equilibrium phase transitions, with well-known examples being, e.g., formation of convection (Bénard) or turbulent (Taylor) vortices (Cross and Hohenberg, 1993). In order for such transitions to occur, three basic requirements have to be satisfied: (i) Dissipation is necessary, to balance the external influx of energy. (ii) The structures may emerge only in systems described by nonlinear equations. (iii) There must be a relevant control parameter entering particular solutions of the equations, which ensures symmetry breaking (viz., transition) above a certain threshold.

10.5.1 *Laning*

One important example of a non-equilibrium phase transition is the formation of lanes – a phenomenon occurring in nature when two species of particles are driven against each other. When the driving forces are strong enough, like-driven particles form "stream lines" and move collectively in lanes. Typically, the lanes exhibit a considerable anisotropic structural order accompanied by an enhancement of their (unidirectional) mobility.

Laning is a truly universal phenomenon – it can be observed in highly populated pedestrian zones (Helbing *et al.*, 2000), as illustrated in Fig. 10.12, but also occurs in different systems of driven particles, such as colloidal dispersions (Leunissen *et al.*, 2005; Dzubiella *et al.*, 2002; Vissers *et al.*, 2011b) and complex plasmas (Morfill *et al.*, 2006; Sütterlin *et al.*, 2009), lattice gases (Schmittmann and Zia, 1998) and molecular ions (Netz, 2003). In other words, this is a ubiquitous generic process of considerable interest in different branches of physics. Laning is an instability which occurs on the particle scale, i.e., the size of the structures formed is comparable to interparticle distances. Therefore particle-resolved experiments, simulations, and theories are needed to understand its origin.

Fig. 10.12   Lane formation in a crowded pedestrian zone. Pedestrians move in lanes in order to maximize their speed.

While steady-state lanes have been studied in detail, the dynamic pathways towards laning are still under debate. Basically, there are two ways to observe laning: One is to start from a completely mixed state in equilibrium and turn the driving force on, as is realized for mixtures that do not phase separate (Dzubiella *et al.*, 2002; Leunissen *et al.*, 2005; Vissers *et al.*, 2011b). The lanes then develop by growing anisotropic structural correlations, most probably via a spinodal instability (Chakrabarti *et al.*,

2003, 2004). The other way to observe laning is to start from a state when two phases are macroscopically separated. In this case particles at the interface are driven against each other, resulting in a RT-like instability (see discussion in Sec. 10.4).

10.5.1.1 *Laning in complex plasmas*

Lane formation can be easily triggered in complex plasmas (Morfill *et al.*, 2006; Sütterlin *et al.*, 2009). As we showed in Sec. 4.2, complex plasmas provide a very important intermediate dynamic regime that is between conventional undamped fluids and fully damped colloidal suspensions: In complex plasmas, the short-time dynamics associated with the interparticle interactions is undamped whereas the large-scale hydrodynamics can be strongly affected by friction. Nevertheless, the mesoscopic appearance of the lane formation in colloids and in complex plasmas is quite similar.

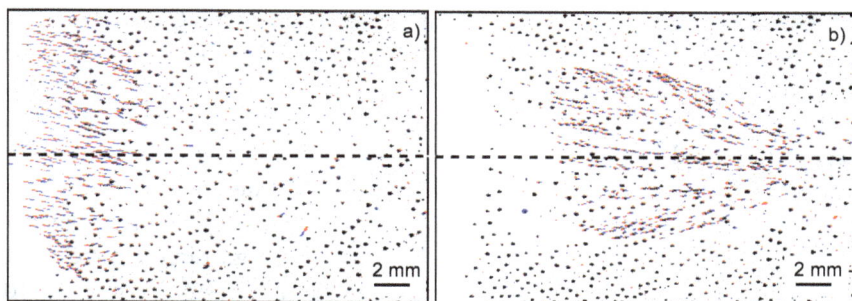

Fig. 10.13   Lane formation in complex plasmas.   A short burst of small (3.4 μm) particles is injected into a cloud of big (9.2 μm) background particles. The experiment is performed in the PK-3 Plus chamber (see Fig. 5.5) under microgravity conditions. Small particles are driven towards the center (chamber midplane is indicated by the horizontal dashed line), stages of (a) initial lane formation and (b) merging of lanes into larger streams are shown. Each figure is a superposition of two consecutive color-coded images (1/50th s apart, red to blue), the time difference between them is $\simeq 1.2$ s. At the stage (b) big particles also form well-defined lanes that can be identified as strings of red/blue dots. From Sütterlin *et al.* (2009).

The experiments by Sütterlin *et al.* (2009) were performed for various combinations of "big" and "small" monodisperse particles, with different neutral gases and pressures (to control the damping rate), and for different rf discharge powers (to control the screening length). First, a stable spheroidal complex plasma cloud of big particles was produced. Then small particles were injected into this cloud. The force field pulls the small particles through the cloud of big particles towards the center, thus making such systems perfectly suited to study lane formation. Figure 10.13 shows a characteristic example of lane formation with 3.4 μm and 9.2 μm particles. When a fraction of individual small particles enters the interface of a cloud formed by big particles, the subsequent penetration is accompanied by a remarkable self-organization sequence: (a) Big particles are pushed collectively by the inflowing cloud of small particles, the latter form "strings" drifting on average along the force field. (b) As the particles approach the center of the chamber, the field decreases and the strings organize themselves into larger "streams". At the later stage, when the field almost vanishes, the streams merge to form a spheroidal droplet with a well-defined surface (shown in Fig. 8.10), indicating the transition to the regime when surface tension plays the primary role. It is noteworthy that during stage (b) big particles also form well defined strings. Small and big particles create an "array" of interpenetrating strings. After the flux of small particles is exhausted, the big-particle strings slowly dissolve. Complementary molecular dynamics simulations revealed remarkable agreement with the experiments.

In this section we restrict ourselves to the discussion of the first two stages illustrated in Fig. 10.13 (the phase separation stage is considered in Sec. 8.3). In order to identify and quantify the string-like structures observed in the experiments and simulations, a suitable *sensitive* order parameter has to be employed. Conventional approaches, e.g., binary correlation or bond-orientational functions, Legendre polynomials, etc. turned out to be too insensitive for the laning characterization. Much more satisfactory results were obtained by implementing an *anisotropic scaling index* method [see paper by Räth *et al.* (2008) for details] which can be used for the local quantification of various anisotropic environments (e.g., it was also used to characterize electrorheological complex plasmas, see Fig. 11.3). For the laning, a "uniaxial vector characterization" has been proposed: Each particle is associated with the unit vector $\mathbf{u}_i$ pointing to a direction of the maximum local anisotropy (determined from the anisotropy of the scaling index). The directions $\mathbf{u}_i$ and $-\mathbf{u}_i$ are equivalent, so that they are defined for the range $-\frac{\pi}{2} \leq \theta_i \leq \frac{\pi}{2}$.

Thus, each particle can now be considered as a uniaxial "molecule" (simple rod) with the direction $\mathbf{u}_i$. Therefore, the global laning on a 2D plane can be characterized with the second-rank tensor $\mathbf{Q} = 2N^{-1}\sum_{i=1}^{N}\mathbf{u}_i\mathbf{u}_i - \mathbf{I}$, analogous to that used to quantify order of the nematic phase. The direction of the global laning, $\langle \mathbf{u} \rangle$, is then the eigenvector ("nematic director") corresponding to the largest eigenvalue of Q, which in turn is the laning order parameter, S. Obviously, $S = 1$ for a perfect alignment and $S = 0$ for a disordered phase, when individual vectors $\mathbf{u}_i$ are uncorrelated. One can also define the global laning angle Θ via the relation $\cos \Theta = \langle \mathbf{u} \rangle \cdot \mathbf{e}_F$ (where $\mathbf{e}_F$ is the unit vector in the direction of the driving field).

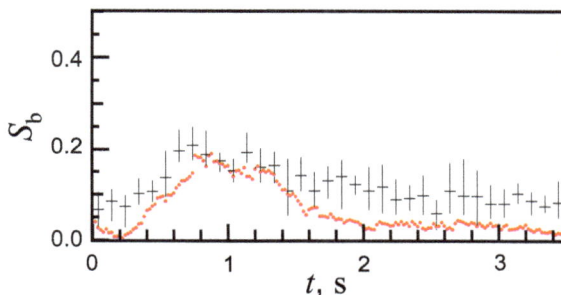

Fig. 10.14   Dynamics of laning. Shown are the evolution of the "nematic" order parameter for big particles, $S_b(t)$, as obtained from the anisotropic scaling index analysis of the experiment presented in Fig. 10.13 (crosses) and molecular dynamics simulation (dots). From Sütterlin *et al.* (2009).

The evolution of the order parameter $S_b(t)$ obtained for big particles by Sütterlin *et al.* (2009) is shown in Fig. 10.14. Initially, there is no anisotropy in the simulation, whereas in the experiment $S_b \simeq 0.05$, due to a weak (transverse) inhomogeneity in the big-particle density. Once the cloud of small particles reaches the region used for the analysis, S_b starts growing due to the increasing number of small particles causing the formation of big-particle strings. Then S_b reaches a maximum and starts falling off, reflecting the onset of string relaxation (after the small particles leave the region). The initial relaxation occurs at a characteristic timescale of ~ 1 s which is an order of magnitude shorter than the self-diffusion timescale for the big particles in the experiment. Note that after this rapid relaxation $S_b(t)$ tends to an intermediate plateau, indicating that the structural relaxation is apparently not complete. This suggests that the ultimate equilibration might involve some metastable states.

10.5.1.2   *Laning in charged colloids*

Oppositely charged colloidal suspensions (Leunissen *et al.*, 2005) are complementary model systems to study the laning instability in electric fields. Typically, the colloidal experiments are performed in 3D capillary cells, and

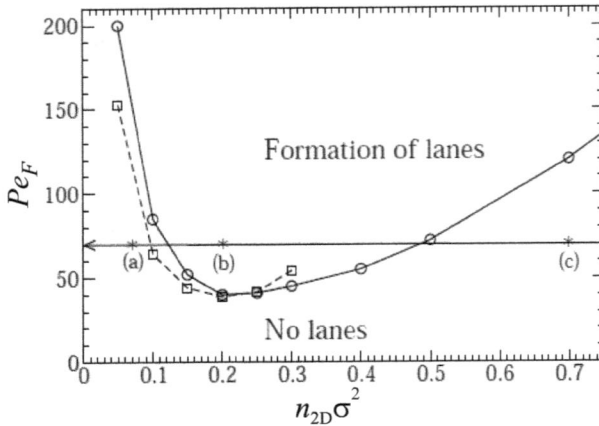

Fig. 10.15   Steady-state phase diagram of laning in 2D hard-sphere system, shown in the parameter space of the normalized density $n_{2D}\sigma^2$ and driving force Pe_F. Computer simulation results (solid line) and data from dynamic density functional theory (dashed line) are shown. Three states along a path of a constant force and increasing density [marked "(a)", "(b)", and "(c)"] demonstrate the reentrant transition to a lane-free state at higher densities. From Chakrabarti *et al.* (2004).

the data obtained are then compared with Brownian dynamics computer simulations.

2D computer simulations for repulsive interparticle interactions have suggested that laning in a plane is a first-order phase transition (Dzubiella *et al.*, 2002), viz., it occurs with a notable hysteresis if the field is increased and decreased. Furthermore, for a fixed driving force and increasing density there is a reentrance of a lane-free state (Chakrabarti *et al.*, 2004). The corresponding steady-state phase diagram is illustrated in Fig. 10.15. The strength of the driving force F_0 is characterized by the Peclet number defined as follows:

$$Pe_F = \frac{F_0\sigma}{2k_BT}, \tag{10.9}$$

(which is the ratio of the drift time $\sigma k_B T/D_0 F_0$ to the diffusive time of a single particle $\tau_D = \sigma^2/D_0$). The instability can be quantitatively described using a dynamical density functional theory with additional phenomenological current term (Chakrabarti *et al.*, 2003). However, while this term can be justified phenomenologically (Vissers, 2010), its microscopic origin is still unclear. Data obtained from a dynamic density functional theory are also presented in Fig. 10.15.

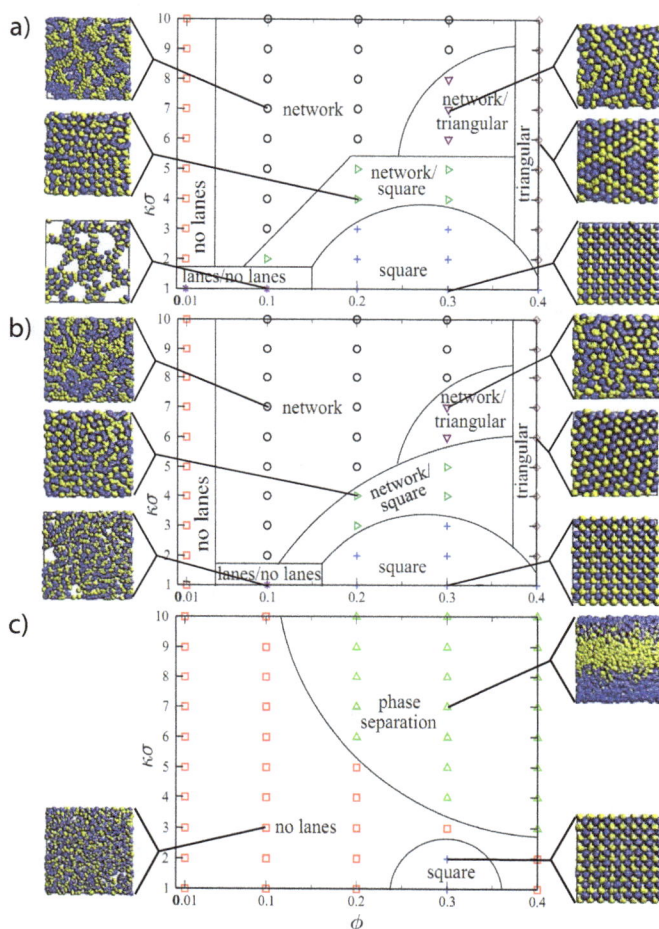

Fig. 10.16   Steady-state phase diagram of laning for oppositely charged hard-core Yukawa particles, shown as a function of the particle packing fraction ϕ and the screening parameter $\kappa\sigma$ (for a fixed driving force of $Pe_F = 75$). Three cases are depicted: (a) no hydrodynamic interactions, (b) driving electric force with the Long-Ajdari mobility tensor, and (c) driving gravitational force with a Rotne-Prager mobility tensor. The snapshots showing projections on a plane perpendicular to the driving force illustrate variety of steady states. From Rex and Löwen (2008).

Comprehensive Brownian dynamics simulations performed by Rex and Löwen (2007) for a 3D mixture of oppositely charged hard-core Yukawa particles revealed a much richer scenario of different steady states. For a fixed driving force, the steady-state diagram is summarized in Fig. 10.16, where the control parameters are the colloid concentration (the total colloid packing fraction ϕ) and the screening parameter $\kappa\sigma$. Apart from a region in the parameter space where no laning occurs (at small ϕ), there is a transition towards laning for any $\kappa\sigma$. The snapshots shown (particle positions projected on a plane perpendicular to the drive) demonstrate impressive diversity of the lateral laning order. At higher $\kappa\sigma$, one can see network structures with a finite spacing, while for lower $\kappa\sigma$ the lanes crystallize into 2D lattices which can exhibit triangular, rhombic, and square structures. Note that for small ϕ and $\kappa\sigma$ chains of lanes are expected to phase separate into a square crystal and a low-density gas.

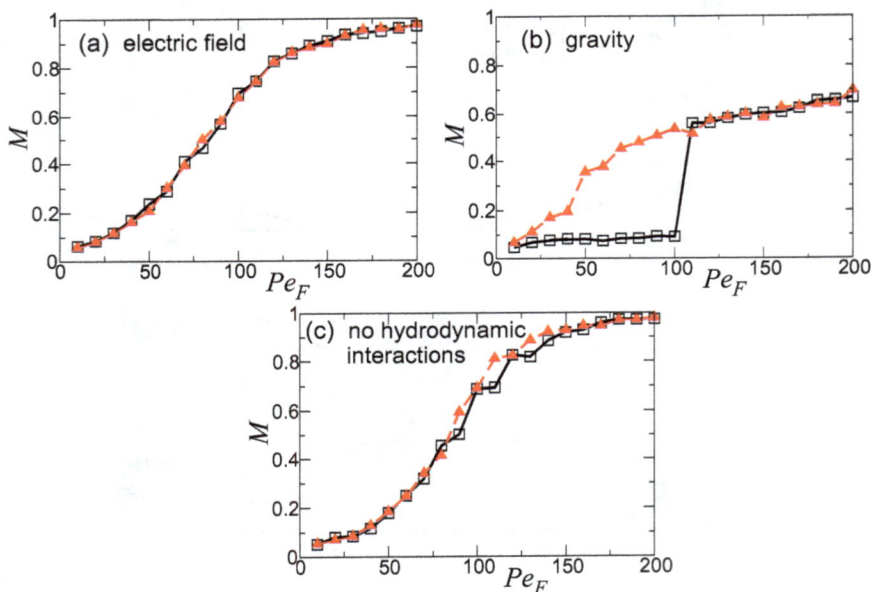

Fig. 10.17   Laning order parameter M (averaged over all particles) versus reduced driving force Pe_F, upon increase (open squares, full line) and decrease (full triangles, dashed line) of the force. Shown are the cases for (a) an electric driving field where hydrodynamic interactions are screened, (b) a gravitational force where hydrodynamic interactions are long-ranged, (c) neglected hydrodynamic interactions. From Löwen (2010).

Since the moving colloidal particles induce a solvent velocity field, there is an issue about the importance of hydrodynamic interactions. One can include these to leading order by employing the long-range pairwise Rotne-Prager tensor [Eq. (3.11)] if the driving field is gravity (see Fig. 10.16c).

For the electric driving force (see Fig. 10.16b), hydrodynamic interactions are described by the screened Long-Ajdari tensor (Long and Ajdari, 2001). In the latter case, the long-range Oseen tensor is exponentially screened because the motion of the counterions is opposed to that of the colloids (which is absent for the gravity-driven case). The results obtained demonstrate (Rex and Löwen, 2008) that hydrodynamic interactions do not destroy the topology of the steady-state diagram shown in Fig. 10.16a. However, the laning driven by gravity (i.e., sedimentation) is connected with macroscopic phase separation, so that the stability range of a lateral lane crystal is strongly suppressed in this case. The same qualitative difference affects the order of the laning transition: Figure 10.17 presents the results of numerical simulations (Löwen, 2010) showing that the transition, which is of the first order for sedimentation (b), becomes continuous (i.e., without any notable hysteresis) when the driving force is electric (a) or hydrodynamic interactions are neglected (c). Note that for all three cases presented here non-hydrodynamic interactions are identical.[6]

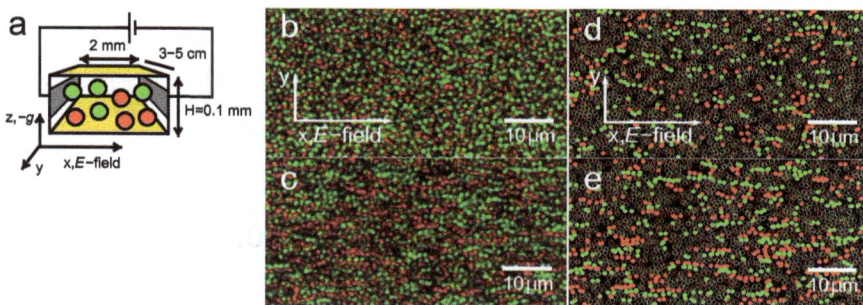

Fig. 10.18   Lane formation in colloidal dispersions. (a) Sketch of the experimental setup. A suspension of oppositely charged colloidal particles (green and red spots) is confined in a millimeter-sized capillary. The electric field E is along the x-axis. Experimental (b,c) and simulation (d,e) snapshots are for $E = 30$ kV/m (upper row) and $E = 110$ kV/m (lower row), a few seconds after the field was turned on. From Vissers *et al.* (2011b).

A more detailed comparison between Brownian dynamics simulations of Yukawa systems and real-space experiments for oppositely charged colloids

---

[6]The laning order parameter M used here is different from the "nematic" order parameter S employed in the previous section (describing laning in complex plasmas): M was calculated by defining a cylindrical shell in the direction of driving force around each particle (of diameter σ and length 3σ). If no oppositely driven particle is found in this cylinder then $M = 1$, otherwise $M = 0$.

was performed recently by Vissers *et al.* (2011b). The confocal microscopy snapshots obtained for two different values of the driving electric force are illustrated in Fig. 10.18 along with the complementary simulation snapshots. Figure 10.19 shows the measured dependence of the (statistically averaged) laning order parameter on the driving electric field, revealing a good quantitative agreement between experiments and simulations. These results demonstrate that even simple Brownian dynamics simulations (neglecting hydrodynamic interactions) can describe the essential physics of laning.

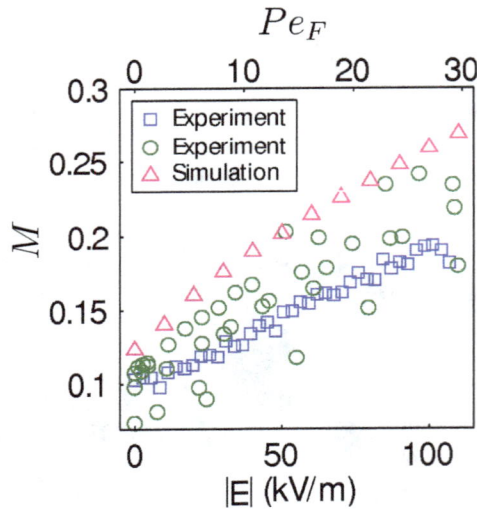

Fig. 10.19   Laning order parameter M (averaged over all particles) as a function of the electric field E (Peclet number Pe_F). Shown are the results of experiments where the field was changing gradually (blue squares) or applied instantaneously (green circles), and the simulation results at constant field (magenta triangles). From Vissers *et al.* (2011b).

Thus, we conclude that the nature of the laning transition depends on the interaction details as well as dimensionality: In the presence of long-range hydrodynamic interactions the transition is of the first order, while for a short-range mobility matrix it is second-order. Such behavior appears to be qualitatively in line with the second-order transition proposed by Corté *et al.* (2008) for oscillatory driven systems. The behavior changes qualitatively in two spatial dimensions (Dzubiella *et al.*, 2002), where laning is found to be of the first-order even for a short-range mobility matrix.

Obviously, this is a topological effect related to the fact that two lanes of particles driven alike cannot merge without destroying an intermediate lane with oppositely moving particles, which makes 2D lanes very stable. Future studies of laning in charged colloidal dispersions require simulations of hydrodynamic interactions and explicit treatment of microions, in order to elucidate the role of the electric dipole moments induced by the applied field. Some modern simulation techniques (Horbach and Frenkel, 2001; Kodama *et al.*, 2004; Hickey *et al.*, 2010; Lobaskin *et al.*, 2007; Yamamoto *et al.*, 2009) which incorporate both hydrodynamics and electrostatics are quite promising in this respect.

10.5.2   *Banding*

Banding (or axial segregation) occurs when driven particles segregate into stripes which are not parallel to the driving field (Wysocki and Löwen, 2009; Aranson and Tsimring, 2006; Löwen, 2010). Apart from periodically shaken granular matter (Mullin, 2000; Sanchez *et al.*, 2004; Tarzia *et al.*, 2005; Ciamarra *et al.*, 2007), this phenomenon has been predicted by Wysocki and Löwen (2009) and observed by Vissers *et al.* (2011a) in oppositely charged colloidal suspensions exposed to an ac electric field. The driving force, which is now time-dependent, can be presented in the following form:

$$\mathbf{F}(t) = \pm\mathbf{F}_0 \cos \omega t, \tag{10.10}$$

where ω denotes the driving frequency and different signs correspond to two species which are driven against each other. In the limit of $\omega \to 0$, one recovers a constant field which was discussed in the preceding section.

Figure 10.20 shows an example of a steady-state diagram in a plane spanned by the normalized driving frequency $\omega\tau_D$ and the Peclet number Pe_F defined in Eq. (10.9), while the particle number density is fixed. Banding occurs for intermediate values of Pe_F and is followed by laning at large Pe_F. On the other hand, for a given Pe_F there is no pattern formation at sufficiently high frequencies. This latter effect becomes clear by taking into account the fact that high frequencies lead to an almost vanishing net force if the timescale of the external drive, ω^{-1}, is much shorter than the particle diffusive time τ_D.

Intuitively, the banding mechanism can be understood as follows (Pooley and Yeomans, 2004): Particles driven alike perform similar harmonic excursions induced by the external field. Hence, if the oppositely charged

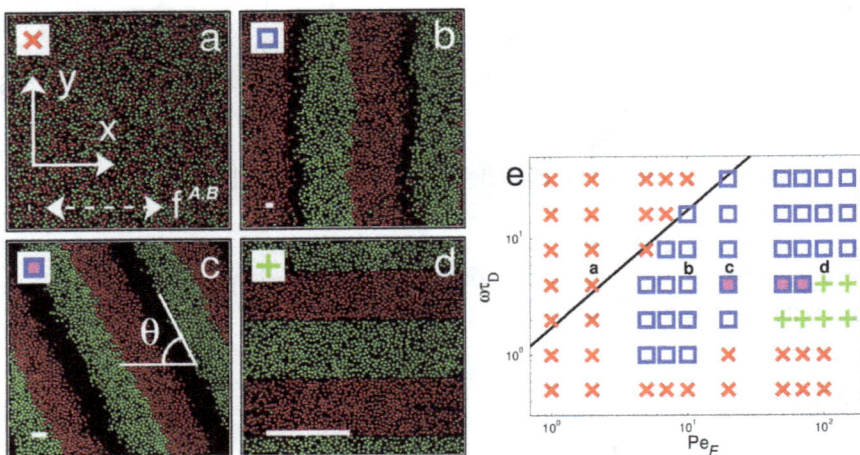

Fig. 10.20   Brownian dynamics simulations of banding. Simulations were performed at fixed driving frequency $\omega\tau_D = 4$ (normalized by the diffusive time τ_D) for different values of the Peclet number Pe_F. The snapshots (a-d) were obtained after 10^4 oscillation periods starting from a fully mixed state. Green and red colors represent oppositely charges species, symbols in the left upper corner identify the states shown in (e). The coordinate frame and the direction of the driving field (indicated by the broken arrow) are shown in (a), the length of the solid bar in the bottom left corner of (a-d) indicates the amplitude of free particle oscillations. For a weak driving force ($Pe_F = 2$, a), a disordered state is observed, for intermediate forces colloids segregate into stripes, perpendicular ($Pe_F = 10$, b) or tilted ($Pe_F = 20$, c) with respect to the field. For a strong driving force ($Pe_F = 110$, d), lanes are formed parallel to the field. (e) Non-equilibrium steady-state phase diagram plotted for the area packing fraction $\phi = 0.4$. The solid line describes a simple theoretical estimate of the disordered-to-segregated phase boundary. From Wysocki and Löwen (2009).

particles arrange themselves into separate bands then there are only two "big collision" events (per band, during one oscillation cycle) involving two oppositely driven bands. By comparing this with the case of oppositely driven lanes, which is accompanied by a continuous friction at the interface between them (not to mention the mixed state, characterized by the mutual volume friction), it is natural to conclude that the band configuration is preferable for minimizing the average dissipation. The formation of bands needs a relatively long time (i.e., many oscillation cycles) when started from a completely mixed configuration, since it is a highly collective process. Finally, although the tilted bands seen in simulations might be an artifacts due to the periodic boundary conditions, similar tilted patterns have been also observed in real-space experiments with oppositely

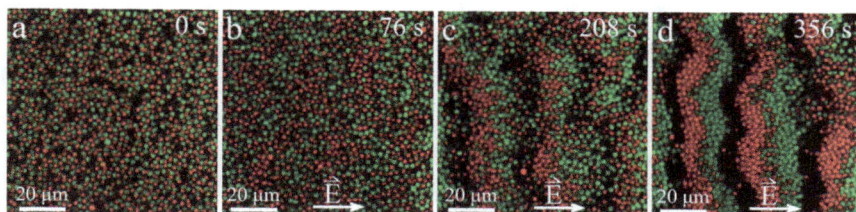

Fig. 10.21   Band formation in experiments with oppositely charged colloids. The results are for colloids of diameter 2.5 μm in an ac electric field (of the amplitude $E = 17.5$ V/mm and frequency 0.02 Hz, switched on at $t = 0$). (a) No field is applied and the mixture is isotropic, (b) local bands start forming, (c) bands become more pronounced, and (d) species are completely separated into distinct bands. From Vissers *et al.* (2011a).

charged suspensions (Vissers *et al.*, 2011a), as illustrated in Fig. 10.21. Hydrodynamic interactions (relevant in the case of gravity-like driving force) may drastically destroy banding, possibly leading to intermittent behavior (Wysocki and Löwen, 2011).

Constructing a microscopic theory of banding is an outstanding unresolved problem. Since banding appears as a secondary instability on top of laning, the theoretical description (even in the mean-field framework) is not easy and probably requires terms beyond a linear stability analysis (Schmittmann and Zia, 1998). It would be a fascinating challenge to realize the oscillatory drive in binary complex plasmas, in order to explore the relevance of band formation for undamped particle dynamics.

10.6   Motion of Particles in Channels

As fluid systems are engineered to smaller and smaller scales, down to the atomic size, the special effects associated with the confinement become increasingly important. The behavior of such systems is a fundamental problem in technology (e.g., lubrication, adhesion, nanofluidics, microchannel spectrometry, surface functionalization, etc.). Obviously, as the system size is reduced, confinement effects become increasingly important. It is inevitable that there will be new physics associated with finite-size effects, due to surface interactions and reduced dimensionality.

There have now been many studies of confined flow systems, e.g., nanoporous materials (ordered or disordered), thin fluid films, microchannels, etc. Amongst the areas of interest are topics such as demixing (segregation) of biological fluid components, flows in nanocapillaries, the effects

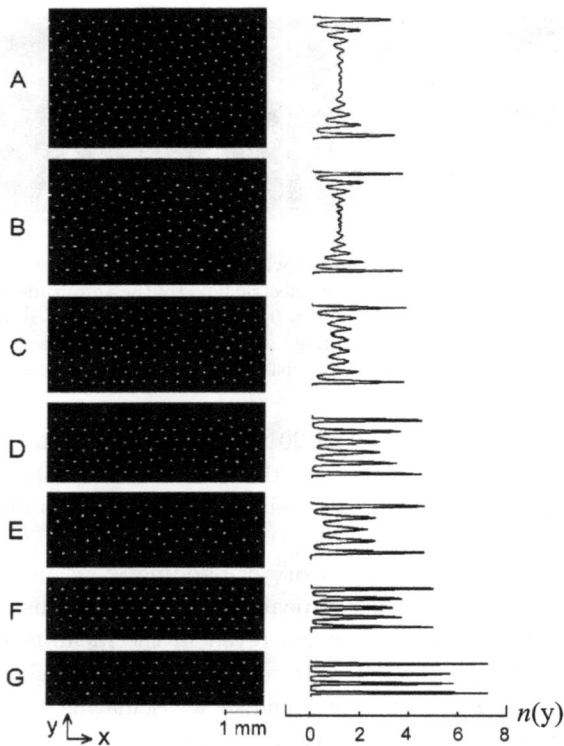

Fig. 10.22   Liquid complex plasmas in narrow channels. The typical snapshots of the 7 μm particle configurations and the corresponding transverse density distributions, $n(y)$, for different experiments with decreasing "number of layers" (measured in units of the interparticle spacing Δ), from 11 to 3. From Teng *et al.* (2003).

of confinement on the fluid structure and on freezing and melting [for a recent review, see Whitby and Quirke (2007) and Alba-Simionesco *et al.* (2006) and references therein]. The optimal way to study the basic (generic) physics is to employ a system where kinetic measurements are possible at all relevant length and time scales. Currently, the only systems capable of satisfying all these requirements are complex plasmas and colloidal dispersions. In Sec. 11.1 we will show that by using external fields one can even "design" the binary interaction potential in these systems. This will provide great opportunities for future basic and applied research, especially for nanofluidics.

In this section we concentrate specifically on the first studies involving driven complex plasmas in narrow channels, where their "kinetic structure" was investigated. All confined flow experiments with complex plasmas have so far been conducted on the ground, i.e., the microparticles are suspended against gravity in the sheath region above a horizontal electrode. The horizontal confinement is either affected by non-conducting glass walls (which then attain floating potential), by conducting segmented electrodes (that can be actively powered and can be used to transport the particles), or by conducting metal channels placed on the horizontal electrode.

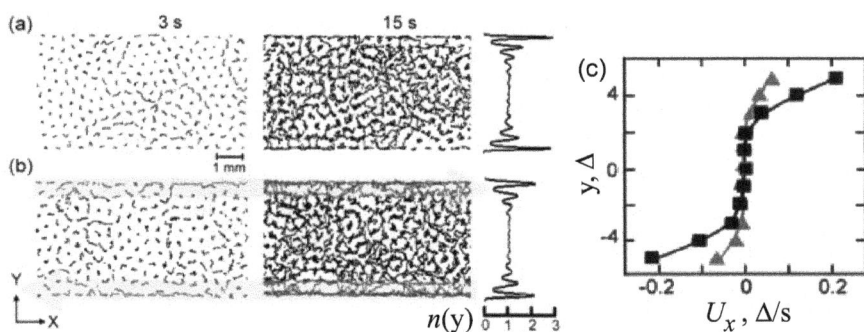

Fig. 10.23   Shear banding in a liquid complex plasma inside a channel. (a) Typical particle trajectories with 3 and 15 s exposure, and the transverse density distribution, $n(y)$, for the shear-free case. (b) The same for the experiment at 100 mW laser power. The gray arrows indicate the positions and directions of two counter-propagating laser beams. (c) The averaged profiles of longitudinal velocity $U_x(y)$ (in units of Δ) for experiments at 50 mW (triangles) and 100 mW (squares). From Chan *et al.* (2004).

Figure 10.22 illustrates the structure of quiescent (quasi-2D) complex plasmas inside narrow channels (Teng *et al.*, 2003). The lateral confinement in this case is produced by two parallel vertical plates which are put on a horizontal rf electrode surface. The transverse density profile $n(y)$ exhibits decaying oscillations from both boundaries, which is a clear manifestation of the confinement-induced freezing. When the "number of layers" exceeds a certain threshold (about 7), the frozen layers near the boundaries sandwich the more disordered isotropic liquid with a flat density profile in the center region. Figure 10.23 demonstrates the individual particle dynamics when the shear flow is induced inside such channels (Chan *et al.*, 2004). Due to the formation of the layered structure, the persistent drive from the external stress along each boundary enhances cage-escape structural rearrangements which cascade into the liquid through many-body interaction. The resulting flow profile consists of two outer shear bands (about three interparticle distances in width) adjacent to the boundaries and a central small-shear zone. The bands have higher level of both longitudinal and transverse velocity fluctuations. This appears very similar to the shear banding observed in glassy materials (such as foams, micelles, dense colloids, and dense granular systems) inside narrow channels (Wearie and Hutzler, 2000).

In a different experiment, converging and diverging ("nano") flows were investigated. One of the interests here was the determination of possible "selection rules" for the flow – e.g., how in detail the system evolves kinetically from N flow lines to $N-1$ flow lines when N becomes small. A second interest was to find out if there was a preferred instantaneous "structure" of the fluid particles during the flow line transitions. An example is shown in Fig. 10.24a. The flow converges by one interparticle spacing Δ over a distance of typically six Δ (i.e., reduction of one flow line), so that the convergence angle is about 10 degrees. The figure shows the following features: (i) The typical structure of the fluid is hexagonal – i.e., the same as the 2D crystalline ground state. (ii) The transition from 4 to 3 flow lines goes via a localized 5/7 dislocation. (iii) The transition from 3 to 2 flow lines goes via alternating jumps ("zipping") of particles from the "central" flow line (which disappears) to the two outer ones. The characteristic "structures" observed are shown schematically in Fig. 10.24b.

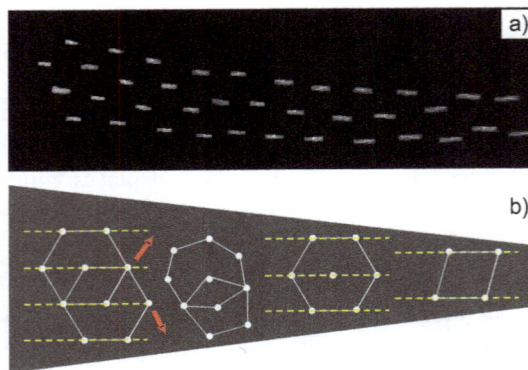

Fig. 10.24   Converging 2D complex plasma flow in the limit of a few flow lines. (a) The convergence of particles of 3.4 μm diameter goes from 4 to 2 lines. The experiment was designed to investigate fluid structure and dynamic selection rules during the (discrete) flow line reduction. (b) Characteristic "fluid structures" observed in different regimes of the converging flow. Courtesy of M. Fink.

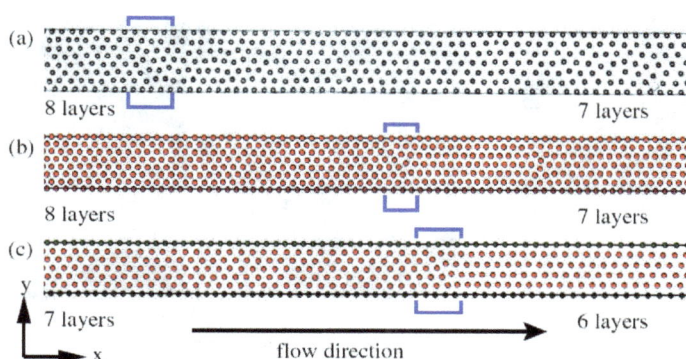

Fig. 10.25 (a) Video microscopy snapshot of colloidal particles moving along a 2D channel (60 μm width, coupling parameter $\Gamma = 72$). (b) Simulation snapshots for a channel with ideal hard walls (45 μm, $\Gamma = 640.5$); (c) the same as in (b) with the particles at the walls kept fixed ($\Gamma = 5026$). The rectangles mark the region of the layer reduction. From Köppl *et al.* (2007).

The results presented in Fig. 10.24 confirm the observations shown in Fig. 10.22 for a plane non-converging channel. As the system becomes smaller (in terms of flow lines), instantaneous snapshots resemble a solid. This is, of course, a consequence of the channel surface which in these experiments is a "slip surface" (i.e., the complex plasma does not have any "wetting" properties). Experiments with rough surfaces (on the scale of the interparticle separation) have not been performed yet. In such a case we would expect surface friction to play a role with associated modification of the flow structure and dynamics.

The flow of confined particles has likewise been studied in colloidal suspensions by real-space methods. Interestingly, experiments carried out with a 2D colloidal suspension flowing through a linear microchannel reveal striking similarity with complex plasma experiments. Figure 10.25 illustrates the experiment performed with super-paramagnetic particles (Köppl *et al.*, 2007) where the flow was generated by gravity, and also shows results of complementary computer simulations. In the experiments, the channel boundaries were smooth on the scale of the neighboring distance between colloidal particles. How the flow depends on the roughness and microstructure of the channel walls certainly needs more exploration.

In three dimensions, the standard setup is a slit geometry of two parallel plates which can move relative to each other (in the longitudinal direction) (Besseling *et al.*, 2008). When the plates do not move we have a quiescent

Fig. 10.26   Structure of a sheared colloidal suspension in a narrow channel. The particle volume fraction is $\phi = 0.61$, one of the plates oscillates in the x-direction at a frequency of 30 Hz, the resulting strain amplitude is $\gamma = 0.38$. (a) A confocal micrograph of a sheared suspension forming hcp layers. (b,c) Micrographs of the suspension in the buckled state. The gap between plates ($\simeq 80~\mu$m) is set slightly below the height commensurate with confinement of four flat hcp layers. Shown is a xy-slice near the upper plate (b) and 1.3 μm below it (c). The y'-direction is aligned with one of the characteristic hcp lattice vectors and forms a 60° angle with the x-axis. From Cohen *et al.* (2004).

confined system discussed in Sec. 7.4. Sheared plates induce a solvent flow, which results in non-equilibrium steady-state configurations of particle. In particular, the effect of shear on layering and crystallization of confined colloids has been explored. For a system consisting of a few layers it was found that shear produces new buckled structures (Cohen *et al.*, 2004). Figure 10.26 shows real-space images of such system. Note that shear flow can align the crystalline layers (Messina and Löwen, 2006) and lead to unusual particle dynamics inside the layers (Das *et al.*, 2002; Shereda *et al.*, 2010), like a collective zigzag motion (Derks *et al.*, 2004, 2009; Wu *et al.*, 2009).

10.7   Summary

It is with driven systems that the complementarity of complex plasmas and colloidal dispersions becomes truly evident. The behavior (for identical direct interactions between particles) is equivalent in (quasi)equilibrium, but upon strong driving (when the relevant Peclet number is large) this equivalence breaks down. In this case, the effects induced by surrounding fluid – dilute gas in complex plasmas and liquid solvent in colloidal dispersions – result in fundamental changes in the dynamics.

Due to their weak damping complex plasmas provide a good analogue for the behavior of atomic and molecular systems, allowing extended access to transport properties. The effect of weak damping is particularly significant when studying shear flows – where the shear-induced heat can play the leading role in determining the rheological properties, and heat transport – the phenomenon which is just impossible to observe in overdamped systems. Conversely, for colloidal dispersions, hydrodynamic coupling and overdamping effects lead to new regimes, where much interest is driven by their importance for applications in flow and processing, with a particular emphasis on rheology.

The remarkable advantage of particle-resolved studies is that they make it possible to identify limits of applicability of continuous description. The examples shown for classical hydrodynamic instabilities demonstrate that the formal continuous approach can actually work down to the scales about the interparticle distance (e.g., like for shear instability observed in complex plasmas), which is similar to the observations on critical scaling discussed in Chap. 6. Furthermore, using particle-resolved studies one can follow a gradual transition from the conventional continuous picture to the "discrete" regime, with the prominent example being the crossover from the Rayleigh-Taylor-like instability to the laning instability.

In fact, the laning should be considered among the most interesting general results obtained on non-equilibrium transitions in driven systems. Unlike equilibrium phase transitions, here the *nature* of the transition depends on details of the particle dynamics. For instance, the laning transition for Brownian particles (in 3D systems) is of the first order, but long-ranged hydrodynamic interactions then make it of the second order. However, in the case of charged colloids (the only colloidal system where it has been realized so far) hydrodynamic screening due to counterion flow destroys the long-ranged nature of the hydrodynamic interactions, and then the transition is of the first order again. The question remains as to the nature of the laning transition observed in numerous experiments with complex plasmas, where the damping is weak – we expect this basic issue to be tackled in the near future.

Chapter 11

Anisotropic Interactions

Although the major qualitative properties of many generic processes have no critical dependence on the particular form of the interparticle interaction, a large number of phenomena only occur when the interaction becomes anisotropic. The most striking feature is the appearance of novel phases: From the liquid crystals introduced by the simplest symmetry breaking to exotic "anisotropic" crystal lattices induced with external fields, the richness in behavior is remarkable. Moreover, the effects of anisotropic interactions are not limited to solids – fluid phases also exhibit string formation, which would not be expected a priori.

To investigate these phenomena in the complete phase space $(\mathbf{r}, \mathbf{v}, t)$ is a very important fundamental step towards a better understanding of self-organization, universality, and scaling in physical systems. In this chapter we summarize different approaches to "tailor" the interparticle interaction in complex plasmas and colloidal dispersions.

11.1   Interactions Tunable by External Fields

To "design" particle systems with prescribed interactions is a topic of broad interest. This fundamental issue can be resolved by employing the so-called "electrorheological effect" which allows us to tune the interparticle interaction to practically any shape – in particular to make it attractive at large distances and change it from isotropic to anisotropic.

11.1.1   Uniaxial fields

In this section we focus on an interesting class of so-called "electrorheological" (ER) fluids which have acquired significant attention in the last decade. "Conventional" ER fluids consist of suspensions of microparticles in (usually) nonconducting fluids with a different dielectric constant (Chen *et al.*, 1992; Dassanayake *et al.*, 2000). The interparticle interaction, and hence the rheology of ER fluids, is determined by an external electric field, which polarizes microspheres and thus induces additional dipole-dipole coupling. The same principle is also employed to produce magnetorheological (MR) fluids, which contain paramagnetic particles polarized in an external magnetic field.

The pair interaction of particles in ER/MR colloids is a superposition of the core isotropic part (determined by the dominating interaction mechanism, see Sec. 3.2) and point-like induced dipole-dipole part $V_{\mathrm{dip}}(\mathbf{r})$ [Eq. (3.14)]. For ER colloids,[1] the dipolar part (in spherical coordinates) is (Hynninen and Dijkstra, 2005; Tao, 1993)

$$V_{\mathrm{dip}}(\mathbf{r}) = -\frac{\mathbf{d}^2}{\varepsilon_{\mathrm{s}} r^3}\left(3\cos^2\theta - 1\right), \tag{11.1}$$

where $\mathbf{d} = \frac{1}{8}\left(\frac{\varepsilon_{\mathrm{r}}-1}{\varepsilon_{\mathrm{r}}+2}\right)\varepsilon_{\mathrm{s}}\sigma^3\mathbf{E}$ is the induced dipole moment determined by the relative dielectric constant $\varepsilon_{\mathrm{r}} = \varepsilon_{\mathrm{p}}/\varepsilon_{\mathrm{s}}$. Typical ac fields used in experiments (Yethiraj and van Blaaderen, 2003; Wen *et al.*, 2008) have frequency $\lesssim$ 1 MHz (to avoid asymmetric effects) and amplitude $E \sim 10^3 - 10^4$ V/cm (ER colloids) and $B \sim 10 - 300$ G (MR colloids).

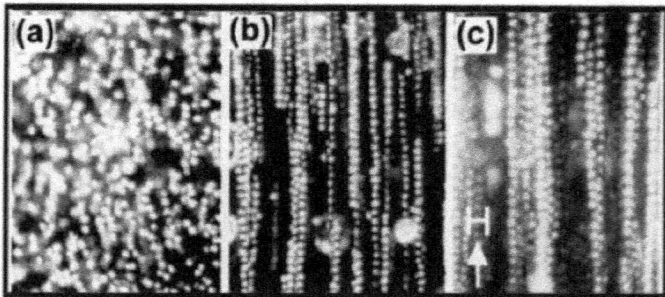

Fig. 11.1   The structural evolution of strings in MR colloids, from no field (a), through a moderate field ($B \simeq 30$ G, b), to a strong field ($B \simeq 200$ G, c). From Wen *et al.* (2008).

The term "electrorheological" (or "magnetorheological") is self-explanatory (Stangroom, 1983; Carlson *et al.*, 1990): At low fields microparticles may be fully disordered and then (provided their concentration is sufficiently low) ER/MR fluids may be just normal Newtonian fluids. At

---

[1]The dipolar part for MR colloids has completely analogous form.

larger fields, however, the situation can change dramatically – due to the increased dipole-dipole attraction particles arrange themselves into strongly coupled chains ("strings", or even "sheets") along the field, as illustrated in Fig. 11.1 (Wen *et al.*, 2008). This naturally changes the rheology – e.g., at low shear stresses ER/MR fluids can behave like elastic solids, whilst at stresses greater than a certain yield stress they are viscous liquids again. ER/MR fluids have a significant industrial application potential – they can be used in hydraulics, photonics, display production, etc. (Stangroom, 1983; Carlson *et al.*, 1990; Yethiraj *et al.*, 2004).

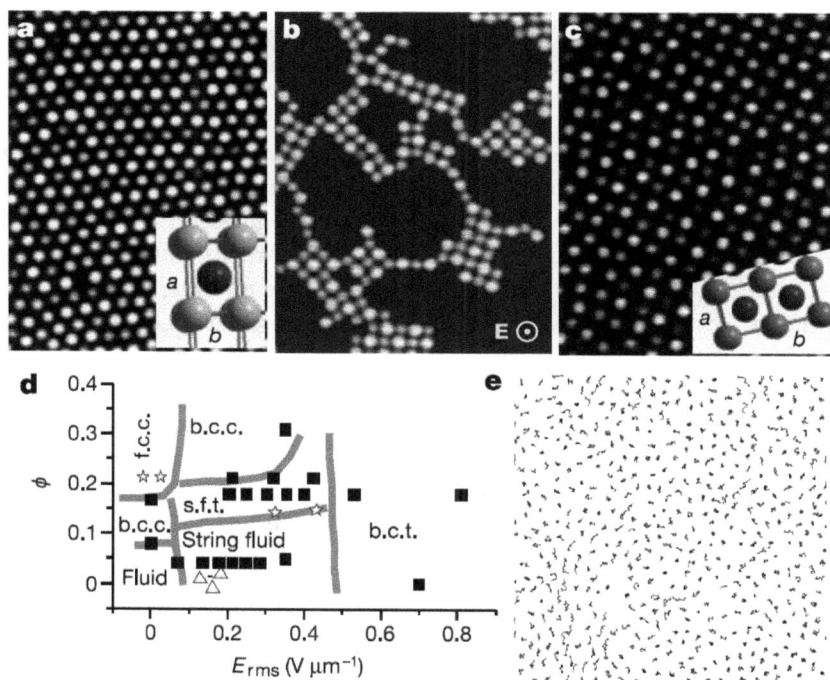

Fig. 11.2   Solid phases of ER colloids. Shown are confocal images (in the plane perpendicular to the field) of (a) body-centered orthorhombic (bco), (b) body-centered tetragonal (bct), and (c) space-filling tetragonal (sft) crystal. (d) The schematic phase diagram of ER colloids. Grey lines indicate approximate phase boundaries, open stars denote two-phase coexistence, and open triangles are in the string fluid. The solid squares indicate states measured in the experiment. (e) Trajectory of strings during 30 s (ordered along the field and diffusing in the transverse direction). From Yethiraj and van Blaaderen (2003).

The electric/magnetic field in ER/MR fluids plays the role of a new degree of freedom which allows us to "tune" the interaction between particles. This makes it possible to create a wealth of new "anisotropic" solid states and makes the phase diagram of such fluids remarkably diverse (Hynninen and Dijkstra, 2005; Yethiraj and van Blaaderen, 2003), as shown in Fig. 11.2.

In ER colloids, the induced dipolar interactions are primarily due to polarization of the microparticles themselves (see Sec. 3.5.4), while in complex plasmas this mechanism is practically negligible, since the typical particle size is much smaller than the interparticle distance. Therefore, the major contribution in complex plasmas is provided by clouds of compensating plasma charges (mostly, excess ions) surrounding negatively charged microparticles (see Sec. 2.2). Without an external field the cloud is spherical ("Debye sphere"), when a field is applied the cloud (which then acquires a fairly complicated shape and is called the "plasma wake") is shifted downstream from the particle, along the field-induced ion drift. In this case the pair interaction between charged microparticles is generally nonreciprocal (i.e., non-Hamiltonian, see Sec. 4.3). The non-reciprocity of the interaction could only be eliminated if the wake potential were an even function of coordinates, i.e., $\varphi(\mathbf{r}) = \varphi(-\mathbf{r})$. A simple "recipe" to create such a reciprocal wake potential is as follows (Ivlev *et al.*, 2008): One has to apply an ac field of a frequency that is (i) much lower than the inverse timescale of the ion response (ion plasma frequency, typically $\sim 10^7$ s^{-1}) and, at the same time, (ii) much higher than the inverse dust response time (dust plasma frequency, typically $\sim 10^2$ s^{-1} or less). Under these conditions, the ions react instantaneously to the field whereas the microparticles do not react at all. The effective interparticle interaction in this case is determined by the *time-averaged* wake potential. The resulting interaction is rigorously reciprocal (Hamiltonian), so that one can directly apply the formalisms of statistical physics to describe ER plasmas.

Quantitatively, the (field-induced) interparticle interaction in ER plasmas can be determined from the linearized dielectric response formalism (see Sec. 2.2.2). For subthermal ion drift the interaction potential is given by Eq. (2.7), which basically represents the far-field asymptotics for the potential expanded into a series over small ion Mach numbers M_T (with the angular dependence of the first three coefficients being proportional to that of the corresponding multipoles, i.e., charge, dipole, quadrupole). Furthermore, all "odd" terms ($\propto M_T^j$ with odd j) are proportional to linear combinations of the odd-order Legendre polynomials whereas "even" terms are combinations of the even-order polynomials. Thus, for an ac field $E(t)$ with $\langle E \rangle_t = 0$, all odd-order terms disappear in the time-averaged potential $\langle \varphi \rangle_t$, which becomes an even function of the coordinates. The effective pair interaction $Q\langle \varphi \rangle_t$ is then a sum of a spherically-symmetric core (represented, e.g., by the Debye-Hückel potential) and a field-induced contribution, with the leading term due to the quadrupole part of the wake. The latter is given by (Ivlev *et al.*, 2008)

$$V_{\text{field}}(\mathbf{r}) = -\left(2 - \frac{\pi}{2}\right) \frac{\langle M_T^2 \rangle_t Q^2 \lambda^2}{r^3} (3\cos^2\theta - 1). \qquad (11.2)$$

The charge-quadrupole interaction is identical to the interaction between two equal and parallel dipoles of magnitude $\simeq 0.65 M_T Q \lambda$ (here and below, the time-averaging brackets are omitted for brevity). This implies that for small M_T the interactions in ER plasmas are *equivalent* to dipolar interactions in conventional ER fluids.

One can compare field-induced interactions in colloids and plasmas in terms of the effective dipole-dipole coupling (Hynninen and Dijkstra, 2005; Tao, 1993). Since the magnitude of the induced dipole is proportional to the volume of the "polarizable sphere", the field necessary to achieve a given coupling in colloids is much larger than that in plasmas. Indeed, colloids of radius a acquire dipoles $\sim a^3 E_{\text{coll}}$ in the field E_{coll}, whereas in plasmas the effective dipoles are $\sim M_T Q \lambda$. The equivalent field for colloids is then readily obtained by comparing Eqs (11.1) and (11.2) (assuming the the mean interparticle distance $\sim a$ for colloids $\sim \lambda$ for dust), which yields $E_{\text{coll}} \sim M_T (a/\lambda)^{1/2} (Q/a^2)$. For typical experimental conditions, the electric field $E \sim 3$ V/cm in plasmas (which corresponds to $M_T \sim 1$) is equivalent to $E_{\text{coll}} \sim 3$ kV/cm in colloids.

ER plasmas were discovered in experiments by Ivlev *et al.* (2008). Particles remained in a disordered fluid state as long as the amplitude of the applied ac field was below a certain threshold. Increasing the field further triggered rearrangement of particles: The fluid became more and more ordered, until eventually well-defined particle strings were formed along the direction of the field. The transition between isotropic and string fluid states was reversible – decreasing the field brought the particles back into their initial isotropic state. The trend to form strings increased with particle size. The MD simulations performed with similar parameters showed remarkable agreement with the experiment.

Figure 11.3 illustrates experiments (Ivlev *et al.*, 2011) performed with complex plasmas in ac and dc electric fields (in otherwise the same conditions), revealing the striking structural difference. To characterize the

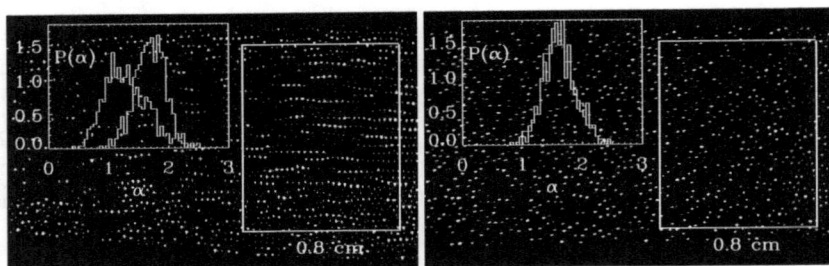

Fig. 11.3   Complex plasmas in external electric fields. The experiment is performed in the PK-4 chamber (see Fig. 5.9) under microgravity conditions. Particles of 6.86 μm diameter illuminated by a laser sheet (somewhat thinner than the interparticle distance, parallel to the applied field) are recorded from the side by a CCD camera. The figure (raw data, single video frames) illustrates the particle string formation observed in the ac mode (left panel) and a practically isotropic distribution in the dc mode (right panel). Both cases represent the same set of experimental parameters corresponding to $M_T \simeq 0.4$. The insets show the corresponding 2D distributions of the anisotropic scaling index α in the longitudinal and transverse directions (left and right curves, respectively). The regions used for the scaling index analysis are marked by rectangular boxes. From Ivlev *et al.* (2011).

particle ordering, an *anisotropic scaling index* was used (see also Sec. 10.5.1). Generally, the scaling index α can be considered as a measure of the local "structural dimensionality" in a particular direction: for an isotropic (uncorrelated) distribution of points on a plane the mean value of α is about two and it does not depend on the direction; when points form aligned chains it tends to unity in the direction of chains. Figure 11.3 shows that while the longitudinal (i.e., along the field, horizontal) and the transverse (vertical) distributions practically coincide in the dc case (inset in the right panel), in the ac case the shape of the longitudinal distribution changes drastically (left panel).

This shows that complex plasmas are more susceptible to ac fields than to dc fields – despite the fact that the attractive forces exerted by wakes in the former case are substantially weaker. We note that the formation of stable particle strings along the field can also be observed in dc fields (see Fig. 4.5b), but the underlying physics in this case is fundamentally different: The attractive interaction between particles in the "dc mode" is governed by nonreciprocal (dipolar and higher odd) parts of $\varphi(\mathbf{r})$, whereas in the "ac mode" this is due to reciprocal (quadrupole and higher even) parts. Therefore, while in the former case the strings are the result of self-organization occurring in a thermodynamically open system, in the latter case they represent the ground state of a particle ensemble that can be studied in the framework of equilibrium statistical physics (Ivlev *et al.*, 2011). The observed distinction is yet another manifestation of the fundamental difference between systems with Hamiltonian and non-Hamiltonian interactions (see discussion in Sec. 4.3).

So far, colloidal suspensions have been the major focus for ER studies, providing a wealth of information (Hynninen and Dijkstra, 2005; Yethiraj and van Blaaderen, 2003). The discovery that complex plasmas also have electrorheological properties adds a new dimension to such research – in terms of time/space scales and for studying new phenomena: A weakly damped system of microparticles in complex plasmas enables us to investigate previously inaccessible rapid elementary processes that govern the dynamic behavior of ER fluids – at the level of individual particles.

11.1.1.1 *Phase diagram of ER/MR systems*

The phase diagram of ER colloids reveals a remarkable variety of crystalline states (Chen *et al.*, 1992; Hynninen and Dijkstra, 2005; Yethiraj and van Blaaderen, 2003). As sketched in Fig. 11.2d, in addition to "isotropic" bcc and fcc lattices, unusual "anisotropic" crystalline states become possible, like body-centered orthorhombic (bco) and body-centered tetragonal (bct). Moreover, the hcp structure can become a ground state in a fairly broad range of phase variables (Hynninen and Dijkstra, 2005). On the other hand, relatively little research has been done on the fluid phase. In particular, the dynamics and details of the phase transition between "isotropic" and "string" fluids is still poorly explored.

So far, the most detailed investigation of the phase behavior of ER/MR systems was carried out by Brandt *et al.* (2009, 2010). In these studies, the generic representation of the interaction potential was used, $V(\mathbf{r}) = \epsilon[V_I(r) - \xi V_A(r)P_2(\theta)]$, where $P_2(\theta)$ is the second Legendre polynomial. The isotropic part $V_I(r)$ was assumed to have the hard-core Yukawa form, Eq. (3.5), while for the anisotropic part $V_A(r)$ different model forms were employed. The focus was put on the effect of the screening parameter characterizing the isotropic interaction, $\kappa\sigma$, and of the relative magnitude of the anisotropy ξ (which is proportional to the squared magnitude of the ac field).

To investigate the solid phases, a variational approach based on the Bogoliubov inequality was employed (Brandt *et al.*, 2009). By increasing $\kappa\sigma$, three distinct regimes of interactions called "soft", "medium", and "hard"

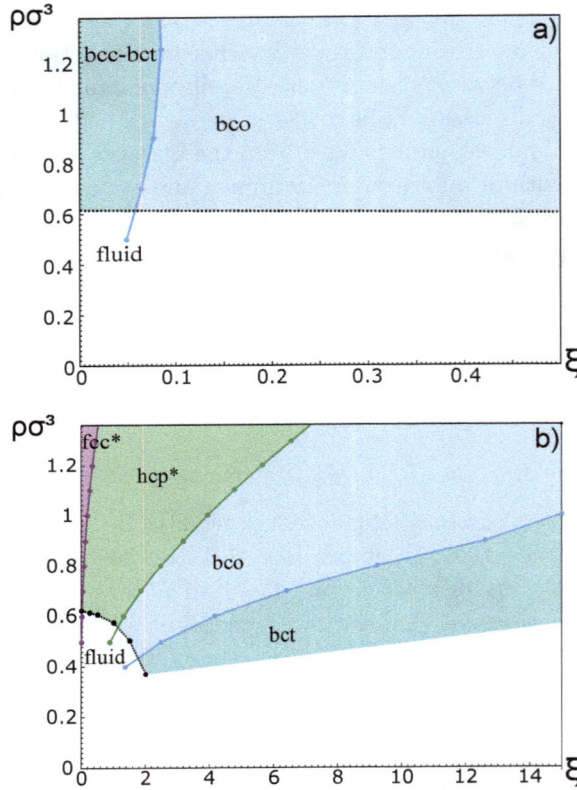

Fig. 11.4   Phase diagram of the ER/MR system with hard-core Yukawa interactions, shown in the parameter space of the normalized density $\rho\sigma^3$ and anisotropic magnitude ξ. The results are for the soft $[\kappa\sigma = 1, \text{a})]$ and medium $[\kappa\sigma = 7, \text{b})]$ interaction regimes, the interaction scale is $\epsilon = 25$ (normalized by $k_B T$). The solid lines indicate boundaries between different crystalline phases, dotted line represents the estimate for fluid-solid transition (calculated using the Ornstein-Zernike theory with the hypernetted-chain closure). From Brandt *et al.* (2009).

were identified. The medium interaction regime presented in Fig. 11.4b offers a rich variety of interesting effects, including a candidate for a lambda line (bco-bct transition). A sequence of solid phases, fcc→hcp→bco→bct, can be obtained for a given density by increasing ξ, and at least three triple points, adjoining fluid and two solid phases, can be found. For the hard interaction regime (not shown), a significant change in the phase balance is observed: The expanding hcp phase practically eliminates the fcc phase at smaller ξ and grows in the neighboring bco phase at larger ξ. When $\kappa\sigma$ increases up to sufficiently large values (corresponding to hard-sphere interactions) the bco phase nearly disappears. In the opposite, soft interaction regime illustrated in Fig. 11.4a, the bcc ground state for $\xi = 0$ turns into a bct phase at infinitesimally small ξ and deforms smoothly upon further increase of the dipolar strength. Due to the high sensitivity to ξ, bct and bco structures with precisely tuneable anisotropic characteristics can be obtained in this regime.

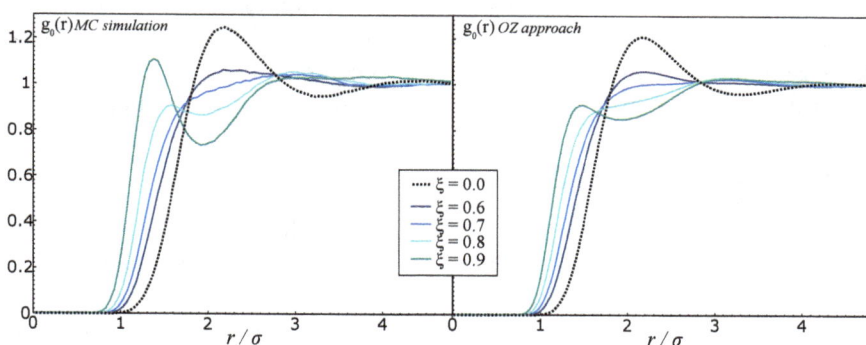

Fig. 11.5   Onset of the string-fluid transition. The figure illustrates the evolution of the zeroth-order (spherical) component of the pair correlation function, $g_0(r)$, in the vicinity of the transition. The left and right panels show the results of the Monte Carlo (MC) simulations and the Ornstein-Zernike (OZ) approach, respectively, performed for the screening parameter $\kappa\sigma = 1$ and normalized density $\rho = 0.1$. The bifurcation of the first peak of $g_0(r)$ into two distinct maxima, which is indicative of the string-fluid transition, is observed in the simulations between $0.825 < \xi < 0.85$, the OZ approach yields $\xi \simeq 0.84$. From Brandt *et al.* (2010).

In order to study fluid phases (Brandt *et al.*, 2010), the OZ approach was extended for ensembles with (aligned) anisotropic interactions (by decomposing anisotropic correlation functions in Legendre polynomials). Using this approach and combining it with the Monte Carlo simulations, two different fluid states representing the weakly anisotropic and string-fluid phases were identified. No threshold for the onset of anisotropy in fluids is found – an anisotropic structure is present at any finite ξ. Such anisotropy is, nevertheless, of a short range and therefore does not change the overall shape of the pair correlation function significantly. However, when ξ reaches a certain critical value the fluid anisotropy becomes significant. The particles line up like pearls on a thread, forming the string-fluid phase characterized by long-range order. This effect is evident in the zeroth-order (spherical) component of the pair correlation function, $g_0(\mathbf{r})$, illustrated in Fig. 11.5: One can see the bifurcation of the first peak, which results in the emergence of two new correlation lengths characterizing the longitudinal and transverse order in string fluids. The occurrence of this bifurcation in simulations practically coincides with the divergence of the structure factor $S(\mathbf{k})$, which manifests the phase transition in the framework of the OZ approach.

The string-fluid transition complements the phase diagram of the ER/MR systems at low densities (Yethiraj and van Blaaderen, 2003). Although such a transition has been observed in very different ER/MR systems ranging from colloidal suspensions to complex plasmas, the detailed experimental research of the string-fluid transition is still ongoing.

11.1.2   *Multiaxial fields*

We showed above that the electric field plays the role of a new degree of freedom in ER/MR colloids and plasmas. In fact, the principle of a uniaxial ac field used to create dipole-dipole interactions can be directly generalized to the method of "tuning" the interaction between particles: One can design new interaction classes – both anisotropic and isotropic – by applying external ac fields with bi- and triaxial polarization (Martin *et al.*, 2004; Kompaneets *et al.*, 2009).

Let us take ER plasmas as an example demonstrating the role of multiaxial fields. We consider the field polarization (and thus the resulting ion drift) of the following form:

$$\frac{\mathbf{u}(t)}{v_{T_i}} = (\mathbf{m}_x \cos\omega_1 t + \mathbf{m}_y \cos\omega_2 t + \mathbf{m}_z \cos\omega_3 t)M_T, \qquad (11.3)$$

where $\mathbf{m}_i$ are the components of a unit vector $\mathbf{m}$ (in a Cartesian coordinate system) which characterizes the anisotropy. To obtain the analytical expression for the field-induced part of the resulting interaction, we assume that the frequencies ω_i are all different and incommensurable, and use the same conditions as for Eq. (11.2), which yields (Kompaneets *et al.*, 2009):

$$V_{\text{field}}(\mathbf{r}) = -\left(2 - \frac{\pi}{2}\right)\frac{M_T^2 Q^2 \lambda^2}{r^3}(3m_x^2 n_x^2 + 3m_y^2 n_y^2 + 3m_z^2 n_z^2 - 1), \quad (11.4)$$

where n_i are the components of the unit vector $\mathbf{n} = \mathbf{r}/r$.

The field-induced part given by Eq. (11.4) dominates at large distances, and various combinations of m_i yield a rich variety of possible anisotropic interactions. One can divide them into (at least) three distinct classes:

(i) *positive* dipolar interactions, (ii) *negative* dipolar interactions, and (iii) *triaxially* anisotropic interactions [we adopted the terminology used for suspensions of magnetic particles (Martin *et al.*, 2004)]. Class (i) corresponds to uniaxial fields discussed in the previous section ($m_x = m_y = 0$ and $m_z = 1$) and the resulting interaction is described by Eq. (11.2). Class (ii) is obtained with biaxial fields of circular polarization ($m_x = m_y = \frac{1}{\sqrt{2}}$ and $m_z = 0$). Remarkably, such modulation induces a dipolar interaction of the same magnitude as in class (i), but reverses its sign! This results in the repulsion between particles along the z-axis and in the attraction in the xy-plane. For these two interaction classes, in dilute particle clouds one obtains the transition from isotropic to "string" (i) and "sheet" (ii) fluids, in sufficiently dense complex plasmas the formation of anisotropic lattices dominated by bco/bct (i) and hcp-like (ii) structures is anticipated (Brandt *et al.*, 2011). Finally, numerous triaxially anisotropic interactions are possible in the general case of $m_x \neq m_y \neq m_z$.

The special case of "isotropically distributed triaxial field with $m_x = m_y = m_z = \frac{1}{\sqrt{3}}$, which results in a spherically symmetric Lennard-Jones-like potential, is considered in Sec. 6.3.

We note that for dilute ER/MR colloids (when the packing fraction is sufficiently small) the field-induced interaction is determined by the pair potential of the form similar to Eq. (11.4). For dense suspensions the local field is strongly affected by the field of neighboring induced dipoles, as discussed in Sec. 3.5.4, resulting in many-body interaction depending on a particular configuration of colloids. This can lead to formation of very intricate anisotropic structures, e.g., "colloidal snowflakes" (Martin *et al.*, 2004).

11.2   Anisotropic Particles

There are various possibilities to prepare particles of anisotropic (i.e., non-spherical) shape [for review, see Glotzer and Solomon (2007)]. In equilibrium, an ensemble of anisotropic particles can exhibit novel liquid-crystalline phases with partial (or "intermediate") order as far as translational and rotational degrees of freedom are concerned. This leads to complex freezing and melting scenarios developing via mesophases.

In this section we summarize some of the recent developments in preparing and observing anisotropic particles. The particle shapes are classified in terms of symmetry and topology ("convexity"), as shown in Fig. 11.6.

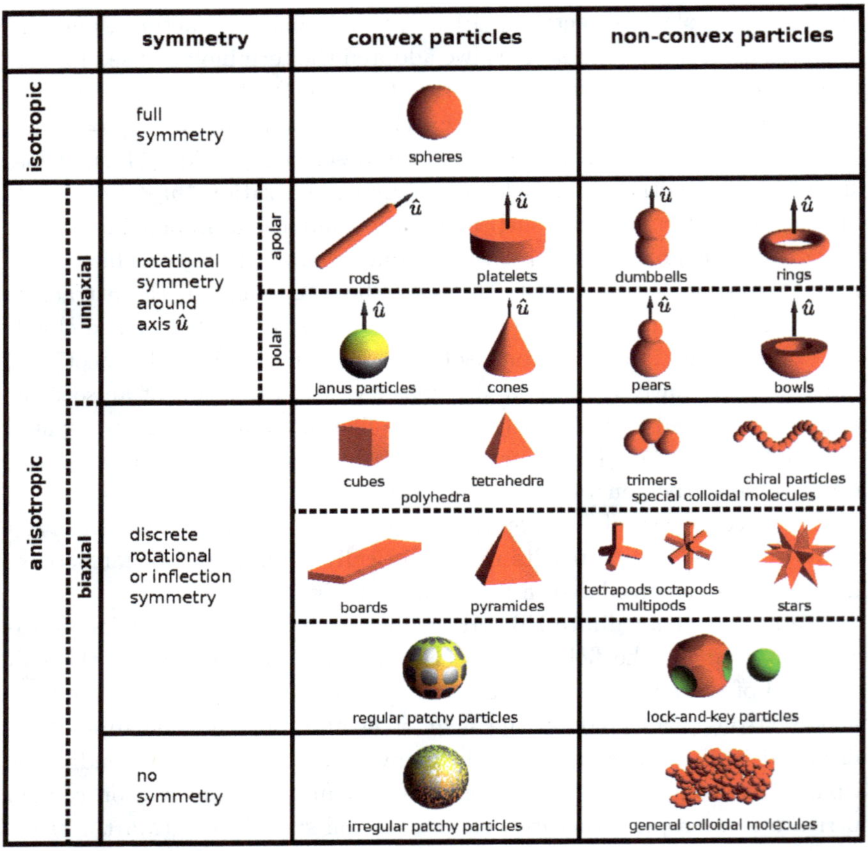

Complex Plasmas and Colloidal Dispersions

	symmetry	convex particles	non-convex particles
isotropic	full symmetry	spheres	
anisotropic / uniaxial / apolar	rotational symmetry around axis $\hat{u}$	rods, platelets	dumbbells, rings
polar		Janus particles, cones	pears, bowls
anisotropic / biaxial	discrete rotational or inflection symmetry	cubes, tetrahedra (polyhedra)	trimers, chiral particles (special colloidal molecules)
		boards, pyramids	tetrapods octapods multipods, stars
		regular patchy particles	lock-and-key particles
	no symmetry	irregular patchy particles	general colloidal molecules

Fig. 11.6   Classification of different particle shapes according to their symmetry and topology. From Wittkowski and Löwen (2012).

Although anisotropic particles can in principle be used both in complex plasmas and colloidal dispersions, the advances achieved so far in synthesizing and performing real-space studies with colloids are far more spectacular. Partially, this is related to the fact that interactions in colloids can be made very short-ranged, so that the actual particle shape becomes crucial in defining distinct thermodynamical configurations. In contrast, complex plasmas are usually characterized by long-range interactions, with the typical screening length being much larger than the relevant particle size, so that "fine details" of the particle shape are simply smeared out and play only a minor role in practice. Nevertheless, as we will show below, ensembles of highly asymmetric rod-like particles result in interesting effects also in complex plasmas.

Figure 11.6 summarizes a "zoo" of major anisotropic classes. The simplest anisotropic shape is an apolar convex *rod* which is rotationally symmetric around its own orientation. Rod-like particles have been added to plasmas (Annaratone *et al.*, 2001; Ivlev *et al.*, 2003), resulting in the formation of various liquid and crystalline structures. A typical example is shown in Fig. 11.7. In the realm of colloids, natural rod-like particles are tobacco-mosaic viruses (TMV) which cause a disease of the tobacco plants. In concentrated TMV suspensions, a variety of liquid and crystalline phases have been observed (Fraden *et al.*, 1989) and predicted (Graf and Löwen, 1999). Several phases observed with silica rod-like colloids are illustrated in Fig. 11.8. Recently, golden nanorods (Roorda *et al.*, 2004) or rods by fused spheres (Zahn *et al.*, 1994; Pantina and Furst, 2005) have been prepared, for which the aspect ratio can be tuned to an arbitrary value.

Fig. 11.7   A monolayer of cylindrical rods of 7.5 μm diameter and 300 μm length levitating in a plasma (top view). The dots are vertically oriented rods. From Annaratone *et al.* (2001).

For infinitely thin hard needles (i.e., for the infinite aspect ratio), the overlap volume of the neighboring rods tends to zero. For such systems, Onsager predicted a phase transition from a disordered (isotropic) to an orientationally ordered (nematic) state, which is driven purely by changes in translational and orientational entropy (Frenkel, 1991). This is a (weakly)

Complex Plasmas and Colloidal Dispersions

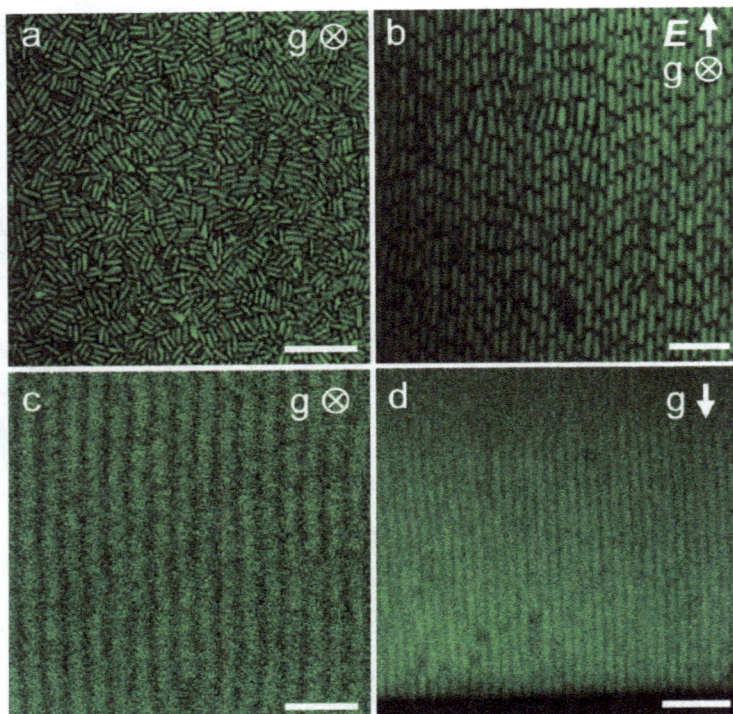

Fig. 11.8   Confocal microscopy images of rods of 640 nm diameter and 2.3 μm length in isotropic phase (a) and paranematic phase (b) induced by an electric field ($\simeq 0.2$ V/μm). For rods of 280 nm diameter and 1.4 μm length, a smectic phase was observed without applying an electric field. The smectic planes are visible in (c) and (d). Scale bars are 10 μm (a,d) and 5 μm (b,c). From Kujik *et al.* (2011).

first-order transition whose coexisting densities can be calculated analytically. For a finite aspect ratio, the phase diagram of hard spherocylinders has been studied in computer simulations, and a wealth of different liquid and crystalline phases was found to be stable in the plane spanned by the aspect ratio and density. These mesophases include isotropic, nematic, and smectic A phases, as well as plastic and differently stacked orientationally ordered crystals (Bolhuis and Frenkel, 1997; Graf *et al.*, 1997). Therefore, similar to hard-sphere systems, hard spherocylinders represent a simple anisotropic model system to study liquid-crystalline phase transitions. It can be realized for sterically-stabilized colloids and for highly screened charged suspensions. In this respect, colloidal rods are very well suited to model molecular liquid crystals (Dogic and Fraden, 2001). In fact, various colloidal rod suspensions have been used to study generic issues of the phase transformation kinetics (Dogic and Fraden, 2001; Huang *et al.*, 2009) and dynamics (van Bruggen *et al.*, 1998; Löwen, 1994a) in liquid crystals.

Platelets are another example of apolar rotationally-symmetric particles presented in Fig. 11.6. However, these particles do not tend to the exact Onsager limit when they become infinitely thin, since the excluded volume of two neighboring plates does not tend to zero. Colloidal particles of such an oblate shape can be prepared (Wierenga *et al.*, 1998). Platelets show complex liquid-crystalline phases similar to those observed with rods. On the other hand, they have much stronger tendency to form columnar stacks (Mourad *et al.*, 2010). In particular, they exhibit an isotropic-nematic transition (van der Beek *et al.*, 2008) with a very small interfacial free energy (van der Beek *et al.*, 2006). In the nematic phase, the orientational motion can be triggered by electric fields (Lapointe *et al.*, 2010). Note that platelet-like models are important to describe real clay suspensions (Dijkstra *et al.*, 1997).

As a next level of complexity, we consider non-convex rotationally-symmetric particle shapes. One of the simplest examples in this class shown in Fig. 11.6 are the *dumbbell* particles modeled as two overlapping hard spheres. This shape is the simplest colloidal "molecule" [analogous to the H_2 molecule (Johnson *et al.*, 2005; Demirörs *et al.*, 2010)]. A transmission electron microscopy image of dumbbell particles is presented in Fig. 11.9. For a small asphericity, hard dumbbells exhibit plastic crystalline phases which are orientationally disordered but translationally ordered (Vega *et al.*, 1992; Marechal and Dijkstra, 2008). There is even a transition between two plastic crystals with different structure (Marechal and Dijkstra, 2008). If the two spheres building the dumbbell are almost touching, an interesting "aperiodic" crystal becomes stable (Wojciechowski *et al.*, 1991; Marechal and Dijkstra, 2008), where the dumbbell centers are non-periodic but the centers of the individual spheres are ordered. These theoretically predicted transitions still need to be confirmed experimentally. However, given the fact that the particle dynamics in dumbbell crystals can be observed in experiments in greatest detail (Gerbode *et al.*, 2010), we believe that this verification is likely in the near future.

Yet another example of apolar non-convex shapes are colloidal *ring*-like particles (Helseth *et al.*, 2005; Li and Yellen, 2010). The principal difference with respect to the shapes discussed above is that ring-like particles can become entangled, which strongly affects dynamics and rheology.

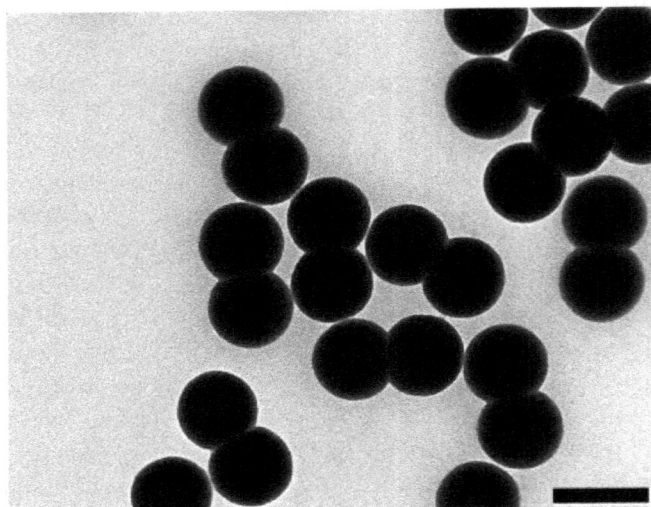

Fig. 11.9   TEM image of silica dumbbells which are formed from seed particles (grown in a microemulsion and aggregated by depletion effects). From Johnson *et al.* (2005).

Polar particles (also illustrated in Fig. 11.6) violate the up-down symmetry. They still have rotationally-symmetric shapes and can either be convex or non-convex. As a typical example, one should mention *Janus* particles which have two different sides. Various types of Janus particles has been prepared (Hong *et al.*, 2006; Ho *et al.*, 2008), some of them serve as self-propelled colloids in a reactive solvent (Erbe *et al.*, 2008), to model microswimmers.

Two examples of polar non-convex particles shown in Fig. 11.6 are *pear*-like particles, which can be synthesized by fusing two spheres of different diameters (Kegel *et al.*, 2006; Hosein *et al.*, 2009), and colloidal *bowls*. Clearly, hard bowls would prefer to stack at high densities, forming columnar (worm-like) aggregates. A wealth of novel liquid-crystalline phases for hard bowls were recently predicted by computer simulations (Marechal *et al.*, 2010). The columnar phase is stable over a broad density range, in agreement with recent real-space experiments (Marechal *et al.*, 2010).

Next, we briefly discuss particle shapes which have only a discrete rotational symmetry. In three dimensions, the simplest convex shape in this respect is a *cube*. Oriented hard cubes (with fixed orientation) show quite unusual second-order (continuous) freezing transition from the disordered

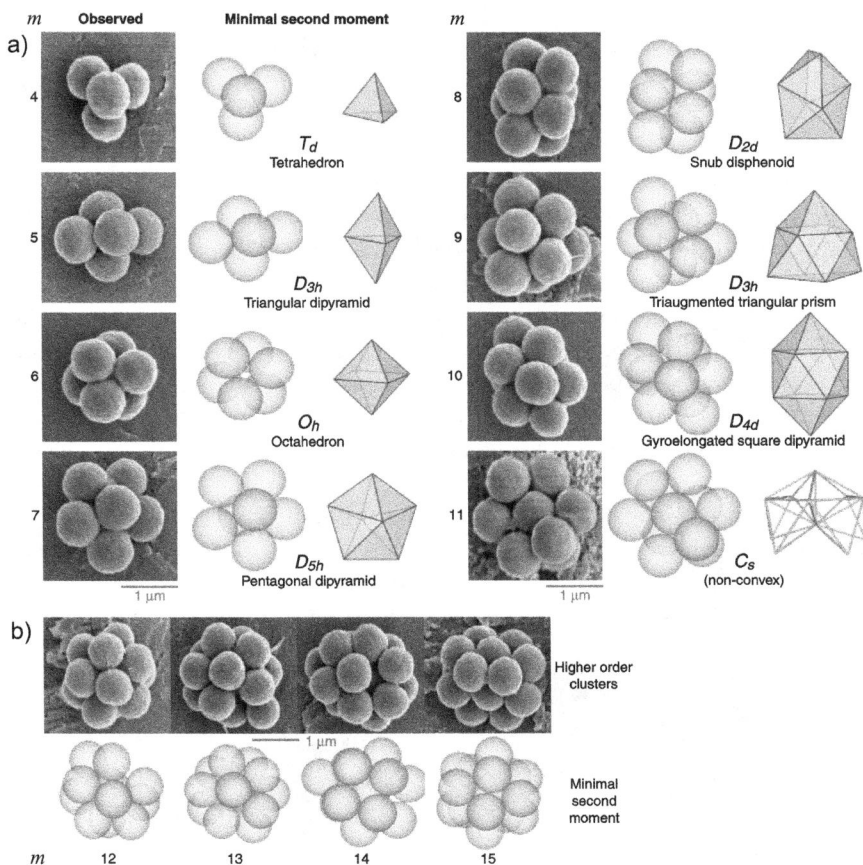

Fig. 11.10   Colloidal cluster configurations consisting of m spherical particles, as determined from the optical and electron microscopy. (a) Left columns show electron micrographs of the clusters; middle columns demonstrate computer renderings of sphere configurations which minimize the second moment of the mass distribution (with the same packing); right columns illustrate the polyhedra formed by drawing lines from the center of each particle to its neighbors, the names of the polyhedra and Schönflies point groups are indicated below. (b) Clusters with $m > 11$ differ increasingly from the minimal second-moment structures, one can observe small variations in the packing from cluster to cluster. From Manoharan *et al.* (2003).

fluid phase (Cuesta and Martínez-Ratón, 1997), while cubes with free orientation exhibit the traditional first-order freezing. Since recently, it is possible to prepare colloidal cubes with "almost" sharp edges (Sevonkaev *et al.*, 2008; Rossi *et al.*, 2011), so that real-space studies now could provide deeper insights into complex freezing scenarios. In particular, it would be interesting to consider cubes with "controllably-rounded" edges – "superballs" – in order to interpolate towards the spherical limit where the freezing transition is known.

More exotic convex shapes are *tetrahedra*, where even the "simplest" close-packed structure was explored only recently (Chen *et al.*, 2010a; Haji-Akbari *et al.*, 2009), or colloidal *boards* characterized by low internal symmetry (van den Pol *et al.*, 2009). *Patchy colloids* with attractive or reactive centers on their surface can be prepared, in particular when the distribution of these patches has some internal (e.g., tetrahedral) symmetry (Cho *et al.*, 2007). To date, most systematic studies have used computer simulations (Zhang and Glotzer, 2004; Sciortino, 2008; Bianchi *et al.*, 2006). In experiments, particles with DNA functionalization can exhibit fascinating self-assembly and produce "colloidal flying carpets" (Geerts and Eiser, 2009).

Non-convex shapes with different internal symmetries lead to a variety of complex particles, such as colloidal *trimers* (Liddell and Summers, 2003), *chiral* particles (Zerrouki *et al.*, 2008; Wensink and Jackson, 2009), and functionalized *lock-and-key* particles (Sacanna *et al.*, 2010).

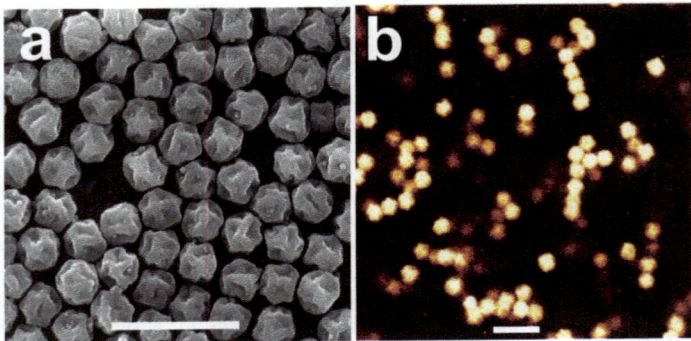

Fig. 11.11   (a) SEM image of polyhedral colloidal "rocks". (b) Confocal image of these rock-shaped particles bond together to form networks which have very low dimension (scale bars are 10 μm). From Rice *et al.* (2011).

And finally, there is an uncountable set of even more complex colloidal particles (Manoharan *et al.*, 2003; Solomon *et al.*, 2010; Quilliet *et al.*, 2008; Kraft *et al.*, 2008; Klix *et al.*, 2009). The general idea is to construct *colloidal molecules* out of monomers: Several experimental techniques have been developed for that, based on emulsification of spherical droplets (Manoharan *et al.*, 2003) and their buckling instabilities (Quilliet *et al.*, 2008), and on self-assembly of colloids in liquid protrusions (Kraft *et al.*, 2008). As an example, Fig. 11.10 shows various colloidal clusters prepared by using the emulsification technique, with quite different internal symmetries. Dispersions of these clusters can then self-assemble into complicated liquid-crystalline phases. Finally, particles having irregular rock-like shape, as illustrated in Fig. 11.11, can assemble themselves into curious linear structures rather than clumps (Rice *et al.*, 2011).

11.3   Summary

We have seen that anisotropic interactions fall mainly into two classes: those which are induced by an external field and those which result from the particle shape. While the latter are predominantly a feature of colloidal dispersions (due to the short-range nature of the interactions), the former illustrate the essential equivalence of colloids and plasmas.

The addition of external fields is perhaps the optimal way to tune interactions in-situ, and thus to arrive at a controllable effective temperature. This has been shown to work both in 2D and 3D electrorheological fluids. The 3D results of particle-resolved studies were among the first to deliver exotic crystal structures which are the most notable – and potentially useful – consequences of anisotropic interactions. Remarkably, although the physical mechanisms governing field-induced interactions in complex plasmas and colloidal dispersions are quite different, the resulting generic form of the interaction potential turns out to be the same.

We have scarcely touched on the explosion in anisotropic colloids, which is only now beginning to feed into systems suitable for particle-resolved studies. Among the most fascinating questions is whether all the richness of crystal structures of the chemical elements – with all the quantum mechanical complexities of their interactions – can be reproduced by hard particles whose phase behavior is governed by entropy alone. Amazingly, the answer appears to be yes – along with more exotic structures not formed by the elements! There seems little doubt that going forward, as new synthesis techniques are developed, we can expect the experimental realization of more and more exotic structures.

Chapter 12

Outlook

In this concluding chapter we present a brief summary of critical open issues and discuss some outstanding problems of interdisciplinary research, where particle-resolved studies of complex plasmas and colloidal dispersions can provide crucial new insights into the understanding of generic phenomena in condensed matter. The summary spans a variety of research topics, mainly in the realm of phase transitions as well as equilibrium and non-equilibrium dynamics. In no way does this list intend to be comprehensive and complete – rather it represents a selection of problems which we believe are important for future research.

Crystallization and melting

The kinetics of crystal nucleation – both homogeneous and heterogeneous – out of the supercooled melt needs better understanding. In particular, this concerns the situation where the thermodynamically stable solid has a structure which is vastly different from that of the liquid. The problem is even more challenging if an external nucleation seed is imposed into a supercooled liquid. The nucleation scenario is expected to be particularly complex if the seed has a structure which is incompatible with the stable crystal. In this case, one has to explore how the final transformation into the stable (bulk) phase evolves.

Furthermore, understanding the role of crystal defects during nucleation and growth is still in its infancy and needs precise real-space data and large-scale simulations. The same is true for the inverse problem of melting, which typically starts at system boundaries and surfaces. Generally, structural heterogeneity is important for melting, e.g., the effect of grain boundaries has received attention (see Chapter 7). However, it is still unclear whether and how the dynamics of crystal defects affects the melting scenario and kinetics.

Another important problem is crystal nucleation out of a solid phase which is not thermodynamically stable and has a structure incompatible with the ground-state crystal. In this case, one might expect nucleation to be suppressed, but this hypothesis needs to be carefully explored. In this context we note that the relation between glass formation and crystallization is only poorly understood at present. The question of how crystallization develops out of a glass is challenging indeed, since both non-equilibrium processes – glass formation and crystal nucleation – are coupled to one another and yet are very slow. Studies of the critical elementary mechanisms triggering one or the other process are clearly needed.

Measurements of homogenous nucleation rates remain a major problem – despite a decade of work (Auer and Frenkel, 2004). In the only system where quantitative data exist that enable a comparison between experiments and simulation – colloidal hard spheres – the nucleation rates differ by many orders of magnitude. It is surely desirable to elucidate the origins of the discrepancy in this most basic system. Particle-resolved studies enable the kind of detailed structural and dynamic information that can provide insights into the mechanisms of nucleation, and can hopefully resolve the conundrum whether the cause lies in experimental inaccuracies or in the assumptions inherent in the predictions.

More fundamentally, the difference between the mechanisms governing crystallization/melting in 2D and 3D systems and, hence, between scenarios of the transition has to be studied. Specifically, the crossover between 2D and 3D crystallization/melting in confined (quasi-2D) systems needs to be further explored (see Sec. 7.4.1). Particular effects to include are the impact of the interaction range on the appearance of an intermediate hexatic phase (which is the prominent feature of 2D melting in many cases, see Sec. 7.1.2) and the role of defects (which is expected to be different in 2D and 3D systems).

Finally, the complementarity of dynamical regimes in colloids and complex plasmas (see Sec. 4.2) should allow us to understand the role of damping in the crystallization process. In Sec. 7.2 we pointed out that the latent heat – which is naturally released in crystallizing complex plasmas and completely absent in colloids – can dramatically alter the process, e.g., change the crystallization speed, morphology, interface structure, etc. Exploring this effect systematically for different degrees of damping is another challenging problem for the future.

Liquid structure and slow dynamics

Although substantial knowledge about the structure of liquids has been accumulated, primarily from real-space studies of colloidal liquids, one still needs to resolve many fundamental issues – even for one-component systems.

First, the essential difference between the structure of coexisting liquid and crystalline phases has to be understood. The relevant questions involve the nature of the critical changes which occur in the (local) structure of solids that give them the ability to flow, and whether there are characteristic patterns in microscopic dynamics associated with the melting transition. These phenomena are expected to be strongly dependent on the interaction type (attractive/repulsive) and range. Both colloids and complex plasmas, where the interaction parameters can be tuned, are perfectly suited for such studies.

One long-standing problem in understanding supercooled liquids is how (if at all) the short-time dynamics influences the long-term evolution, e.g., whether short-time dynamical signatures (such as the Bose peaks) have an impact on the long-time behavior. A comparative study of this problem with colloids and complex plasmas is particularly important. Other major issues of slow dynamics include the role of particle polydispersity and its influence on glass formation, and (possibly general) conclusions about the difference between 2D and 3D glasses.

The fundamental problem of a self-consistent description of liquids certainly requires feedback from particle-resolved studies. The "straightforward" truncation of high-order correlations resulting in different (inconsistent) closures has well-known drawbacks. Careful studies of the liquid structure at the particle level might suggest certain generic form(s) of the local order (e.g., icosahedral clusters in supercooled liquids, see Sec. 9.5) which can be used as a "basis" to describe higher correlations. Most likely, if such generic structures do exist, they should also depend on the interaction type and range.

Liquid-vapor phase transition, binary mixtures

The universality of phenomena occurring near the critical point (be it for a liquid-vapor transition or phase separation in binary mixtures) is generally well understood by now. Nevertheless, even here there are certain issues to be studied more carefully: For example, how large is the vicinity where the corresponding critical scaling is preserved (e.g., how significant is the discreteness effect, see discussion in Sec. 6.3), or how the crossover to the mean-field behavior occurs for different classes of interaction (other than extensively studied inverse power-law potentials). Note that long-range interactions which can be realized in complex plasmas and colloidal dispersions provide a fascinating opportunity to perform particle-resolved studies of an experimental system which is truly mean-field. What is less clear here is whether similar criticality concepts are applicable to non-equilibrium situations. For example, the influence of shear fields and non-conservative interactions (see Sec. 4.3) on criticality needs better understanding both in terms of fundamental theory and experiments.

Far away from the critical point, the fundamental problem of a gradual crossover from the regime of nucleation to the spinodal decomposition requires careful in-depth investigation. Despite substantial efforts to resolve this issue it remains poorly understood, even the appropriate concept of how to treat the problem still needs to be developed (see discussion in Sec. 8.3.3). We believe that careful particle-resolved studies can help us to identify the leading microscopic processes which govern the crossover, and elucidate the role of the interaction range in this complex problem.

As regards the specific processes occurring in binary mixtures (sufficiently far away from the critical point), the full kinetics of fluid demixing – from the micro- to macroscale – is not completely understood. In particular, the influence of the interaction range(s) on different demixing regimes (see Sec. 8.3.4) should be further explored. The nonadditivity parameter governing the cross-interaction in binary mixtures is expected to strongly affect the structuring and the associated scaling behavior, and systematic particle-resolved studies would be helpful in identifying the general trends. And, of course, the whole spectrum of problems associated with crystallization and melting in binary mixtures is very broad (see discussion in Sec. 8.2). A complex scenario of nucleation, which may strongly depend on the detailed microstructure of the stable crystalline phase, is expected in this case.

Driven systems

As we pointed out in the introduction to Chap. 10, the "discreteness problem" is one of the central issues of fluid mechanics. This concerns the question as to what is the smallest scale at which the conventional hydrodynamic description breaks down. Of primary relevance here would be comprehensive kinetic investigations of the onset and nonlinear development of hydrodynamic instabilities. Individual particle observations can clarify numerous important aspects – e.g., whether (and under what conditions) the coarse-grained concept is still adequate on interparticle distance scales and what is the dynamics and structure of fluid interfaces in this case, are there any microscopic origins of instabilities (in particular, what are the corresponding "critical" trajectories of individual particles) and whether "discrete" instability patterns differ from the hydrodynamic ones, etc. In Sec. 10.4 we showed that while some of these questions have already been partially addressed for the RT-like instabilities (which were demonstrated to have close interconnections with laning, see Sec. 10.5), the respective studies of shear instabilities still have to be carried out.

Generally, non-equilibrium phase behavior of driven systems discussed in Sec. 10.5 needs to be properly classified. This task is expected to be fairly complicated, since the relevant parameter space in non-equilibrium systems is much larger than that in equilibrium. The characteristic examples showing a rich variety of possible steady-state phases in non-equilibrium systems are laning and banding phenomena (illustrated in Figs 10.16 and 10.20, respectively). We note that so far there have been no self-consistent (microscopic) theories proposed for their description – this can be considered as a truly outstanding theoretical challenge. Furthermore, banding is a dynamical phenomenon and therefore complementary studies with complex plasmas (where no experiments have been carried out so far) would allow us to explore the effect of damping. Since banding is apparently triggered as a result of a higher-order instability (on top of laning), many subtle processes can come into play in this case. For instance, it would be important to identify the role of short- and long-term relaxation processes and (possible) metastable states.

We stress that the non-equilibrium phase transitions tackled to date only represent the "tip of the iceberg" – steady state. As we move further away from equilibrium, we can expect far richer behavior. The tools to apply 3D time-resolved particle tracking have recently been developed for driven systems. The work done so far on sheared colloidal suspensions is clearly the start of studies of non-equilibrium phenomena with unprecedented detail.

Tunable interactions, anisotropic and active particles

We expect that the ability to tune the interparticle interactions will be broadly utilized in the future. In particular, this is related to anisotropic interactions and anisotropic particle shapes discussed in Chapter 11.

New interaction classes can be designed by applying external ac fields with bi- and triaxial polarization, leading to novel phase behavior. For instance, "negative dipole" interactions can be created (by using biaxial rotating fields), to favor sheet-like structures (instead of chain-like structures in uniaxial ac fields). Properties of such 2D fluids promise to be very interesting (e.g., as a convenient model system to study 2D phase transitions), but this needs to be explored systematically. Furthermore, while both electric and magnetic ac fields have been used to control and tune interactions between colloids, in complex plasmas only electric manipulation has been probed so far. The use of magnetic fields (or their combinations with ac electric fields) is expected to provide new possibilities in controlling the particle charge and the interaction anisotropy (in the regime of strongly magnetized electrons).

Scenarios of the equilibrium freezing in a system of anisotropic particles is another fascinating unresolved problem. For example, there are some indications that for hard tetrahedra a stable quasi-crystalline phase may intervene at intermediate densities (Haji-Akbari *et al.*, 2009) (while for other complex shapes the scenario might be different). The question of whether there are new fundamental principles governing freezing in such systems needs to be clarified and, hence, further particle-resolved studies in this direction are certainly necessary. Finally, non-convex particle shapes which are able to "swallow" smaller particles can serve as lock-and-key colloids or as living particle which "annihilate" and "create" other particles. Such "living" systems need a better understanding, both in terms of their single-particle and many-body properties.

Yet another "hot topic" is associated with the possibility to functionalize particles (currently, this topic is much more advanced for colloids than for complex plasmas, see Sec. 11.2). Much effort was recently extended on the study of self-propelled ("active") colloids which have their own intrinsic "motors". For example, a catalytic chemical reaction at the cap of the "Janus" particles can drive their motion. Unlike passive particles, this internal drive is non-Hamiltonian and therefore represents a genuine non-equilibrium situation. The ability to functionalize particles to almost "living" entities bears a large potential for future studies. Such self-propelled particles can serve as a model system for real microswimmers, like bacteria. While the basics of single microswimmers are well understood, the collective processes (like swarming and turbulence) remain unexplored at the individual-particle scale. Self-propelled particles can also be observed in experiments with complex plasmas. The shape of such particles is believed to be irregular, and the driving force is then exerted in the presence of plasma (ion) flow (Krasheninnikov, 2010). However, systematic studies of possible mechanism(s) driving self-propulsion of individual particles in complex plasmas still need to be carried out.

Appendix A

Symbols

Major Notations

a	particle radius
A_H	Hamacker constant
$B, \mathbf{B}$	magnetic field
c	concentration
$c(\mathbf{r})$	direct correlation function
$C_\mathrm{l,t}$	longitudinal (l), transverse (t) acoustic velocity
$d, \mathbf{d}$	dipole moment
D, D	diffusion constant, tensor
D_0	self-diffusion constant
e	elementary charge
$E, \mathbf{E}$	electric field
E_a	activation energy, energy barrier
E_c	defect core energy
$f^{(N)}(t, \mathbf{p}_1, \mathbf{r}_1, \ldots \mathbf{p}_N, \mathbf{r}_N)$	N-particle distribution function
$f(v)$	(isotropic) velocity distribution function
f_k	Debye-Waller factor
$F, \mathbf{F}$	force
F	Helmholtz free energy
$F(\mathbf{k}, t)$	intermediate scattering function (ISF)
g	gravitational acceleration
$g(r)$	pair correlation (radial distribution) function
$g_{G,6}(r)$ or $g_{G,6}(t)$	translational (G), bond-orientational (6) correlation functions
G	shear modulus
$G(\mathbf{r}, t)$	van Hove correlation function

271

$h(r) = g(r) - 1$	total correlation function
I	electron/ion current; radiation intensity
$k, \mathbf{k}$	wave number
k_B	Boltzmann constant
ℓ_{ij}	mean free path for species i and j
ℓ_g	gravitational length
L	characteristic length
$L(t)$	stochastic Langevin force
m	mass; cluster size; magnetic moment
M	laning order parameter
M	mobility tensor
n	volume number density
$n_{e,i,n,d}$	electron, ion, neutral, dust number density
n_{2D}	areal number density
$n_\mathbf{k}$	number density in Fourier space
n_r	relative refractive index
p	pressure
$p, \mathbf{p}$	momentum
Q	particle electric charge
Q^*	far-field charge
$Q_{SR,LR}$	short-range (SR), long-range (LR) charge for the double-Yukawa potential
R_0	Coulomb radius
R_g	polymer radius of gyration
S	entropy; "nematic" order parameter
$S(\mathbf{k}), S_\mathbf{k}$	static structure factor
T	kinetic temperature
T_{cr}	critical temperature
$T_{e,i,n,d}$	electron, ion, neutral, dust temperature
$u, \mathbf{u}$	velocity
$U, \mathbf{U}$	characteristic velocity
$v, \mathbf{v}$	velocity
$v_T = \sqrt{T/m}$	thermal velocity
V	volume
$V(r)$	pair potential/energy
$Z = Q/e$	charge number

α	characteristic exponent; parabolic confinement strength
β	thermal expansion
γ	shear strain; surface tension
$\dot{\gamma}$	shear rate
Υ	velocity gradient tensor
δr^2	mean squared displacement
$\Delta = n^{-1/3}$	mean interparticle distance (volume)
$\epsilon_{\rm Y}$	contact Yukawa potential
ε	dielectric constant
$\varepsilon(\omega, \mathbf{k})$	plasma permitivity
ζ	Stokes (viscous) damping rate
η	dynamic viscosity
Θ	characteristic temperature difference
$\kappa = \lambda^{-1}$	inverse screening length
λ	(effective) screening length
$\lambda_{\rm B}$	Bjerrum length
$\lambda_{\rm D}$	Debye screening length
$\lambda_{\rm SR,LR}$	short-range (SR), long-range (LR) screening length for the double-Yukawa potential
μ	mobility; chemical potential
$\nu \ (= \nu_{dn}, \zeta)$	relevant damping rate
ν_{dn}	Epstein (free molecular) damping rate
ξ	structural correlation length
$\rho = mn$	mass density
σ	hard-sphere diameter
σ_{ij}	collision cross section for species i and j
$\Sigma, \boldsymbol{\Sigma}$	shear stress, tensor
τ_i	characteristic relaxation time of process i
$\tau_D = \sigma^2/D_0$	diffusive timescale
$\varphi(r)$	electrostatic potential
$\Phi_{\mathbf{k}}(t) = F(\mathbf{k}, t)/S(\mathbf{k})$	density correlator
χ	thermal conductivity; effective magnetic susceptibility per particle
ψ	order parameter
$\psi_{G,6}$	translational (G), bond-orientational (6) order parameter

274   *Complex Plasmas and Colloidal Dispersions*

$\Psi(t, \mathbf{r}_1, \ldots \mathbf{r}_N)$ coordinate part of the N-particle distribution function

ω wave frequency

$\omega_\mathrm{p} = \sqrt{4\pi Q^2 n/m}$ plasma frequency

Ω Smoluchowski operator

Ω_conf confinement frequency

Ω_E Einstein frequency

Major Similarity (Dimensionless) Numbers

$h = L/\sigma - 1$ scaled slit width for hard spheres

$Kn = (\sigma_{nn} n_n a)^{-1}$ Knudsen number for neutral-neutral collisions

$\tilde{L} = \kappa L$ reduced slit width for Yukawa particles

$\tilde{m}$ magnetic moment ratio (asymmetry) for binary systems

$M_T = u_i/v_{T_i}$ ion "thermal" Mach number

$P_\mathrm{H} = |Z| n_d/n_e$ Havnes parameter

$Pe = \dot{\gamma}\sigma^2/D_0$ Peclet number (shear flow)

$Pe_F = F_0\sigma/2k_\mathrm{B}T$ Peclet number (driven flow)

$Pr = \eta/\rho\chi$ Prandtl number

$q = 2R_\mathrm{g}/\sigma$ polymer-colloid size ratio

$\tilde{Q}$ charge asymmetry for binary systems

$Ra = g\rho\beta\Theta L^3/\eta\chi$ Rayleigh number

$\tilde{S}$ size asymmetry for binary systems

$Wi = \dot{\gamma}\tau_{\alpha,\beta}$ Weissenberg number for α-, β-relaxation

x_i composition of species i

$z = |eQ|/k_\mathrm{B}T_e a$ dimensionless dust charge

$\tilde{\alpha} = \alpha/Q^2\kappa^3$ normalized parabolic confinement strength for Yukawa particles

$\alpha_\mathrm{fr} = \sqrt{2\nu/\chi}\, L$ friction number

$\beta_{ij} = |Q_i Q_j|/\lambda k_\mathrm{B}T$ scattering parameter for species i and j

γ_L modified Lindemann parameter for 2D systems

$\Gamma = Q^2/k_\mathrm{B}T\Delta$ electrostatic coupling parameter (strength)

$\Gamma^{(s)} = \Gamma e^{-\kappa_\mathrm{p}}$ screened coupling parameter

δ interaction nonadditivity parameter

$\epsilon_Q = Q^*_{\mathrm{LR}}/Q^*_{\mathrm{SR}}$	charge ratio for the double-Yukawa potential
$\eta_{\mathrm{hw}} = n_{2\mathrm{D}}L^2$	normalized density of Yukawa particles in hard-wall slit confinement
$\eta_{\mathrm{par}} = n_{2\mathrm{D}}/\kappa^2$	normalized density of Yukawa particles in parabolic confinement
$\kappa_{\mathrm{p}} = \Delta/\lambda$	plasma screening parameter
$\Lambda = \lambda_{\mathrm{LR}}/\lambda_{\mathrm{SR}}$	screening length ratio for the double-Yukawa potential
$\nu = Q^*/Q$	renormalizing charge ratio
ξ	relative magnitude of the field-induced interaction anisotropy
$\tilde{\rho} = n_{2\mathrm{D}}\sigma^3/L$	normalized density of hard spheres in hard-wall slit confinement
$\tau = T_e/T_i$	electron-to-ion temperature ratio
$\phi = \frac{\pi}{6}\sigma^3 n$	packing (volume) fraction
$\phi_{2\mathrm{D}} = \frac{\pi}{4}\sigma^2 n_{2\mathrm{D}}$	areal fraction

Bibliography

Aarts, D. G. A. L., Dullens, R. P. A. and Lekkerkerker, H. N. W. (2005). Interfacial dynamics in demixing systems with ultralow interfacial tension, *New J. Phys.* **7**, p. 40.

Abraham, F. F. and Gao, H. J. (2000). How fast can cracks propagate? *Phys. Rev. Lett.* **84**, pp. 3113–3116.

Adam, G. and Gibbs, J. H. (1965). On the temperature dependence of relaxation phenomena in glass-forming liquids, *J. Chem. Phys.* **43**, pp. 139–146.

Alba-Simionesco, C., Coasne, B., Dosseh, G., Dudziak, G., Gubbins, K. E., Radhakrishnan, R. and Sliwinska-Bartkowiak, M. (2006). Effects of confinement on freezing and melting, *J. Phys.: Condens. Matter* **18**, pp. R15–R68.

Alder, B. J. and Wainwright, T. E. (1957). Phase transition for a hard sphere system, *J. Chem. Phys.* **27**, pp. 1208–1209.

Aleksandrov, A. F., Bogdankevich, L. S. and Rukhadze, A. A. (1984). *Principles of plasma electrodynamics* (Springer, New York).

Alexander, S., Chaikin, P. M., Grant, P., Morales, G. J., Pincus, P. and Hone, D. (1984). Charge renormalisation, osmotic-pressure, and bulk modulus of colloidal crystals – theory, *J. Chem. Phys.* **80**, pp. 5776–5781.

Allen, J. E. (1992). Probe theory - the orbital motion approach, *Phys. Scripta* **45**, pp. 497–503.

Al'pert, Y. L., Gurevich, A. V. and Pitaevskii, L. P. (1965). *Space physics with artificial satellites* (Consultants Bureau, New York).

Alsayed, A. M., Islam, M. F., Zhang, J., Collings, P. J. and Yodh, A. G. (2005). Premelting at defects within bulk colloidal crystals, *Science* **309**, pp. 1207–1210.

Amini, M. and Laird, B. B. (2006). Kinetic coefficient for hard-sphere crystal growth from the melt, *Phys. Rev. Lett.* **97**, p. 216102.

Andelman, D. (1995). *Handbook of biological physics* (Elsevier).

Anderson, V. J. and Lekkerkerker, H. N. W. (2002). Insights into phase transition kinetics from colloid science, *Nature* **416**, pp. 811–815.

Angell, C. A. (1988). Structural instability and relaxation in liquid and glassy phases near the fragile liquid limit, *J. Non-Cryst. Solids* **102**, pp. 205–221.

Annaratone, B. M., Antonova, T., Goldbeck, D. D., Thomas, H. M. and Morfill, G. E. (2004). Complex-plasma manipulation by radiofrequency biasing, *Plasma. Phys. Controlled Fusion* **46**, pp. B495–B509.

Annaratone, B. M., Khrapak, A. G., Ivlev, A. V., Söllner, G., Bryant, P., Sütterlin, R. K., Konopka, U., Yoshino, K., Zuzic, M., Thomas, H. M. *et al.* (2001). Levitation of cylindrical particles in the sheath of an RF plasma, *Phys. Rev. E* **63**, p. 36406.

Antonova, T., Annaratone, B. M., Goldbeck, D. D., Yaroshenko, V. V., Thomas, H. M. and Morfill, G. E. (2006). Measurement of the interaction force among particles in three-dimensional plasma clusters, *Phys. Rev. Lett.* **96**, p. 115001.

Apolinario, S. W. S., Partoens, B. and Peeters, F. M. (2006). Inhomogeneous melting in anisotropically confined two-dimensional clusters, *Phys. Rev. E* **74**, p. 31107.

Aranson, I. S. and Tsimring, L. S. (2006). Patterns and collective behavior in granular media: theoretical concepts, *Rev. Mod. Phys.* **78**, pp. 641–692.

Archer, A. J. and Evans, R. (2004). Dynamical density functional theory and its application to spinodal decomposition, *J. Chem. Phys.* **121**, pp. 4246–4254.

Archer, A. J. and Wilding, N. B. (2007). Phase behavior of a fluid with competing attractive and repulsive interactions, *Phys. Rev. E* **76**, p. 031501.

Arp, O., Block, D., Klindworth, M. and Piel, A. (2005). Confinement of Coulomb balls, *Phys. Plasmas* **12**, p. 122102.

Arp, O., Block, D., Piel, A. and Melzer, A. (2004). Dust Coulomb balls: three-dimensional plasma crystals, *Phys. Rev. Lett.* **93**, p. 165004.

Asakura, S. and Oosawa, F. (1954). On interaction between 2 bodies immersed in a solution of macromolecules, *J. Chem. Phys.* **22**, pp. 1255–1256.

Asakura, S. and Oosawa, F. (1958). Interaction between particles suspended in solutions of macromolecules, *J. Poly. Sci.* **33**, pp. 183–192.

Ashkin, A. (2000). History of optical trapping and manipulation of small-neutral particles, atoms, and molecules, *IEEE J. Sel. Top. Quantum Electron.* **6**, pp. 841–856.

Ashkin, A. and Dziedzic, J. M. (1987). Optical trapping and manipulation of viruses and bacteria, *Science* **235**, pp. 1517–1520.

Asinovskii, E. I., Kirillin, A. V. and Markovets, V. V. (2006). Plasma coagulation of microparticles on cooling of glow discharge by liquid helium, *Phys. Lett. A* **350**, pp. 126–128.

Assoud, L., Ebert, F., Keim, P., Messina, R., Maret, G. and Löwen, H. (2009a). Crystal nuclei and structural correlations in two-dimensional colloidal mixtures: experiment versus simulation, *J. Phys.: Condens. Matter* **21**, p. 464114.

Assoud, L., Ebert, F., Keim, P., Messina, R., Maret, G. and Löwen, H. (2009b). Ultrafast quenching of binary colloidal suspensions in an external magnetic field, *Phys. Rev. Lett.* **102**, p. 238301.

Assoud, L., Messina, R. and Löwen, H. (2007). Stable crystalline lattices in two-dimensional binary mixtures of dipolar particles, *Europhys. Lett.* **80**, p. 48001.

Assoud, L., Messina, R. and Löwen, H. (2008). Binary crystals in two-dimensional two-component Yukawa mixtures, *J. Chem. Phys.* **129**, p. 164511.

Assoud, L., Messina, R. and Löwen, H. (2010). Ionic mixtures in two dimensions: From regular to empty crystals, *Europhys. Lett.* **89**, p. 36001.

Auer, S. and Frenkel, D. (2004). Quantitative prediction of crystal-nucleation rates for spherical colloids: a computational approach, *Ann. Rev. Phys. Chem.* **55**, pp. 333–361.

Barkan, A., D'Angelo, N. and Merlino, R. L. (1994). Charging of dust grains in a plasma, *Phys. Rev. Lett.* **73**, pp. 3093–3096.

Barkan, A., Merlino, R. L. and D'Angelo, N. (1995). Laboratory observation of the dust-acoustic wave mode, *Phys. Plasmas* **2**, pp. 3563–3565.

Barker, J. A. and Henderson, D. (1976). What is liquid - understanding states of matter, *Rev. Mod. Phys.* **48**, pp. 587–671.

Barnes, M. S., Keller, J. H., Forster, J. C., ONeill, J. A. and Coultas, D. K. (1992). Transport of dust particles in glow-discharge plasmas, *Phys. Rev. Lett.* **68**, pp. 313–316.

Baroody, E. M. (1962). Classical scattering by some important repulsive potentials, *Phys. Fluids* **5**, pp. 925–932.

Bartlett, P. and Campbell, A. I. (2005). Three-dimensional binary superlattices of oppositely charged colloids, *Phys. Rev. Lett.* **95**, p. 128302.

Bartlett, P., Ottewill, R. H. and Pusey, P. N. (1992). Superlattice formation in binary mixtures of hard-sphere colloids, *Phys. Rev. Lett.* **68**, pp. 3801–3804.

Bartlett, P. and Pusey, P. N. (1993). Freezing of binary mixtures of hard-sphere colloids, *Physica A* **194**, pp. 415–423.

Batchelor, G. K. (1977). Effect of Brownian-motion on bulk stress in a suspension of spherical-particles, *J. Fluid. Mech.* **83**, pp. 97–117.

Baumgartl, J., Arauz-Lara, J. L. and Bechinger, C. (2006). Like-charge attraction in confinement: myth or truth? *Soft Matter* **2**, pp. 631–635.

Baumgartl, J., Dietrich, J., Dobnikar, J., Bechinger, C. and von Grünberg, H. H. (2008). Phonon dispersion curves of two-dimensional colloidal crystals: the wavelength-dependence of friction, *Soft Matter* **4**, pp. 2199–2206.

Baumgartl, J., Zvyagolskaya, M. and Bechinger, C. (2007). Tailoring of phononic band structures in colloidal crystals, *Phys. Rev. Lett.* **99**, p. 205503.

Baumgartner, H., Block, D. and Bonitz, M. (2009). Structure and phase transitions of Yukawa balls, *Contrib. Plasma Phys.* **49**, pp. 281–302.

Bayer, M., Brader, J. M., Ebert, F., Fuchs, M., Lange, E., Maret, G., Schilling, R., Sperl, M. and Wittmer, J. P. (2007). Dynamic glass transition in two dimensions, *Phys. Rev. E* **76**, p. 011508.

Beatus, T., Tlusty, T. and Bar-Ziv, R. (2006). Phonons in a one-dimensional microfluidic crystal, *Nature Physics* **2**, pp. 743–748.

Bechinger, C., Brunner, M. and Leiderer, P. (2001). Phase behavior of two-dimensional colloidal systems in the presence of periodic light fields, *Phys. Rev. Lett.* **86**, pp. 930–933.

Bedanov, V. M., Gadiyak, G. V. and Lozovik, Y. E. (1985). On a modified Lindemann-like criterion for 2D melting, *Phys. Lett. A* **109**, pp. 289–291.

Bender, J. and Wagner, N. J. (1996). Reversible shear thickening in monodisperse and bidisperse colloidal dispersions, *J. Rheol.* **40**, pp. 899–916.

Benito, D. C., Carberry, D. M., Simpson, S. H., Gibson, G. M., Padgett, M. J., Rarity, J. G., Miles, M. J. and Hanna, S. (2008). Constructing 3D crystal templates for photonic band gap materials using holographic optical tweezers, *Optics Express* **16**, pp. 13005–13015.

Bernal, J. D. (1959). Geometrical approach to the structure of liquids, *Nature* **183**, pp. 141–147.

Berthier, L., Biroli, G., Bouchaud, J.-P., Cipelletti, L. and van Saarloos, W. (eds.) (2011). *Dynamical heterogeneities in glasses, colloids and granular materials* (Oxford University Press).

Besseling, R., Isa, L., Balletsa, P., Petekidis, G., Cates, M. E. and Poon, W. C. K. (2010). Shear banding and flow-concentration coupling in colloidal glasses, *Phys. Rev. Lett.* **105**, p. 268301.

Besseling, R., Isa, L., Weeks, E. R. and Poon, W. C. K. (2008). Quantitative imaging of colloidal flows, *Adv. Coll. Interf. Sci.* **146**, pp. 1–17.

Besseling, R., Weeks, E. R., Schofield, A. B. and Poon, W. C. K. (2007). Three-dimensional imaging of colloidal glasses under steady shear, *Phys. Rev. Lett.* **99**, p. 028301.

Bianchi, E., Largo, J., Tartaglia, P., Zaccarelli, E. and Sciortino, F. (2006). Phase diagram of patchy colloids: towards empty liquids, *Phys. Rev. Lett.* **97**, p. 168301.

Binder, K. (1977). Theory for dynamics of clusters. 2. Critical diffusion in binary-systems and kinetics of phase separation, *Phys. Rev. B* **15**, pp. 4425–4447.

Binder, K. (1987). Theory of 1st-order phase-transitions, *Rep. Prog. Phys.* **50**, pp. 783–859.

Bitzer, F., Palberg, T., Löwen, H., Simon, R. and Leiderer, P. (1994). Dynamical test of interaction potentials for colloidal suspensions, *Phys. Rev. E* **50**, pp. 2821–2826.

Bocquet, L., Trizac, E. and Aubouy, M. (2002). Effective charge saturation in colloidal suspensions, *J. Chem. Phys.* **117**, pp. 8138–8152.

Boda, D. and Henderson, D. (2008). The effects of deviations from Lorentz-Berthelot rules on the properties of a simple mixture, *Molecular Physics* **106**, pp. 2367–2370.

Bolhuis, P. and Frenkel, D. (1997). Tracing the phase boundaries of hard spherocylinders, *J. Chem. Phys.* **106**, pp. 666–687.

Bonitz, M., Block, D., Arp, O., Golubnychiy, V., Baumgartner, H., Ludwig, P., Piel, A. and Filinov, A. (2006). Structural properties of screened Coulomb balls, *Phys. Rev. Lett.* **96**, p. 75001.

Bonitz, M., Henning, C. and Block, D. (2010). Complex plasmas: a laboratory for strong correlations, *Rep. Prog. Phys.* **73**, p. 066501.

Bouchaud, J.-P. and Biroli, G. (2004). On the Adam-Gibbs-Kirkpatrick-Thirumalai-Wolynes scenario for the viscosity increase in glasses, *J. Chem. Phys.* **121**, pp. 7347–7354.

Bouchoule, A. (1999). *Dusty plasmas: physics, chemistry and technological impacts in plasma processing* (Wiley, Chichester).

Brader, J. M. (2010). Nonlinear rheology of colloidal dispersions, *J. Phys.: Condens. Matter* **22**, p. 363101.

Brader, J. M., Cates, M. E. and Fuchs, M. (2008). First-principles constitutive equation for suspension rheology, *Phys. Rev. Lett.* **101**, p. 138301.

Brader, J. M., Voigtmann, T., Cates, M. E. and Fuchs, M. (2007). Dense colloidal suspensions under time-dependent shear, *Phys. Rev. Lett.* **98**, p. 058301.

Brader, J. M., Voigtmann, T., Fuchs, M., Larson, R. G. and Cates, M. E. (2009). Glass rheology: from mode-coupling theory to a dynamical yield criterion, *Proc. Nat. Acad. Sci.* **106**, pp. 15186–15191.

Brambilla, G., El Masri, D., Pierno, M., Berthier, L., Cipelletti, L., Petekidis, G. and Schofield, A. B. (2009). Probing the equilibrium dynamics of colloidal hard spheres above the mode-coupling glass transition, *Phys. Rev. Lett.* **102**, p. 085703.

Brandt, P. C., Ivlev, A. V. and Morfill, G. E. (2009). Solid phases in electro- and magnetorheological systems, *J. Chem. Phys.* **130**, p. 204513.

Brandt, P. C., Ivlev, A. V. and Morfill, G. E. (2010). String-fluid transition in systems with aligned anisotropic interactions, *J. Chem. Phys.* **132**, p. 234709.

Brandt, P. C., Ivlev, A. V. and Morfill, G. E. (2011). Liquid-vapor transition in fluid monoloayers with tunable interactions, (to be submitted to Phys. Rev. Lett.).

Bray, A. J. (2002). Theory of phase-ordering kinetics, *Advances in Physics* **51**, pp. 481–587.

Brujic, J., Song, C., Wang, P., Briscoe, C., Marty, G. and Makse, H. A. (2007). Measuring the coordination number and entropy of a 3D jammed emulsion packing by confocal microscopy, *Phys. Rev. Lett.* **98**, p. 248001.

Brunner, M. and Bechinger, C. (2002). Phase behavior of colloidal molecular crystals on triangular light lattices, *Phys. Rev. Lett.* **88**, p. 248302.

Brunner, M., Bechinger, C., Strepp, W., Lobaskin, V. and von Grünberg, H. H. (2002). Density-dependent pair interactions in 2D, *Europhys. Lett.* **58**, pp. 926–965.

Bryant, G., Williams, S. R., Qian, L., Snook, I. K., Perez, E. and Pincet, F. (2002). How hard is a colloidal 'hard-sphere' interaction? *Phys. Rev. E* **66**, p. 060501.

Bubeck, R., Bechinger, C., Neser, S. and Leiderer, P. (1999). Melting and reentrant freezing of two-dimensional colloidal crystals in confined geometry, *Phys. Rev. Lett.* **82**, pp. 3364–3367.

Butler, S. and Harrowell, P. (1991). The origin of glassy dynamics in the 2D facilitated kinetic Ising model, *J. Chem. Phys.* **95**, pp. 4454–4464.

Cacciuto, A., Auer, S. and Frenkel, D. (2004). Onset of heterogeneous crystal nucleation in colloidal suspensions, *Nature* **428**, pp. 404–406.

Cacciuto, A. and Frenkel, D. (2005). Simulation of colloidal crystallization on finite structured templates, *Phys. Rev. E* **72**, p. 041604.

Cahn, J. W. and Hilliard, J. E. (1958). Free energy of a nonuniform system. 1. Interfacial free energy, *J. Chem. Phys.* **28**, pp. 258–267.

Cahn, J. W. and Hilliard, J. E. (1959). Free energy of a nonuniform system. 3. Nucleation in a two-component incompressible fluid, *J. Chem. Phys.* **31**, pp. 688–699.

Cairns, R. J. R., Ottewill, R. H., Osmond, D. W. J. and Wagstaff, I. (1976). Studies on the preparation and properites in lattices in nonpolar media, *J. Coll. Interf. Sci.* **54**, pp. 45–51.

Campbell, A. I., Anderson, V. J., van Duijneveldt, J. S. and Bartlett, P. (2005). Dynamical arrest in attractive colloids: the effect of long-range repulsion, *Phys. Rev. Lett.* **94**, p. 208301.

Carlson, J., Sprecher, A. and Conrad, H. (eds.) (1990). *Electrorheological fluids* (Technomic, Lancaster).

Castaldo, C., Ratynskaia, S. V., Pericoli, V., de Angelis, U., Rypdal, K., Pieroni, L., Giovannozzi, E., Maddaluno, G., Marmolino, C., Rufoloni, A., Tuccillo, A., Kretschmer, M. and Morfill, G. E. (2007). Diagnostics of fast dust particles in tokamak edge plasmas, *Nuclear Fusion* **47**, pp. L5–L9.

Cavagna, A. (2009). Supercooled liquids for pedestrians, *Phys. Rep.* **476**, pp. 51–124.

Chaikin, P. M. and Lubensky, T. C. (1986). *Principles of condensed matter physics* (Cambridge, Cambridge), 4nd ed.

Chakrabarti, J., Dzubiella, J. and Löwen, H. (2003). Dynamical instability in driven colloids, *Europhys. Lett.* **61**, pp. 415–421.

Chakrabarti, J., Dzubiella, J. and Löwen, H. (2004). Reentrance effect in the lane formation of driven colloids, *Phys. Rev. E* **70**, p. 012401.

Chakrabarti, J., Krishnamurthy, H. R., Sood, A. K. and Sengupta, S. (1995). Reentrant melting in laser field modulated colloidal suspensions, *Phys. Rev. Lett.* **75**, pp. 2232–2235.

Chan, C. L., Woon, W. Y. and I, L. (2004). Shear banding in mesoscopic dusty plasma liquids, *Phys. Rev. Lett.* **93**, p. 220602.

Chandler, D. and Garrahan, J.-P. (2010). Dynamics on the way to forming glass: bubbles in space-time, in S. Leone, P. Cremer, J. Groves, M. Johnson and G. Richmond (eds.), *Annual Review of Physical Chemistry* **61**, pp. 191–217.

Chandrasekhar, S. (1961). *Hydrodynamic and hydromagnetic stability* (Oxford University Press, Oxford, UK).

Chang, K. (2008). The nature of glass remains anything but clear, *New York Times* July 29 2008.

Charbonneau, P. and Reichman, D. (2005). Mode-coupling theory, *J. Stat. Mech: Theory and Experiment* **2005**, pp. 1–22.

Chaudhuri, M., Khrapak, S. A. and Morfill, G. E. (2008). Ion drag force on a small grain in highly collisional weakly anisotropic plasma: effect of plasma production and loss mechanisms, *Phys. Plasmas* **15**, p. 053703.

Chaudhuri, M., Kompaneets, R. and Morfill, G. E. (2010). On the possibility of collective attraction in complex plasmas, *Phys. Plasmas* **17**, p. 063705.

Chayes, J. T. and Chayes, L. (1984). On the validity of the inverse conjecture in classical density functional theory, *J. Stat. Phys.* **36**, pp. 471–488.

Chen, E. R., Engel, M. and Glotzer, S. C. (2010a). Dense crystalline dimer packings of regular tetrahedra, *Discrete & Computational Geometry* **44**, pp. 253–280.

Chen, K., Ellenbroek, W. G., Zhang, Z., Chen, D. T. N., Yunker, P., Henkes, S., Brito, C., Dauchot, O., van Saarloos, W., Liu, A. J. and Yodh, A. G.

(2010b). Low-frequency vibrations of soft colloidal glasses, *Phys. Rev. Lett.* **105**, p. 025501.

Chen, T. J., Zitter, R. N. and Tao, R. (1992). Laser diffraction determination of the crystalline-structure of an electrorheological fluid, *Phys. Rev. Lett.* **68**, pp. 2555–2558.

Cho, Y. S., Yi, G. R., Kim, S. H., Jeon, S. J., Elsesser, M. T., Yu, H. K., Yang, S. M. and Pine, D. J. (2007). Particles with coordinated patches or windows from oil-in-water emulsions, *Chem. Mater.* **19**, pp. 3183–3193.

Chowdhury, A., Ackerson, B. J. and Clark, N. A. (1985). Laser-induced freezing, *Phys. Rev. Lett.* **55**, pp. 833–836.

Chu, J. H. and I, L. (1994). Direct observation of Coulomb crystals and liquids in strongly coupled RF dusty plasmas, *Phys. Rev. Lett.* **72**, pp. 4009–4012.

Chui, S. T. (1982). Grain-boundary theory of melting in 2 dimensions, *Phys. Rev. Lett.* **48**, pp. 933–935.

Chui, S. T. (1983). Grain-boundary theory of melting in 2 dimensions, *Phys. Rev. B* **28**, pp. 178–194.

Chung, P. M., Talbot, L. and Touryan, K. J. (1975). *Electric probes in stationary and flowing plasmas: theory and application* (Springer, New York).

Ciamarra, M. P., Coniglio, A., De Martino, D. and Nicodemi, M. (2007). Shear- and vibration-induced order-disorder transitions in granular media, *Eur. Phys. J. E* **24**, pp. 411–415.

Cianci, G. C., Courtland, R. E. and Weeks, E. R. (2006). Correlations of structure and dynamics in an aging colloidal glass, *Solid State Communications* **139**, pp. 599–604.

Cipelletti, L. and Ramos, L. (2005). Slow dynamics in glassy soft matter, *J. Phys.: Condens. Matter* **17**, pp. R253–R285.

Clark, N. A. and Ackerson, B. J. (1980). Observation of the coupling of concentration fluctuations to steady-state shear-flow, *Phys. Rev. Lett.* **44**, pp. 1005–1008.

Cohen, I., Mason, T. G. and Weitz, D. A. (2004). Shear-induced configurations of confined colloidal suspensions, *Phys. Rev. Lett.* **93**, p. 046001.

Corté, L., Chaikin, P. M., Gollub, J. P. and Pine, D. J. (2008). Random organization in periodically driven systems, *Nature Physics* **4**, pp. 420–424.

Couëdel, L., Nosenko, V., Ivlev, A. V., Zhdanov, S. K., Thomas, H. M. and Morfill, G. E. (2010). Direct observation of mode-coupling instability in two-dimensional plasma crystals, *Phys. Rev. Lett.* **104**, p. 195001.

Couëdel, L., Zhdanov, S. K., Ivlev, A. V., Nosenko, V., Thomas, H. M. and Morfill, G. E. (2011). Wave mode coupling due to plasma wakes in two-dimensional plasma crystals: in-depth view, *Phys. Plasmas* **18**, p. 083707.

Crocker, J. C. and Grier, D. G. (1995). Methods of digital video microscopy for colloidal studies, *J. Coll. Interf. Sci.* **179**, pp. 298–310.

Cross, M. C. and Hohenberg, P. C. (1993). Pattern-formation outside of equilibrium, *Rev. Mod. Phys.* **65**, pp. 851–1112.

Cuesta, J. A. and Martínez-Ratón, Y. (1997). Fundamental measure theory for mixtures of parallel hard cubes. I. General formalism, *J. Chem. Phys.* **107**, pp. 6379–6389.

284   *Complex Plasmas and Colloidal Dispersions*

Cui, B., Lin, B. and Rice, S. A. (2001). Dynamical heterogeneity in a dense quasi-two-dimensional colloidal liquid, *J. Chem. Phys.* **114**, pp. 9142–9156.

Cui, C. and Goree, J. (1994). Fluctuations of the charge on a dust grain in a plasma, *IEEE Trans. Plasma Sci.* **22**, pp. 151–158.

Curtin, W. A. (1987). Density-functional theory of the solid-liquid interface, *Phys. Rev. Lett.* **59**, pp. 1228–1231.

Dagotto, E. (2005). Complexity in strongly correlated electronic systems, *Science* **309**, pp. 257–262.

Das, M., Ramaswamy, S. and Ananthakrishna, G. (2002). Melting-freezing cycles in a relatively sheared pair of crystalline monolayers, *Europhys. Lett.* **60**, pp. 636–642.

Dassanayake, U., Fraden, S. and van Blaaderen, A. (2000). Structure of electrorheological fluids, *J. Chem. Phys.* **112**, pp. 3851–3858.

Daugherty, J. E., Porteous, R. K., Kilgore, M. D. and Graves, D. B. (1992). Sheath structure around particles in low-pressure discharges, *J. Appl. Phys.* **72**, pp. 3934–3942.

Davidchak, R. L. and Laird, B. B. (2000). Direct calculation of the hard-sphere crystal/melt interfacial free energy, *Phys. Rev. Lett.* **85**, pp. 4751–4754.

de Villeneuve, V. W. A., Dullens, R. P. A., Aarts, D. G. A. L., Groeneveld, E., Scherff, J. H., Kegel, W. K. and Lekkerkerker, H. N. W. (2005). Colloidal hard-sphere crystal growth frustrated by large spherical impurities, *Science* **309**, pp. 1231–1233.

Debenedetti, P. G. and Stillinger, F. H. (2001). Supercooled liquids and the glass transition, *Nature* **410**, pp. 259–267.

Demirörs, A. F., Johnson, P. M., van Kats, C. M., van Blaaderen, A. and Imhof, A. (2010). Directed self-assembly of colloidal dumbbells with an electric field, *Langmuir* **26**, pp. 14466–14471.

Derks, D., Wisman, H., van Blaaderen, A. and Imhof, A. (2004). Confocal microscopy of colloidal dispersions in shear flow using a counter-rotating cone-plate shear cell, *J. Phys.: Condens. Matter* **16**, pp. S3917–S3927.

Derks, D., Wu, Y. L., van Blaaderen, A. and Imhof, A. (2009). Dynamics of colloidal crystals in shear flow, *Soft Matter* **5**, pp. 1060–1065.

Dhont, J. K. G. (2003). *An introduction to the dynamics of colloids* (Elsevier).

Dhont, J. K. G. and Kang, K. (2010). Electric-field-induced polarization and interactions of uncharged colloids in salt solutions, *Eur. Phys. J. E* **33**, pp. 51–68.

Di Cicco, A., Trapananti, A., Faggioni, S. and Filipponi, A. (2003). Is there icosahedral ordering in liquid and undercooled metals? *Phys. Rev. Lett.* **91**, p. 135505.

Dibble, C. J., Kogan, M. and Solomon, M. J. (2006). Structure and dynamics of colloidal depletion gels: Coincidence of transitions and heterogeneity, *Phys. Rev. E.* **74**, p. 041403.

Dijkstra, M., Hansen, J.-P. and Madden, P. A. (1997). Statistical model for the structure and gelation of smectite clay suspensions, *Phys. Rev. E* **55**, pp. 3044–3053.

Dijkstra, M., van Roij, R. and Evans, R. (1999). Phase behaviour and structure of model colloid-polymer mixtures, *J. Phys.: Condens. Matter* **11**, pp. 10079–10106.

Dinsmore, A. D., Prasad, V., Wong, I. Y. and Weitz, D. A. (2006). Microscopic structure and elasticity of weakly aggregated colloidal gels, *Phys. Rev. Lett.* **96**, p. 185502.

Dinsmore, A. D. and Weitz, D. A. (2002). Direct imaging of three-dimensional structure and topology of colloidal gels, *J. Phys.: Condens. Matter* **14**, pp. 7581–7597.

Dogic, Z. and Fraden, S. (2001). Development of model colloidal liquid crystals and the kinetics of the isotropic–smectic transition, *Phil. Trans.* **359**, pp. 997–1014.

Doi, M. and Edwards, S. F. (1986). The theory of polymer dynamics, Oxford Univeristy Press.

Donko, Z., Goree, J., Hartmann, P. and Kutasi, K. (2006). Shear viscosity and shear thinning in two-dimensional Yukawa liquids, *Phys. Rev. Lett.* **96**, p. 145003.

Donko, Z., Hartmann, P., Kalman, G. J. and Golden, K. I. (2003). Simulations of strongly coupled charged particle systems: static and dynamical properties of classical bilayers, *J. Phys. A* **36**, pp. 5877–5885.

Dosho, S., Ise, N., Ito, K., Iwai, S., Kitano, H., Matsuoka, H., Nakamura, H., Okamura, H., Ono, T., Sogami, I. S., Ueno, Y., Yoshida, H. and Yoshiyama, T. (1993). Recent study of polymer latex dispersions, *Langmuir* **9**, pp. 394–411.

Dullens, R. P. A., Aarts, D. G. A. L. and Kegel, W. K. (2006). Dynamic broadening of the crystal-fluid interface of colloidal hard spheres, *Phys. Rev. Lett.* **97**, p. 228301.

Dünweg, B. and Ladd, A. J. C. (2009). *Advanced computer simulation approaches for soft matter sciences III* (Springer, Berlin).

Dzubiella, J., Hoffmann, G. P. and Löwen, H. (2002). Lane formation in colloidal mixtures driven by an external field, *Phys. Rev. E* **65**, p. 021402.

Ebert, F., Keim, P. and Maret, G. (2008). Local crystalline order in a 2D colloidal glass former, *Eur. Phys. J. E* **26**, pp. 161–168.

Ediger, M. D. (2000). Spatially heterogeneous dynamics in supercooled liquids, *Ann. Rev. Phys. Chem.* **51**, pp. 99–128.

Edwards, S. F. and Doi, M. (1986). *The theory of polymer dynamics* (Oxford University Press, Oxford, UK).

el Kader, A. A. and Earnshaw, J. C. (1999). Shear-induced changes in two-dimensional foam, *Phys. Rev. Lett.* **82**, pp. 2610–2613.

Eldridge, M. D., Madden, P. A. and Frenkel, D. (1993). Entropy-driven formation of a superlattice in a hard-sphere binary mixture, *Nature* **365**, pp. 35–37.

Eldridge, M. D., Madden, P. A., Pusey, P. N. and Bartlett, P. (1995). Binary hard-sphere mixtures: a comparison between computer simulation and experiment, *Molecular Physics* **84**, pp. 395–420.

Elliot, J. R. and Hu, L. G. (1999). Vapor-liquid equilibria of square-well spheres, *J. Chem. Phys.* **110**, pp. 3043–3048.

286   *Complex Plasmas and Colloidal Dispersions*

Else, D., Kompaneets, R. and Vladimirov, S. V. (2009). Instability of the ionization-absorption balance in a complex plasma at ion time scales, *Phys. Rev. E* **80**, p. 016403.

Epstein, P. (1924). On the resistance experienced by spheres in their motion through gases, *Phys. Rev.* **23**, pp. 710–733.

Erbe, A., Zientara, M., Baraban, L., Kreidler, C. and Leiderer, P. (2008). Various driving mechanisms for generating motion of colloidal particles, *J. Phys.: Condens. Matter* **20**, p. 404215.

Ernst, M. H., Hauge, E. H. and Vanleeuw, J. M. (1970). Asymptotic time behavior of correlation functions, *Phys. Rev. Lett.* **25**, pp. 1254–1256.

Evans, D. J. and Morriss, G. P. (1990). *Statistical mechanics of nonequilibrium liquids* (Academic, New York).

Evans, R. (1966). *Crystal chemistry* (Cambridge Univ. Press, London and New York).

Federici, G., Skinner, C. H., Brooks, J. N., Coad, J. P., Grisolia, C., Haasz, A. A., Hassanein, A., Philipps, V., Pitcher, C. S., Roth, J., Wampler, W. R. and Whyte, D. G. (2001). Plasma-material interactions in current tokamaks and their implications for next step fusion reactors, *Nuclear Fusion* **41**, pp. 1967–2137.

Filion, L. (2011). *Self-assembly in colloidal hard-sphere systems*, Ph.D. thesis, University of Utrecht.

Filion, L. and Dijkstra, M. (2009). Prediction of binary hard-sphere crystal structures, *Phys. Rev. E* **79**, p. 46714.

Filippov, A. V., Zagorodny, A. G., Momot, A. I., Pal, A. F. and Starostin, A. N. (2007). Charge screening in a plasma with an external ionization source, *JETP* **104**, pp. 147–161.

Fischer, E. W. (1993). Light-scattering and dielectric studies on glass-forming liquids, *Physica A* **201**, pp. 183–206.

Fisher, M. E. (1974). Renormalization group in theory of critical behavior, *Rev. Mod. Phys.* **46**, pp. 597–616.

Fontecha, A. B., Palberg, T. and Schöpe, H. J. (2007). Construction and stability of a close-packed structure observed in thin colloidal crystals, *Phys. Rev. E* **76**, p. 050402(R).

Fontecha, A. B., Schöpe, H. J., König, H., Palberg, T., Messina, R. and Löwen, H. (2005). A comparative study on the phase behaviour of highly charged colloidal spheres in a confining wedge geometry, *J. Phys.: Condens. Matter* **17**, pp. S2779–S2786.

Fornleitner, J., F. Lo Verso, F., Kahl, G. and Likos, C. N. (2008). Genetic algorithms predict formation of exotic ordered configurations for two-component dipolar monolayers, *Soft Matter* **4**, pp. 480–484.

Fortini, A. and Dijkstra, M. (2006). Phase behaviour of hard spheres confined between hard plates: manipulation of colloidal crystal structures by confinement, *J. Phys.: Condens. Matter* **18**, pp. L371–L378.

Fortov, V. E., Iabukov, I. and Khrapak, A. G. (2006). *Physics of strongly coupled plasma* (Oxford University Press, Oxford, UK).

Fortov, V. E., Ivlev, A. V., Khrapak, S. A., Khrapak, A. G. and Morfill, G. E.
(2005). Complex (dusty) plasmas: current status, open issues, perspectives,
Phys. Rep. **421**, pp. 1–103.

Fortov, V. E., Khrapak, A. G., Khrapak, S. A., Molotkov, V. I., Nefedov, A. P.,
Petrov, O. F. and Torchinskii, V. M. (2000). Mechanism of dust-acoustic
instability in a direct current glow discharge plasma, *Phys. Plasmas* **7**,
pp. 1374–1380.

Fortov, V. E., Khrapak, A. G., Khrapak, S. A., Molotkov, V. I. and Petrov, O. F.
(2004a). Dusty plasmas, *Phys. Usp.* **47**, pp. 447–492.

Fortov, V. E. and Morfill, G. E. (eds.) (2010). *Complex and dusty plasmas: from
laboratory to space, Ch. 1* (CRC Press, Boca Raton).

Fortov, V. E., Nefedov, A. P., Petrov, O. F., Samarian, A. A. and Chernyshev,
A. V. (1996a). Particle ordered structures in a strongly coupled classical
thermal plasma, *Phys. Rev. E* **54**, pp. R2236–R2239.

Fortov, V. E., Nefedov, A. P., Torchinskii, V. M., Molotkov, V. I., Khrapak,
A. G., Petrov, O. F. and Volykhin, K. F. (1996b). Crystallization of a
dusty plasma in the positive column of a glow discharge, *JETP Lett.* **64**,
pp. 92–98.

Fortov, V. E., Nefedov, A. P., Vaulina, O. S., Lipaev, A. M., Molotkov, V. I.,
Samarian, A. A., Nikitski, V. P., Ivanov, A. I., Savin, S. F., Kalmykov,
A. V., Solov'ev, A. Y. and Vinogradov, P. V. (1998). Dusty plasma induced
by solar radiation under microgravitational conditions: an experiment on
board the MIR orbiting space station, *JETP* **87**, pp. 1087–1097.

Fortov, V. E., Petrov, O. F., Molotkov, V. I., Poustylnik, M. Y., Torchinskii,
V. M., Khrapak, A. G. and Chernyshev, A. V. (2004b). Large-amplitude
dust waves excited by the gas-dynamic impact in a DC glow discharge
plasma, *Phys. Rev. E* **69**, p. 016402.

Fortov, V. E., Vasilyak, L. M., Vetchinin, S. P., Zimnukhov, V. S., Nefedov,
A. P. and Polyakov, D. N. (2002). Plasma-dust structures at cryogenic
temperatures, *Dokl. Phys.* **47**, pp. 21–24.

Fortov, V. E., Vaulina, O. S., Petrov, O. F., Molotkov, V. I., Lipaev, A. M.,
Torchinskii, V. M., Thomas, H. M., Morfill, G. E., Khrapak, S. A., Semenov, Y. P., Ivanov, A. I., Krikalev, S. K., Kaleri, A. Y., Zaletin, S. V.
and Gidzenko, Y. P. (2003). Transport of microparticles in weakly ionized
gas-discharge plasmas under microgravity conditions, *Phys. Rev. Lett.* **90**,
p. 245005.

Fortov, V. E., Vaulina, O. S., Petrov, O. F., Vasiliev, M. N., Gavrikov, A. V.,
Shakhova, I., Vorona, N. A., Khrustalyov, Y. V., Manohin, A. A. and
Chernyshev, A. V. (2007). Experimental study of the heat transport pro-
cesses in dusty plasma fluid, *Phys. Rev. E* **75**, p. 026403.

Fraden, S., Maret, G., Caspar, D. L. D. and Meyer, R. B. (1989). Isotropic-
nematic phase transition and angular correlations in isotropic suspensions
of Tobacco Mosaic Virus, *Phys. Rev. Lett.* **63**, pp. 2068–2071.

Frank, F. C. (1952). Supercooling of liquids, *Proc. R. Soc. Lond. A.* **215**, pp. 43–
46.

Franzrahe, K. and Nielaba, P. (2009). Controlled structuring of binary hard-disk mixtures via a periodic, external potential, *Phys. Rev. E* **79**, p. 051505.

Frenkel, D. (1991). *Liquids, freezing and the glass transition* (North-Holland, Amsterdam).

Frenkel, D. (1999). Entropy-driven phase transitions, *Physica A* **263**, pp. 26–28.

Frenkel, D. (2006). Colloidal encounters: a matter of attraction, *Science* **314**, pp. 768–769.

Fuchs, M. and Ballauff, M. (2005). Flow curves of dense colloidal dispersions: schematic model analysis of the shear-dependent viscosity near the colloidal glass transition, *J. Chem. Phys.* **122**, p. 094707.

Fuchs, M. and Cates, M. E. (2002). Theory of nonlinear rheology and yielding of dense colloidal suspensions, *Phys. Rev. Lett.* **89**, p. 248304.

Fuchs, M. and Cates, M. E. (2003). Schematic models for dynamic yielding of sheared colloidal glasses, *Faraday Discuss.* **123**, pp. 267–286.

Fuchs, M. and Cates, M. E. (2009). A mode coupling theory for Brownian particles in homogeneous steady shear flow, *J. Rheol.* **53**, pp. 957–1000.

Fuchs, M., Götze, W. and Mayr, M. R. (1998). Asymptotic laws for tagged-particle motion in glassy systems, *Phys. Rev. E* **58**, pp. 3384–3399.

Furukawa, H. (1984). Dynamic-scaling theory for phase-separating unmixing mixtures - growth-rates of droplets and scaling properties of auto-correlation functions, *Physica A* **123**, pp. 497–515.

Gasser, U. (2009). Crystallization in three- and two-dimensional colloidal suspensions, *J. Phys.: Condens. Matter* **21**, p. 203101.

Gasser, U., Weeks, E. R., Schofield, A. B., Pusey, P. N. and Weitz, D. A. (2001). Real-space imaging of nucleation and growth in colloidal crystallization, *Science* **292**, pp. 258–262.

Gavrikov, A. V., Shakhova, I., Ivanov, A. I., Petrov, O. F., Vorona, N. A. and Fortov, V. E. (2005). Experimental study of laminar flow in dusty plasma liquid, *Phys. Lett. A* **336**, pp. 378–383.

Geerts, N. and Eiser, E. (2009). Flying colloidal carpets, *Soft Matter* **6**, pp. 664–669.

Gerbode, S. J., Agarwal, U., Ong, D. C., Liddell, C. M., Escobedo, F. A. and Cohen, I. (2010). Glassy dislocation dynamics in 2D colloidal dimer crystals, *Phys. Rev. Lett.* **105**, p. 078301.

Ghosh, A., Chikkadi, V. K., Schall, P., Kurchan, J. and Bonn, D. (2010). Density of states of colloidal glasses, *Phys. Rev. Lett.* **104**, p. 248305.

Gilbert, S. L., Bollinger, J. J. and Wineland, D. J. (1988). Shell-structure phase of magnetically confined strongly coupled plasmas, *Phys. Rev. Lett.* **60**, pp. 2022–2025.

Ginzburg, V. L. (1996). Radiation by uniformly moving sources (Vavilov-Cherenkov effect, transition radiation, and other phenomena), *Phys. Usp.* **39**, pp. 973–982.

Gleiter, H. (1989). Nanocrystalline materials, *Prog. Mater. Sci.* **33**, pp. 223–315.

Gleiter, H. (2000). Nanostructured materials: basic concepts and microstructure, *Acta Materialia* **48**, pp. 1–29.

Glotzer, S. C. and Solomon, M. J. (2007). Anisotropy of building blocks and their assembly into complex structures, *Nature Materials* **6**, pp. 557–562.

Goertz, C. K. (1992). Dusty plasmas in the solar system, *Rev. Geophys.* **27**, pp. 271–292.

Goldoni, G. and Peeters, F. M. (1998). Stability, dynamical properties, and melting of a classical bilayer Wigner crystal, *Phys. Rev. B* **53**, pp. 4591–4603.

Gompper, G., Ihle, T., Kroll, D. and Winkler, R. (2009). Multi-particle collision dynamics: a particle-based mesoscale simulation approach to the hydrodynamics of complex fluids, *Advanced Computer Simulation Approaches for Soft Matter Sciences III* , pp. 1–87.

Goree, J. (1994). Charging of particles in a plasma, *Plasma Sources Sci. Technol.* **3**, pp. 400–406.

Goree, J., Morfill, G. E., Tsytovich, V. N. and Vladimirov, S. V. (1999). Theory of dust voids in plasmas, *Phys. Rev. E* **59**, pp. 7055–7067.

Götze, I. O., Brader, J. M., Schmidt, M. and Löwen, H. (2003). Laser-induced condensation in colloid-polymer mixtures, *Molecular Physics* **101**, pp. 1651–1658.

Götze, W. (1991). *Aspects of structural glass transitions* (North-Holland, Amsterdam).

Götze, W. (1999). Recent tests of the mode-coupling theory for glassy dynamics, *J. Phys.: Condens. Matter* **11**, pp. A1–A45.

Götze, W. (2009). *Complex dynamics of glass-forming liquids: a mode-coupling theory* (Oxford University Press, Oxford, UK).

Götze, W. and Sjogren, L. (1992). Relaxation processes in supercooled liquids, *Rep. Prog. Phys.* **55**, pp. 241–376.

Graf, H. and Löwen, H. (1999). Phase diagram of Tobacco Mosaic Virus solutions, *Phys. Rev. E* **59**, pp. 1932–1942.

Graf, H., Löwen, H. and Schmidt, M. (1997). *Cell theory for the phase diagram of hard spherocylinders* (Springer), pp. 177–179.

Grandner, S. and Klapp, S. H. L. (2008). Freezing of charged colloids in slit pores, *J. Chem. Phys.* **129**, p. 244703.

Grier, D. G. (2003). A revolution in optical manipulation, *Nature* **424**, pp. 810–816.

Groenewold, J. and Kegel, W. K. (2001). Anomalously large equilibrium clusters of colloids, *J. Phys. Chem. B.* **105**, pp. 11702–11709.

Grün, E., Morfill, G. E. and Mendis, D. A. (1984). *Dust-magnetosphere interactions* (Univ. Arizona Press, Tucson), pp. 275–332.

Gumbsch, P. and Gao, H. J. (1999). Dislocations faster than the speed of sound, *Science* **283**, pp. 965–968.

Hahn, H. S., Mason, E. A. and Smith, F. J. (1971). Quantum transport cross sections for ionized gases, *Phys. Fluids* **14**, p. 278.

Haji-Akbari, A., Engel, M., Keys, A. S., Zheng, X., Petschek, R. G., Palffy-Muhoray, P. and Glotzer, S. C. (2009). Disordered, quasicrystalline and crystalline phases of densely packed tetrahedra, *Nature* **462**, pp. 773–777.

Halperin, B. I. and Nelson, D. R. (1978). Theory of two-dimensional melting, *Phys. Rev. Lett.* **41**, pp. 121–124.

Hamaguchi, S., Farouki, R. T. and Dubin, D. H. E. (1997). Triple point of Yukawa systems, *Phys. Rev. E* **56**, pp. 4671–4682.

Hamaker, H. C. (1937). The London-van der Waals attraction between spherical particles, *Physica* **4**, pp. 1058–1072.

Hamanaka, T. and Onuki, A. (2007). Heterogeneous dynamics in polycrystal and glass in a binary mixture with changing size dispersity and composition, *Phys. Rev. E* **75**, p. 041503.

Han, Y., Ha, N. Y., Alsayed, A. M. and Yodh, A. G. (2008). Melting of two-dimensional tunable-diameter colloidal crystals, *Phys. Rev. E* **77**, p. 041406.

Hanley, H. J. M., Rainwater, J. C., Clark, N. A. and Ackerson, B. J. (1983). Liquid structure under shear - comparison between computer-simulations and colloidal suspensions, *J. Chem. Phys.* **79**, pp. 4448–4458.

Hansen, J.-P. and Löwen, H. (2000). Effective interactions between electric double layers, *Ann. Rev. Phys. Chem.* **51**, pp. 209–242.

Hansen, J.-P. and MacDonald, I. R. (2006). *Theory of simple liquids* (Academic, London).

Hansen, J.-P. and Schiff, D. (1973). Influence of interatomic repulsion on structure of liquids at melting, *Molecular Physics* **25**, pp. 1281–1290.

Hansen, J.-P. and Verlet, L. (1969). Phase transitions of Lennard-Jones system, *Phys. Rev.* **184**, pp. 151–161.

Hansen, J.-P. and Victor, J. M. (1985). Spinodal decomposition and the liquid vapor equilibrium in charged colloidal dispersions, *J. Chem. Soc., Faraday Trans. II* **81**, pp. 43–61.

Harada, Y. and Asakura, T. (1996). Radiation forces on a dielectric sphere in the Rayleigh scattering regime, *Optics Communications* **124**, pp. 529–541.

Hargis, P. J., Greenberg, K. E., Miller, P. A., Gerardo, J. B., Torczynski, J. R., Riley, M. E., Hebner, G. A., Roberts, J. R., Olthoff, J. K., Whetstone, J. R., Vanbrunt, R. J., Sobolewski, M. A., Anderson, H. M., Splichal, M. P., Mock, J. L., Bletzinger, P., Garscadden, A., Gottscho, R. A., Selwyn, G. S., Dalvie, M., Heidenreich, J. E., Butterbaugh, J. W., Brake, M. L., Passow, M. L., Pender, J., Lujan, A., Elta, M. E., Graves, D. B., Sawin, H. H., Kushner, M. J., Verdeyen, J. T., Horwath, R. and Turner, T. R. (1994). The gaseous electronics conference radiofrequency reference cell - a defined parallel-plate radiofrequency system for experimental and theoretical-studies of plasma-processing discharges, *Rev. Sci. Instrum.* **65**, pp. 140–154.

Hartquist, T. W., Havnes, O. and Morfill, G. E. (1992). The effects of dust on the dynamics of astronomical and space plasmas, *Fund. Cosmic Phys.* **15**, pp. 107–142.

Havnes, O., Morfill, G. E. and Goertz, C. K. (1984). Plasma potential and grain charges in a dust cloud embedded in a plasma, *J. Geophys. Res.* **89**, pp. 999–1003.

Hayakawa, H. and Ichiki, K. (1995). Statistical-theory of sedimentation of disordered suspensions, *Phys. Rev. E* **51**, pp. R3815–R3818.

Hayashi, Y. and Tachibana, K. (1994). Observation of Coulomb-crystal formation from carbon particles grown in a methane plasma, *Jpn. J. Appl. Phys.* **33**, pp. L804–L806.

Helbing, D., Farkas, I. J. and Vicsek, T. (2000). Freezing by heating in a driven mesoscopic system, *Phys. Rev. Lett.* **84**, pp. 1240–1243.

Helseth, L. E., Muruganathan, R. M., Zhang, Y. and Fischer, T. M. (2005). Colloidal rings in a liquid mixture, *Langmuir* **21**, pp. 7271–7275.

Henderson, D. and Leonard, P. J. (1971). *Physical chemistry: an advanced treatise*, Vol. 8B (Academic Press, New York).

Henderson, R. L. (1974). Uniqueness theorem for fluid pair correlation-functions, *Phys. Lett. A* **49**, pp. 197–198.

Heni, M. and Löwen, H. (2000). Surface freezing on patterned substrates, *Phys. Rev. Lett.* **85**, pp. 3668–3671.

Henrich, O., Weysser, F., Cates, M. E. and Fuchs, M. (2009). Hard discs under steady shear: comparison of Brownian dynamics simulations and mode coupling theory, *Phil. Trans. A* **367**, pp. 5033–5050.

Hermes, M., Vermolen, E. C. M., Leunissen, M. E., Vossen, D. L. J., van Oostrum, P. D. J., Dijkstra, M. and van Blaaderen, A. (2011). Nucleation of colloidal crystals on configurable seed structures, *Soft Matter* **7**, pp. 4623–4628.

Hernandez-Guzman, J. and Weeks, E. R. (2009). The equilibrium intrinsic crystal-liquid interface of colloids, *Proc. Nat. Acad. Sci.* **106**, pp. 15198–15202.

Hickey, O. A., Holm, C., Harden, J. L. and Slater, G. W. (2010). Implicit method for simulating electrohydrodynamics of polyelectrolytes, *Phys. Rev. Lett.* **105**, p. 148301.

Ho, C. C., Chen, W. S., Shie, T. Y., Lin, J. N. and Kuo, C. (2008). Novel fabrication of Janus particles from the surfaces of electrospun polymer fibers, *Langmuir* **24**, pp. 5663–5666.

Hoffmann, N., Ebert, F., Likos, C. N., Löwen, H. and Maret, G. (2006a). Partial clustering in binary two-dimensional colloidal suspensions, *Phys. Rev. Lett.* **97**, pp. 078301–078301.

Hoffmann, N., Likos, C. N. and Löwen, H. (2006b). Microphase structuring in two-dimensional magnetic colloid mixtures, *J. Phys.: Condens. Matter* **18**, pp. 10193–10193.

Hong, L., Cacciuto, A., Luijten, E. and Granick, S. (2006). Clusters of charged Janus spheres, *Nano Lett.* **6**, pp. 2510–2514.

Hoogenboom, J. P., van Langen-Suurling, A. K., Romijn, J. and van Blaaderen, A. (2003a). Hard-sphere crystals with hcp and non-close-packed structure grown by colloidal epitaxy, *Phys. Rev. Lett.* **90**, p. 138301.

Hoogenboom, J. P., Vergeer, P. and van Blaaderen, A. (2003b). A real-space analysis of colloidal crystallization in a gravitational field at a flat bottom wall, *J. Chem. Phys.* **119**, pp. 3371–3383.

Hoover, W. G. (1985). Canonical dynamics - equilibrium phase-space distributions, *Phys. Rev. A* **31**, pp. 1695–1697.

Hoover, W. G. and Ree, F. H. (1968). Melting transition and communal entropy for hard spheres, *J. Chem. Phys.* **49**, pp. 3609–3618.

Horbach, J. and Frenkel, D. (2001). Lattice-Boltzmann method for the simulation of transport phenomena in charged colloids, *Phys. Rev. E* **64**, p. 061507.

Horowitz, C. J., Pérez-Garcia, M. A. and Piekarewicz, J. (2004). Neutrino-"pasta" scattering: the opacity of nonuniform neutron-rich matter, *Phys. Rev. C* **69**, p. 045804.

292   *Complex Plasmas and Colloidal Dispersions*

Hosein, I. D., John, B. S., Lee, S. H., Escobedo, F. A. and Liddell, C. M. (2009).
    Rotator and crystalline films via self-assembly of short-bond-length colloidal
    dimers, *J. Mater. Chem.* **19**, pp. 344–349.
Hou, L. J., Wang, Y. N. and Miskovic, Z. L. (2003). Induced potential of a dust
    particle in a collisional radio-frequency sheath, *Phys. Rev. E* **68**, p. 016410.
Huang, F., Rotstein, R., Fraden, S., Kasza, K. E. and Flynn, N. T. (2009).
    Phase behavior and rheology of attractive rod-like particles, *Soft Matter* **5**,
    pp. 2766–2771.
Huang, Y.-H. and Lin, I. (2007). Memory and persistence correlation of mi-
    crostructural fluctuations in two-dimensional dusty plasma liquids, *Phys.
    Rev. E* **76**, p. 016403.
Huse, D. A. (1986). Corrections to late-stage behavior in spinodal decomposition
    - Lifshitz-Slyozov scaling and Monte-Carlo simulations, *Phys. Rev. B* **34**,
    pp. 7845–7850.
Hynninen, A.-P., Christova, C. G., van Roij, R., van Blaaderen, A. and Dijkstra,
    M. (2006a). Prediction and observation of crystal structures of oppositely
    charged colloids, *Phys. Rev. Lett.* **96**, p. 138308.
Hynninen, A.-P. and Dijkstra, M. (2005). Phase diagram of dipolar hard and soft
    spheres: manipulation of colloidal crystal structures by an external field,
    Phys. Rev. Lett. **94**, p. 138303.
Hynninen, A.-P., Fortini, A. and Dijkstra, M. (2009). Phase behavior and struc-
    ture of colloidal suspensions in bulk, confinement, and external fields, *Struc-
    ture and Functional Properties of Colloidal Systems* , pp. 165–199.
Hynninen, A.-P., Leunissen, M. E., van Blaaderen, A. and Dijkstra, M. (2006b).
    CuAu structure in the restricted primitive model and oppositely charged
    colloids, *Phys. Rev. Lett.* **96**, p. 018303.
Hynninen, A.-P., Thijssen, J. H. J., Vermolen, E. C. M., Dijkstra, M. and
    van Blaaderen, A. (2007). Self-assembly route for photonic crystals with
    a bandgap in the visible region, *Nature Materials* **6**, pp. 202–205.
Ilett, S. M., Orrock, A., Poon, W. C. K. and Pusey, P. N. (1995). Phase behaviour
    of a model colloid-polymer mixture, *Phys. Rev. E* **52**, pp. 1344–1352.
Ishihara, O. and Vladimirov, S. V. (1997). Wake potential of a dust grain in a
    plasma with ion flow, *Phys. Plasmas* **4**, pp. 69–74.
Ivanov, Y. and Melzer, A. (2007). Particle positioning techniques for dusty plasma
    experiments, *Rev. Sci. Instrum.* **78**, p. 033506.
Ivlev, A. V., Khrapak, A. G., Khrapak, S. A., Annaratone, B. M., Morfill, G. E.
    and Yoshino, K. (2003). Rodlike particles in gas discharge plasmas: theo-
    retical model, *Phys. Rev. E* **68**, p. 026403.
Ivlev, A. V., Konopka, U. and Morfill, G. E. (2000). Influence of charge variation
    on particle oscillations in the plasma sheath, *Phys. Rev. E* **62**, pp. 2739–
    2744.
Ivlev, A. V., Konopka, U., Morfill, G. E. and Joyce, G. (2003). Melting of mono-
    layer plasma crystals, *Phys. Rev. E* **68**, p. 026405.
Ivlev, A. V. and Morfill, G. E. (2001). Anisotropic dust lattice modes, *Phys. Rev.
    E* **63**, p. 016409.

Ivlev, A. V., Morfill, G. E., Thomas, H. M., Räth, C., Joyce, G., Huber, P., Kompaneets, R., Fortov, V. E., Lipaev, A. M., Molotkov, V. I., Reiter, T., Turin, M. and Vinogradov, P. V. (2008). First observation of electrorheological plasmas, *Phys. Rev. Lett.* **100**, p. 095003.

Ivlev, A. V., Steinberg, V., Kompaneets, R., Höfner, H., Sidorenko, I. and Morfill, G. E. (2007a). Non-Newtonian viscosity of complex-plasma fluids, *Phys. Rev. Lett.* **98**, p. 145003.

Ivlev, A. V., Thoma, M. H., Räth, C., Joyce, G. and Morfill, G. E. (2011). Complex plasmas in external fields: the role of non-Hamiltonian interactions, *Phys. Rev. Lett.* **106**, p. 155001.

Ivlev, A. V., Zhdanov, S. K., Khrapak, S. A. and Morfill, G. E. (2005). Kinetic approach for the ion drag force in a collisional plasma, *Phys. Rev. E* **71**, p. 016405.

Ivlev, A. V., Zhdanov, S. K. and Morfill, G. E. (2007b). Free thermal convection in complex plasma with background-gas friction, *Phys. Rev. Lett.* **99**, p. 135004.

Ivlev, A. V., Zhdanov, S. K., Thomas, H. M. and Morfill, G. E. (2009). Fluid phase separation in binary complex plasmas, *Europhys. Lett.* **85**, p. 45001.

Jenkins, M. C. and Egelhaaf, S. U. (2008). Confocal microscopy of colloidal particles: towards reliable, optimum coordinates, *J. Coll. Interf. Sci.* **136**, pp. 65–92.

Jenkins, M. C., Haw, M. D., Barker, G. C., Poon, W. C. K. and Egelhaaf, S. U. (2010). Finding bridges in packings of colloidal spheres, *Soft Matter* **7**, pp. 684–690.

Jiang, K., Hou, L. J., Ivlev, A. V., Li, Y. F., Du, C. R., Thomas, H. M., Morfill, G. E. and Sütterlin, R. K. (2011). Initial stages in phase separation of binary complex plasmas: numerical experiments, *Europhys. Lett.* **93**, p. 55001.

Johnson, P. M., van Kats, C. M. and van Blaaderen, A. (2005). Synthesis of colloidal silica dumbbells, *Langmuir* **21**, pp. 11510–11517.

Jonsson, H. and Andersen, H. C. (1988). Icosahedral ordering in the Lennard-Jones liquid and glass, *Phys. Rev. Lett.* **60**, pp. 2295–2298.

Juan, W. T., Chen, M. H. and I, L. (2001). Nonlinear transports and microvortex excitations in sheared quasi-two-dimensional dust Coulomb liquids, *Phys. Rev. E* **64**, p. 016402.

Juan, W. T., Huang, Z. H., Hsu, J. W., Lai, Y. J. and I, L. (1998). Observation of dust Coulomb clusters in a plasma trap, *Phys. Rev. E* **58**, pp. R6947–R6950.

Kadau, K., Germann, T. C., Hadjiconstantinou, N. G., Lomdahl, P. S., Dimonte, G., Holian, B. L. and Alder, B. J. (2004). Nanohydrodynamics simulations: an atomistic view of the Rayleigh-Taylor instability, *Proc. Nat. Acad. Sci.* **101**, pp. 5851–5855.

Kawasaki, T., Araki, T. and Tanaka, H. (2007). Correlation between dynamic heterogeneity and medium-range order in two-dimensional glass-forming liquids, *Phys. Rev. Lett.* **99**, p. 215701.

Kaya, D., Green, N. L., Maloney, C. E. and Islam, M. F. (2010). Normal modes and density of states of disordered colloidal solids, *Science* **329**, pp. 656–658.

Kegel, W. K., Breed, D., Elsesser, M. T. and Pine, D. J. (2006). Formation of anisotropic polymer colloids by disparate relaxation times, *Langmuir* **22**, pp. 7135–7136.

Kegel, W. K. and van Blaaderen, A. (2001). Direct observation of dynamic heterogeneities in colloidal hard-sphere suspensions, *Science* **287**, pp. 290–293.

Keim, P., Maret, G., Herz, U. and von Grünberg, H. H. (2004). Harmonic lattice behavior of two-dimensional colloidal crystals, *Phys. Rev. Lett.* **92**, p. 215504.

Kennedy, R. V. and Allen, J. E. (2003). The floating potential of spherical probes and dust grains. II: Orbital motion theory, *J. Plasma Phys.* **69**, pp. 485–506.

Khodataev, Y. K., Khrapak, S. A., Nefedov, A. P. and Petrov, O. F. (1998). Dynamics of the ordered structure formation in a thermal dusty plasma, *Phys. Rev. E* **57**, pp. 7086–7092.

Khrapak, S. A., Ivlev, A. V. and Morfill, G. E. (2001). Interaction potential of microparticles in a plasma: role of collisions with plasma particles, *Phys. Rev. E* **64**, p. 046403.

Khrapak, S. A., Ivlev, A. V. and Morfill, G. E. (2004). Momentum transfer in complex plasmas, *Phys. Rev. E* **70**, p. 056405.

Khrapak, S. A., Ivlev, A. V. and Morfill, G. E. (2010a). Shielding of a test charge: role of plasma production and loss balance, *Phys. Plasmas* **17**, p. 042107.

Khrapak, S. A., Ivlev, A. V., Morfill, G. E. and Thomas, H. M. (2002). Ion drag force in complex plasmas, *Phys. Rev. E.* **66**, p. 046414.

Khrapak, S. A., Ivlev, A. V., Morfill, G. E. and Zhdanov, S. K. (2003). Scattering in the attractive Yukawa potential in the limit of strong interaction, *Phys. Rev. Lett.* **90**, p. 225002.

Khrapak, S. A., Klumov, B. A., Huber, P., Molotkov, V. I., Lipaev, A. M., Naumkin, V. N., Thomas, H. M., Ivlev, A. V., Morfill, G. E., Petrov, O. F., Fortov, V. E., Malentschenko, Y. and Volkov, S. (2011). Freezing and melting of 3D complex plasma structures under microgravity conditions driven by neutral gas pressure manipulation, *Phys. Rev. Lett* **106**, p. 205001.

Khrapak, S. A. and Morfill, G. E. (2001). Waves in two component electron-dust plasma, *Phys. Plasmas* **8**, pp. 2629–2634.

Khrapak, S. A. and Morfill, G. E. (2006). Grain surface temperature in noble gas discharges: refined analytical model, *Phys. Plasmas* **13**, p. 104506.

Khrapak, S. A. and Morfill, G. E. (2008). A note on the binary interaction potential in complex (dusty) plasmas, *Phys. Plasmas* **15**, p. 084502.

Khrapak, S. A. and Morfill, G. E. (2009). Basic processes in complex (dusty) plasmas: charging, interactions, and ion drag force, *Contrib. Plasma Phys.* **49**, pp. 148–168.

Khrapak, S. A., Morfill, G. E., Khrapak, A. G. and D'yachkov, L. G. (2006). Charging properties of a dust grain in collisional plasmas, *Phys. Plasmas* **13**, p. 052114.

Khrapak, S. A., Nefedov, A. P., Petrov, O. F. and Vaulina, O. S. (1999). Dynamical properties of random charge fluctuations in a dusty plasma with different charging mechanisms, *Phys. Rev. E* **59**, pp. 6017–6022.

Khrapak, S. A., Ratynskaia, S. V., Zobnin, A. V., Usachev, A. D., Yaroshenko, V. V., Thoma, M. H., Kretschmer, M., Höfner, H., Morfill, G. E., Petrov, O. F. *et al.* (2005). Particle charge in the bulk of gas discharges, *Phys. Rev. E* **72**, p. 016406.

Khrapak, S. A., Thomas, H. M. and Morfill, G. E. (2010b). Multiple phase transitions associated with charge cannibalism effect in complex (dusty) plasmas, *Europhys. Lett.* **91**, p. 25001.

Kim, S. H. and Karilla, S. J. (1991). *Microhydrodynamics, principles and selected applications* (Butterworth-Heinemann, Boston, MA).

Kirchner, H. O. K., Michot, G. and Schweizer, J. (2002). Fracture toughness of snow in shear under friction, *Phys. Rev. E* **66**, p. 027103.

Kirkpatrick, T. R., Thirumalai, D. and Wolynes, P. G. (1989). Scaling concepts for the dynamics of viscous liquids near an ideal glassy state, *Phys. Rev. A* **40**, pp. 1045–1054.

Kirkwood, J. E. (1981). *Phase transformations in solids* (Wiley, New York).

Kirkwood, J. G., Buff, F. P. and Green, M. S. (1949). The statistical mechanical theory of transport processes. 3. The coefficients of shear and bulk viscosity of liquids, *J. Chem. Phys.* **17**, pp. 988–994.

Kittel, C. (1961). *Introduction to solid state physics* (Wiley, New York).

Klix, C. L., Murata, K., Tanaka, H., Williams, S., Malins, A. and Royall, C. P. (2009). The role of weak charging in metastable colloidal clusters, *ArXiv cond-mat.soft*, URL http://arxiv.org/abs/0905.3393.

Klix, C. L., Royall, C. P. and Tanaka, H. (2010). Structural and dynamical features of multiple metastable glassy states in a colloidal system with competing interactions, *Phys. Rev. Lett.* **104**, p. 165702.

Knapek, C. A., Samsonov, D., Zhdanov, S. K., Konopka, U. and Morfill, G. E. (2007). Recrystallization of a 2D plasma crystal, *Phys. Rev. Lett.* **98**, p. 015004.

Kodama, H., Takeshita, K., Araki, T. and Tanaka, H. (2004). Fluid particle dynamics simulation of charged colloidal suspensions, *J. Phys.: Condens. Matter* **16**, pp. L115–L123.

Kollmann, M., Hund, R., Rinn, B., Nägele, G., Zahn, K., König, H., Maret, G., Klein, R. and Dhont, J. K. G. (2002). Structure and tracer-diffusion in quasi two-dimensional and strongly asymmetric magnetic colloidal mixtures, *Europhys. Lett* **58**, pp. 919–925.

Kompaneets, R. (2007). *Complex plasmas: interaction potentials and non-Hamiltonian dynamics*, Ph.D. thesis, Ludwig-Maximilians-Universität München, available at http://edoc.ub.uni-muenchen.de/7380.

Kompaneets, R., Konopka, U., Ivlev, A. V., Tsytovich, V. N. and Morfill, G. E. (2007). Potential around a charged dust particle in a collisional sheath, *Phys. Plasmas* **14**, p. 052108.

Kompaneets, R., Morfill, G. E. and Ivlev, A. V. (2009). Design of new binary interaction classes in complex plasmas, *Phys. Plasmas* **16**, p. 043705.

Kompaneets, R., Vladimirov, S. V., Ivlev, A. V. and Morfill, G. E. (2008). Reciprocal interparticle attraction in complex plasmas with cold ion flows, *New J. Phys.* **10**, p. 063018.

König, H. (2005). Elementary triangles in a 2D binary colloidal glass former, *Europhys. Lett.* **71**, pp. 838–844.

König, H., Hund, R., Zahn, K. and Maret, G. (2005). Experimental realisation of a model glass former in 2D, *Eur. Phys. J. E* **18**, pp. 287–293.

Köppl, M., Henseler, P., Erbe, A., Nielaba, P. and Leiderer, P. (2007). Layer reduction in driven 2D-colloidal systems through microchannels, *Phys. Rev. Lett.* **97**, pp. 208302–208302.

Kosterlitz, J. M. and Thouless, D. J. (1973). Ordering, metastability and phase-transitions in 2 dimensional systems, *J. Phys. C - Solid State Physics* **6**, pp. 1181–1203.

Koumakis, N., Schofield, A. B. and Petekidis, G. (2008). Effects of shear induced crystallization on the rheology and ageing of hard sphere glasses, *Soft Matter* **4**, pp. 2008–2018.

Kraft, D. J., Vlug, W. S., van Kats, C. M., van Blaaderen, A., Imhof, A. and Kegel, W. K. (2008). Self-assembly of colloids with liquid protrusions, *J. Am. Chem. Soc.* **131**, pp. 1182–1186.

Kranendonk, W. G. T. and Frenkel, D. (1991). Computer simulation of solid-liquid coexistence in binary hard sphere mixtures, *Molecular Physics* **72**, pp. 679–697.

Krasheninnikov, S. I. (2010). On the dynamics of nonspherical dust grain in plasma, *Phys. Plasmas* **17**, p. 033703.

Kremer, K., Robbins, M. O. and Grest, G. S. (1986). Phase-diagram of Yukawa systems – model for charge-stabilized colloids, *Phys. Rev. Lett.* **57**, pp. 2694–2697.

Kroll, M., Harms, S., Block, D. and Piel, A. (2008). Digital in-line holography of dusty plasmas, *Phys. Plasmas* **15**, p. 063703.

Kroll, M., Schablinski, J., Block, D. and Piel, A. (2010). On the influence of wakefields on three-dimensional particle arrangements, *Phys. Plasmas* **17**, p. 013702.

Kujik, A., van Blaaderen, A. and Imhof, A. (2011). Synthesis of monodisperse, rodlike silica colloids with tunable aspect ratio, *J. Am. Chem. Soc.* **133**, pp. 2346–2349.

Lacevic, N., Starr, F. W., Schroder, T. B. and Glotzer, S. C. (2003). Spatially heterogeneous dynamics investigated via a time-dependent four-point density correlation function, *J. Chem. Phys.* **119**, pp. 7372–7387.

Lai, Y. J. and I, L. (2002). Avalanche excitations of fast particles in quasi-2D cold dusty-plasma liquids, *Phys. Rev. Lett.* **89**, p. 155002.

Lampe, M., Gavrishchaka, V., Ganguli, G. and Joyce, G. (2001a). Effect of trapped ions on shielding of a charged spherical object in a plasma, *Phys. Rev. Lett.* **86**, pp. 5278–5281.

Lampe, M., Goswami, R., Sternovsky, Z., Robertson, S., Gavrishchaka, V., Ganguli, G. and Joyce, G. (2003). Trapped ion effect on shielding, current flow, and charging of a small object in a plasma, *Phys. Plasmas* **10**, pp. 1500–1513.

Lampe, M., Joyce, G. and Ganguli, G. (2001b). Analytic and simulation studies of dust grain interaction and structuring, *Phys. Scripta* **T89**, pp. 106–111.

Lampe, M., Joyce, G., Ganguli, G. and Gavrishchaka, V. (2000). Interactions between dust grains in a dusty plasma, *Phys. Plasmas* **7**, pp. 3851–3861.

Landau, L. D. and Lifshitz, E. M. (1978). *Statistical physics. Part I* (Pergamon, Oxford).

Landau, L. D. and Lifshitz, E. M. (1986). *Theory of elasticity* (Pergamon, Oxford).

Landau, L. D. and Lifshitz, E. M. (1987). *Fluid mechanics* (Pergamon, Oxford).

Langer, J. S. (1974). Metastable states, *Physica* **73**, pp. 61–72.

Langer, J. S., Baron, M. and Miller, H. D. (1975). New computational method in theory of spinodal decomposition, *Phys. Rev. A* **11**, pp. 1417–1429.

Langmuir, I., Found, G. and Dittmer, A. F. (1924). A new type of electric discharge: the streamer discharge, *Science* **60**, pp. 392–394.

Lapenta, G. (2000). Linear theory of plasma wakes, *Phys. Rev. E* **62**, pp. 1175–1181.

Lapenta, G. (2002). Nature of the force field in plasma wakes, *Phys. Rev. E* **66**, p. 026409.

Lapointe, C. P., Hopkins, S., Mason, T. G. and Smalyukh, I. I. (2010). Electrically driven multiaxis rotational dynamics of colloidal platelets in nematic liquid crystals, *Phys. Rev. Lett.* **105**, p. 178301.

Laurati, M., Egelhaaf, S. U. and Petekidis, G. (2011). Nonlinear rheology of colloidal gels with intermediate volume fraction, *J. Rheol.* **55**, pp. 673–706.

Lebowitz, J. L. and Penrose, O. (1966). Rigorous treatment of van der Waals-Maxwell theory of liquid-vapor transition, *J. Math. Phys.* **7**, pp. 98–114.

Lekkerkerker, H. N. W., Poon, W. C. K., Pusey, P. N., Stroobants, A. and Warren, P. B. (1992). Phase-behavior of colloid plus polymer mixtures, *Europhys. Lett.* **20**, pp. 559–564.

Lemons, D. S., Murillo, M. S., Daughton, W. and Winske, D. (2000). Two-dimensional wake potentials in sub- and supersonic dusty plasmas, *Phys. Plasmas* **7**, pp. 2306–2313.

Leunissen, M. E., Christova, C. G., Hynninen, A.-P., Royall, C. P., Campbell, A. I., Imhof, A., Dijkstra, M., van Roij, R. and van Blaaderen, A. (2005). Ionic colloidal crystals of oppositely charged particles, *Nature* **437**, pp. 235–240.

Li, K. H. and Yellen, B. B. (2010). Magnetically tunable self-assembly of colloidal rings, *App. Phys. Lett.* **97**, p. 083105.

Liddell, C. M. and Summers, C. J. (2003). Monodispersed ZnS dimers, trimers, and tetramers for lower symmetry photonic crystal lattices, *Adv. Mater.* **15**, pp. 1715–1719.

Lieberman, M. A. and Lichtenberg, A. J. (1994). *Principles of plasma discharges and materials processing* (Wiley, New York).

Lifshitz, E. M. and Pitaevskii, L. P. (1981). *Physical kinetics* (Pergamon, Oxford).

Lifshitz, I. M. and Slyozov, V. V. (1961). The kinetics of precipitation from supersaturated solid solutions, *J. Phys. Chem. Solids* **19**, pp. 35–50.

Likos, C. N. and Henley, C. L. (1993). Complex alloy phases for binary hard-disc mixtures, *Phil. Mag. B* **68**, pp. 85–113.

Lindemann, F. A. (1910). The calculation of molecular vibration frequencies, *Physik. Z.* **11**, pp. 609–612.

Lipaev, A. M., Khrapak, S. A., Molotkov, V. I., Morfill, G. E., Fortov, V. E., Ivlev, A. V., Thomas, H. M., Khrapak, A. G., Naumkin, V. N., Ivanov, A. I. et al. (2007). Void closure in complex plasmas under microgravity conditions, *Phys. Rev. Lett.* **98**, p. 265006.

Lipaev, A. M., Molotkov, V. I., Nefedov, A. P., Petrov, O. F., Torchinskii, V. M., Fortov, V. E., Khrapak, A. G. and Khrapak, S. A. (1997). Ordered structures in a nonideal dusty glow-discharge plasma, *JETP* **85**, pp. 1110–1118.

Liu, B. and Goree, J. (2008). Superdiffusion and non-Gaussian statistics in a driven-dissipative 2D dusty plasma, *Phys. Rev. Lett.* **100**, p. 055003.

Liu, B., Goree, J., Nosenko, V. and Boufendi, L. (2003). Radiation pressure and gas drag forces on a melamine-formaidehyde microsphere in a dusty plasma, *Phys. Plasmas* **10**, pp. 9–20.

Lobaskin, V., Dünweg, B., Medebach, M., Palberg, T. and Holm, C. (2007). Electrophoresis of colloidal dispersions in the low-salt regime, *Phys. Rev. Lett.* **98**, p. 176105.

Long, D. and Ajdari, A. (2001). A note on the screening of hydrodynamic interactions, in electrophoresis, and in porous media, *Eur. Phys. J. E* **4**, pp. 29–32.

Lorenz, N. J. and Palberg, T. (2010). Melting and freezing lines for a mixture of charged colloidal spheres with spindle-type phase diagram, *J. Chem. Phys.* **133**, p. 104501.

Lorenz, N. J., Schöpe, H. J., Reiber, H., Palberg, T., Wette, P., Klassen, I., Holland-Moritz, D., Herlach, D. and Okubo, T. (2009). Phase behaviour of deionized binary mixtures of charged colloidal spheres, *J. Phys.: Condens. Matter* **21**, p. 464116.

Löwen, H. (1994a). Brownian dynamics of hard spherocylinders, *Phys. Rev. E* **50**, pp. 1232–1242.

Löwen, H. (1994b). Melting, freezing and colloidal suspensions, *Phys. Rep.* **237**, pp. 249–324.

Löwen, H. (1996). Dynamical criterion for two-dimensional freezing, *Phys. Rev. E* **53**, pp. R29–R32.

Löwen, H. (1998). The apparent mass in sedimentation profiles of charged suspensions, *J. Phys.: Condens. Matter* **10**, pp. L479–L485.

Löwen, H. (2010). Particle-resolved instabilities in colloidal dispersions, *Soft Matter* **6**, pp. 3133–3142.

Löwen, H. and Allahyarov, E. (1998). The role of effective triplet interactions in charged colloidal suspensions, *J. Phys.: Condens. Matter* **10**, pp. 4147–4160.

Löwen, H. and Kramposthuber, G. (1993). Optimal effective pair potential for charged colloids, *Europhys. Lett.* **23**, pp. 673–678.

Löwen, H., Palberg, T. and Simon, R. (1993). Dynamic criterion for freezing of colloidal liquids, *Phys. Rev. Lett.* **70**, pp. 1557–1560.

Lu, P. J., Sims, P. A., Oki, H., Macarthur, J. B. and Weitz, D. A. (2007). Target-locking acquisition with real-time confocal (tarc) microscopy, *Optics Express* **15**, pp. 8702–8712.

Lu, P. J., Zaccarelli, E., Ciulla, F., Schofield, A. B., Sciortino, F. and Weitz, D. A. (2008). Gelation of particles with short-range attraction, *Nature* **435**, pp. 499–504.

Lyklema, J. (2005). *Fundamentals of interface and colloid science, volume IV: Particulate colloids* (Elsevier).

Mangold, K., Birk, J., Leiderer, P. and Bechinger, C. (2004). Binary colloidal systems in two-dimensional circular cavities: structure and dynamics, *Phys. Chem. Chem. Phys* **6**, pp. 1623–1626.

Manoharan, V. N., Elsesser, M. T. and Pine, D. J. (2003). Dense packing and symmetry in small clusters of microspheres, *Science* **301**, pp. 483–487.

Manzano, F. R., Bonet, E., Rodriguez, I. and Meseguer, F. (2009). Layering transitions in colloidal crystal thin films between 1 and 4 monolayers, *Soft Matter* **5**, pp. 4279–4282.

March, N. H. and Tosi, M. P. (2002). *Introduction to liquid state physics* (World Scientific, London).

Marcus, A. H., Lin, B. and Rice, S. A. (1996). Self-diffusion in dilute quasi-two-dimensional hard sphere suspensions: evanescent wave light scattering and video microscopy studies, *Phys, Rev. E* **53**, pp. 1765–1776.

Marcus, A. H. and Rice, S. A. (1997). Phase transitions in a confined quasi-two-dimensional colloid suspension, *Phys. Rev. E* **55**, pp. 637–656.

Marechal, M. and Dijkstra, M. (2008). Stability of orientationally disordered crystal structures of colloidal hard dumbbells, *Phys. Rev. E* **77**, p. 061405.

Marechal, M., Kortschot, R. J., Demirörs, A. F., Imhof, A. and Dijkstra, M. (2010). Phase behavior and structure of a new colloidal model system of bowl-shaped particles, *Nano Lett.* **10**, pp. 1907–1911.

Marr, D. W. and Gast, A. P. (1993). Planar density-functional approach to the solid-fluid interface of simple liquids, *Phys. Rev. E* **47**, pp. 1212–1221.

Marro, J., Bortz, A. B., Kalos, M. H. and Lebowitz, J. L. (1975). Time evolution of a quenched binary alloy. 2. Computer-simulation of a three-dimensional model system, *Phys. Rev. B* **12**, pp. 2000–2011.

Martin, J., Venturini, E., Gulley, G. L. and Williamson, J. (2004). Using triaxial magnetic fields to create high susceptibility particle composites, *Phys. Rev. E* **69**, p. 021508.

Martin, S., Bryant, G. and van Megen, W. (2005). Crystallization kinetics of polydisperse colloidal hard spheres. II. Binary mixtures, *Phys. Rev. E* **71**, p. 021404.

Matsoukas, T. and Russell, M. (1997). Fokker-Planck description of particle charging in ionized gases, *Phys. Rev. E* **55**, pp. 991–994.

Meijer, E. J. and Frenkel, D. (1991). Melting line of Yukawa system by computer-simulation, *J. Chem. Phys.* **94**, pp. 2269–2271.

Melandso, F. and Goree, J. (1995). Polarized supersonic plasma-flow simulation for charged bodies such as dust particles and spacecraft, *Phys. Rev. E* **52**, pp. 5312–5326.

Meller, A. and Stavans, J. (1992). Glass transition and phase diagrams of strongly interacting binary colloidal mixtures, *Phys. Rev. Lett.* **68**, pp. 3646–3649.

Melzer, A., Klindworth, M. and Piel, A. (2001). Normal modes of 2D finite clusters in complex plasmas, *Phys. Rev. Lett.* **87**, p. 115002.

Melzer, A., Schweiger, V. A., Schweigert, I. V., Homann, A., Peters, S. and Piel, A. (1996). Structure and stability of the plasma crystal, *Phys. Rev. E* **54**, pp. R46–R49.

Melzer, A., Schweigert, V. A. and Piel, A. (1999). Transition from attractive to repulsive forces between dust molecules in a plasma sheath, *Phys. Rev. Lett.* **83**, pp. 3194–3197.

Merlino, R. L., Barkan, A., Thompson, C. and D'Angelo, N. (1998). Laboratory studies of waves and instabilities in dusty plasmas, *Phys. Plasmas* **5**, pp. 1607–1614.

Messina, R. (2009). Electrostatics in soft matter, *J. Phys.: Condens. Matter* **21**, p. 113102.

Messina, R. and Löwen, H. (2003). Reentrant transitions in colloidal or dusty plasma bilayers, *Phys. Rev. Lett.* **91**, p. 146101.

Messina, R. and Löwen, H. (2006). Confined colloidal bilayers under shear: steady state and relaxation back to equilibrium, *Phys. Rev. E* **73**, p. 011405.

Mikhael, J., Roth, J., Helden, L. and Bechinger, C. (2008). Archimedean-like tiling on decagonal quasicrystalline surfaces, *Nature* **454**, pp. 501–504.

Miloch, W. J., Trulsen, J. and Pecseli, H. L. (2008). Numerical studies of ion focusing behind macroscopic obstacles in a supersonic plasma flow, *Phys. Rev. E* **77**, p. 056408.

Minsky, M. (1957). Microscopy apparatus, *US patent 3013467*.

Molotkov, V. I., Petrov, O. F., Pustyl'nik, M. Y., Torchinskii, V. M., Fortov, V. E. and Khrapak, A. G. (2004). Dusty plasma of a DC glow discharge: methods of investigation and characteristic features of behavior, *High Temp.* **42**, pp. 827–841.

Montgomery, D., Joyce, G. and Sugihara, R. (1968). Inverse 3rd power law for shielding of test particles, *Plasma Phys.* **10**, p. 681.

Morfill, G. E. and Ivlev, A. V. (2009). Complex plasmas: an interdisciplinary research field, *Rev. Mod. Phys.* **81**, pp. 1353–1404.

Morfill, G. E., Khrapak, S. A., Ivlev, A. V., Klumov, B. A., Rubin-Zuzic, M. and Thomas, H. M. (2004). From fluid flows to crystallization: new results from complex plasmas, *Phys. Scripta* **T107**, pp. 59–64.

Morfill, G. E., Konopka, U., Kretschmer, M., Rubin-Zuzic, M., Thomas, H. M., Zhdanov, S. K. and Tsytovich, V. N. (2006). The "classical tunnelling effect" – observations and theory, *New J. Phys.* **8**, pp. 7–26.

Morfill, G. E., Rubin-Zuzic, M., Rothermel, H., Ivlev, A. V., Klumov, B. A., Thomas, H. M., Konopka, U. and Steinberg, V. (2004). Highly resolved fluid flows: "liquid plasmas" at the kinetic level, *Phys. Rev. Lett.* **92**, p. 175004.

Morfill, G. E., Thomas, H. M., Konopka, U., Rothermel, H., Zuzic, M., Ivlev, A. V. and Goree, J. (1999). Condensed plasmas under microgravity, *Phys. Rev. Lett.* **83**, pp. 1598–1601.

Mourad, M. C. D., Petukhov, A. V., Vroege, G. J. and Lekkerkerker, H. N. W. (2010). Lyotropic hexagonal columnar liquid crystals of large colloidal Gibbsite platelets, *Langmuir* **26**, pp. 14182–14187.

Mu, Y., Houk, A. and Song, X. Y. (2005). Anisotropic interfacial free energies of the hard-sphere crystal-melt interfaces, *J. Phys. Chem. B* **109**, pp. 6500–6504.

Mukhija, D. and Solomon, M. J. (2007). Translational and rotational dynamics of colloidal rods by direct visualization with confocal microscopy, *J. Coll. Interf. Sci.* **314**, pp. 98–106.

Mullin, T. (2000). Coarsening of self-organized clusters in binary mixtures of particles, *Phys. Rev. Lett.* **84**, pp. 4741–4744.

Murayama, M., Howe, J. M., Hidaka, H. and Takaki, S. (2002). Atomic-level observation of disclination dipoles in mechanically milled, nanocrystalline Fe, *Science* **295**, pp. 2433–2435.

Murray, C. A. and Grier, D. G. (1996). Video microscopy of monodisperse colloidal systems, *Physical Chemistry* **47**, pp. 421–462.

Murray, C. A., Sprenger, W. O. and Wenk, R. A. (1990a). Comparison of melting in three and two dimensions: microscopy of colloidal spheres, *Phys. Rev. B* **42**, pp. 688–703.

Murray, C. A. and van Winkle, D. H. (1987). Experimental observation of two-stage melting in a classical two-dimensional screened Coulomb system, *Phys. Rev. Lett.* **58**, pp. 1200–1203.

Murray, C. A., van Winkle, D. H. and Wenk, R. A. (1990b). Digital imaging studies of submicron colloidal spheres confined into a single layer between two smooth glass plates: two-dimensional melting, *Phase Transitions* **21**, pp. 93–126.

Murray, M. J. and Sanders, J. V. (1980). Close-packed structures of spheres of two different sizes II. The packing densities of likely arrangements, *Phil. Mag. A* **42**, pp. 721–740.

Nambu, M., Vladimirov, S. V. and Shukla, P. K. (1995). Attractive forces between charged particulates in plasmas, *Phys. Lett. A* **203**, pp. 40–42.

Nefedov, A. P., Morfill, G. E., Fortov, V. E., Thomas, H. M., Rothermel, H., Hagl, T., Ivlev, A. V., Zuzic, M., Klumov, B. A., Lipaev, A. M., Molotkov, V. I., Petrov, O. F., Gidzenko, Y. P., Krikalev, S. K., Shepherd, W., Ivanov, A. I., Roth, M., Binnenbruck, H., Goree, J. and Semenov, Y. P. (2003). PKE-Nefedov: plasma crystal experiments on the international space station, *New J. Phys.* **5**, p. 33.

Nefedov, A. P., Petrov, O. F. and Fortov, V. E. (1997). Quasicrystalline structures in strongly coupled dusty plasmas, *Phys. Usp.* **40**, pp. 1163–1173.

Nefedov, A. P., Petrov, O. F., Khodataev, Y. K. and Khrapak, S. A. (1999). Dynamics of formation of ordered structures in a thermal plasma with macroparticles, *JETP* **88**, pp. 460–464.

Nefedov, A. P., Petrov, O. F., Molotkov, V. I. and Fortov, V. E. (2000). Formation of liquidlike and crystalline structures in dusty plasmas, *JETP Lett.* **72**, pp. 218–226.

Nefedov, A. P., Vaulina, O. S., Petrov, O. F., Molotkov, V. I., Torchinskii, V. M., Fortov, V. E., Chernyshev, A. V., Lipaev, A. M., Ivanov, A. I., Kaleri, A. Y., Semenov, Y. P. and Zaletin, S. V. (2002). The dynamics of macroparticles in a direct current glow discharge plasma under microgravitation conditions, *JETP* **95**, pp. 673–681.

Nelson, D. R. (2002). *Defects and geometry in condensed matter physics* (Cambridge University Press, Cambridge).

Nelson, D. R. and Halperin, B. I. (1979). Dislocation-mediated melting in 2 dimensions, *Phys. Rev. B* **19**, pp. 2457–2484.

Neser, S., Bechinger, C., Leiderer, P. and Palberg, T. (1997). Finite-size effects on the closest packing of hard spheres, *Phys. Rev. Lett.* **79**, pp. 2348–2351.

Netz, R. R. (2003). Conduction and diffusion in two-dimensional electrolytes, *Europhys. Lett.* **63**, pp. 616–622.

Nosenko, V. and Goree, J. (2004). Shear flows and shear viscosity in a two-dimensional Yukawa system (dusty plasma), *Phys. Rev. Lett.* **93**, p. 155004.

Nosenko, V., Goree, J. and Piel, A. (2006). Laser method of heating monolayer dusty plasmas, *Phys. Plasmas* **13**, p. 032106.

Nosenko, V., Morfill, G. E. and Rosakis, P. (2011). Direct experimental measurement of the speed-stress relation for dislocations in a plasma crystal, *Phys. Rev. Lett.* **106**, p. 155002.

Nosenko, V., Zhdanov, S. K., Ivlev, A. V., Knapek, C. A. and Morfill, G. E. (2009). 2D melting of plasma crystals: equilibrium and nonequilibrium regimes, *Phys. Rev. Lett.* **103**, p. 015001.

Nosenko, V., Zhdanov, S. K., Ivlev, A. V., Morfill, G. E., Goree, J. and Piel, A. (2008). Heat transport in a two-dimensional complex (dusty) plasma at melting conditions, *Phys. Rev. Lett.* **100**, p. 025003.

Nosenko, V., Zhdanov, S. K. and Morfill, G. E. (2007). Supersonic dislocations observed in a plasma crystal, *Phys. Rev. Lett.* **99**, p. 025002.

Nugent, C. R., Edmond, K. V., Patel, H. V. and Weeks, E. R. (2007). Colloidal glass transition observed in confinement, *Phys. Rev. Lett.* **99**, p. 025702.

Nunomura, S., Goree, J., Hu, S., Wang, X., Bhattacharjee, A. and Avinash, K. (2002). Phonon spectrum in a plasma crystal, *Phys. Rev. Lett.* **89**, p. 035001.

Nunomura, S., Samsonov, D., Zhdanov, S. K. and Morfill, G. E. (2006). Self-diffusion in a liquid complex plasma, *Phys. Rev. Lett.* **96**, p. 015003.

Nunomura, S., Zhdanov, S. K., Samsonov, D. and Morfill, G. E. (2005). Wave spectra in solid and liquid complex (dusty) plasmas, *Phys. Rev. Lett.* **94**, p. 045001.

Ohnesorge, R., Löwen, H. and Wagner, H. (1994). Density-functional theory of crystal fluid interfaces and surface melting, *Phys. Rev. E* **50**, pp. 4801–4809.

Ohta, H. and Hamaguchi, S. (2000). Molecular dynamics evaluation of self-diffusion in Yukawa systems, *Phys. Plasmas* **7**, pp. 4506–4514.

Ohtsuka, T., Royall, C. P. and Tanaka, H. (2008). Local structure and dynamics in colloidal fluids and gels, *Europhys. Lett.* **84**, p. 46002.

Okubo, T. and Fujita, H. (1996). Phase diagram of alloy crystal in the exhaustively deionized suspensions of binary mixtures of colloidal spheres, *Colloid & Polymer Science* **274**, pp. 368–374.

Onuki, A. (2004). *Phase transition dynamics* (Cambridge University Press, Cambridge).

Ott, T. and Bonitz, M. (2009). Is diffusion anomalous in two-dimensional Yukawa liquids? *Phys. Rev. Lett.* **103**, p. 195001.

Oğuz, E. C., Messina, R. and Löwen, H. (2009a). Crystalline multilayers of the confined Yukawa system, *Europhys. Lett.* **86**, p. 28002.

Oğuz, E. C., Messina, R. and Löwen, H. (2009b). Multilayered crystals of macroions under slit confinement, *J. Phys.: Condens. Matter* **21**, p. 424110.

Oğuz, E. C., Messina, R. and Löwen, H. (2011). Buckling in confining potentials, (submitted).

Padding, J. T. and Louis, A. A. (2004). Hydrodynamic and Brownian fluctuations in sedimenting suspensions, *Phys. Rev. Lett.* **93**, p. 220601.

Palberg, T. (1999). Crystallization kinetics of repulsive colloidal spheres, *J. Phys.: Condens. Matter* **11**, pp. R323–R360.

Pantina, J. P. and Furst, E. M. (2005). Elasticity and critical bending moment of model colloidal aggregates, *Phys. Rev. Lett.* **94**, p. 138301.

Parthe, E. (1961). Space filling of crystal structures, *Z. Kristallogr* **115**, pp. 52–79.

Pawley, J. (2006). *Handbook of biological confocal microscopy* (Springer, New York.).

Peeters, F. M., Schweigert, V. A. and Bedanov, V. M. (1995). Classical two-dimensional atoms, *Physica B* **212**, pp. 237–244.

Peng, Y., Wang, Z., Alsayed, A. M., Yodh, A. G. and Han, Y. (2010). Melting of colloidal crystal films, *Phys. Rev. Lett.* **104**, p. 205703.

Penrose, O. and Lebowitz, J. L. (1971). Rigorous treatment of metastable states in the van der Waals-Maxwell theory, *J. Stat. Phys.* **3**, pp. 211–241.

Perrin, J. (1913). *Les atomes* (Felix Alcan).

Petekidis, G., Vlassopoulos, D. and Pusey, P. N. (2003). Yielding and flow of colloidal glasses, *Faraday Discuss.* **123**, pp. 287–302.

Pham, K. N., Petekidis, G., Vlassopoulos, D., Egelhaaf, S. U., Poon, W. C. K. and Pusey, P. N. (2008). Yielding behavior of repulsion- and attraction-dominated colloidal glasses, *J. Rheol.* **52**, pp. 649–676.

Pham, K. N., Puertas, A. M., Bergenholtz, J., Egelhaaf, S. U., Moussaid, A., Pusey, P. N. and Schofield, A. B. (2002). Multiple glassy states in a simple model system, *Science* **296**, pp. 104–106.

Piazza, R., Bellini, T. and Degiorgio, V. (1993). Equilibrium sedimentation profiles of screened charged colloids: a test of the hard sphere equation of state, *Phys. Rev. Lett.* **25**, pp. 4267–4270.

Piazza, R. and Parola, A. (2008). Thermophoresis in colloidal suspensions, *J. Phys.: Condens. Matter* **20**, p. 153102.

Pieper, J. B., Goree, J. and Quinn, R. A. (1996). Three-dimensional structure in a crystallized dusty plasma, *Phys. Rev. E* **54**, pp. 5636–5640.

Pieranski, P., Strzelecki, L. and Pansu, B. (1983). Thin colloidal crystals, *Phys. Rev. Lett.* **50**, pp. 900–903.

Pooley, C. M. and Yeomans, J. M. (2004). Stripe formation in differentially forced binary systems, *Phys. Rev. Lett.* **93**, p. 118001.

Prigogine, I. (1980). *From being to becoming* (Freeman, San Francisco).

Puertas, A. M., Fuchs, M. and Cates, M. E. (2002). Comparative simulation study of colloidal gels and glasses, *Phys. Rev. Lett.* **88**, p. 098301.

Puri, S. and Wadhawan, V. (eds.) (2009). *Kinetics of phase transitions, ch. 1* (CRC Press, Boca Raton).

Pusey, P. N. (1991). *Liquids, freezing and the glass transition* (North-Holland, Amsterdam).

Pusey, P. N. (2005). Freezing and melting: action at grain boundaries, *Science* **309**, pp. 1198–1199.

Pusey, P. N. and van Megen, W. (1986). Phase behaviour of concentrated suspensions of nearly hard colloidal spheres, *Nature* **320**, pp. 340–342.

Qian, T., Wang, X.-P. and Sheng, P. (2006). A variational approach to moving contact line hydrodynamics, *J. Fluid. Mech.* **564**, pp. 333–360.

Quilliet, C., Zoldesi, C., Riera, C., van Blaaderen, A. and Imhof, A. (2008). Anisotropic colloids through non-trivial buckling, *Eur. Phys. J. E* **27**, pp. 13–20.

Raizer, Y. P. (1991). *Gas discharge physics* (Springer-Verlag, Berlin).

Ramiro-Manzano, F., Bonet, E., Rodriguez, I. and Meseguer, F. (2007). Layering transitions in confined colloidal crystals: the hcp-like phase, *Phys. Rev. E* **76**, p. 050401(R).

Ramiro-Manzano, F., Meseguer, F., Bonet, E. and Rodriguez, I. (2006). Faceting and commensurability in crystal structures of colloidal thin films, *Phys. Rev. Lett.* **97**, p. 028304.

Ramsteiner, I. B., Weitz, D. A. and Spaepen, F. (2010). Stiffness of the crystal-liquid interface in a hard-sphere colloidal system measured from capillary fluctuations, *Phys. Rev. E* **82**, p. 041603.

Rasa, M. and Philipse, A. P. (2004). Evidence for a macroscopic electric field in the sedimentation profiles of charged colloids, *Nature* **429**, pp. 857–860.

Räth, C., Monetti, R., Bauer, J., Sidorenko, I., Müller, D., Matsuura, M., Lochmueller, E. M., Zysset, P. and Eckstein, F. (2008). Strength through structure: visualization and local assessment of the trabecular bone structure, *New J. Phys.* **10**, p. 125010.

Ratynskaia, S. V., Rypdal, K., Knapek, C. A., Khrapak, S. A., Milovanov, A. V., Ivlev, A. V., Rasmussen, J. J. and Morfill, G. E. (2006). Superdiffusion and viscoelastic vortex flows in a two-dimensional complex plasma, *Phys. Rev. Lett.* **96**, p. 105010.

Raveche, H. J., Mountain, R. D. and Streett, W. B. (1974). Freezing and melting properties of Lennard-Jones system, *J. Chem. Phys.* **61**, pp. 1970–1984.

Reichhardt, C. and Olson, C. J. (2002). Novel colloidal crystalline states on two-dimensional periodic substrates, *Phys. Rev. Lett.* **88**, p. 248301.

Reichman, D., Rabani, E. and Geissler, P. L. (2005). Comparison of dynamical heterogeneity in hard-sphere and attractive glass formers, *J. Phys. Chem. B* **109**, pp. 14654–14658.

Reinke, D., Stark, H., von Grünberg, H. H., Schofield, A. B., Maret, G. and Gasser, U. (2007). Noncentral forces in crystals of charged colloids, *Phys. Rev. Lett.* **98**, p. 038301.

Rex, M. and Löwen, H. (2007). Lane formation in oppositely charged colloids driven by an electric field: chaining and two-dimensional crystallization, *Phys. Rev. E* **75**, p. 051402.

Rex, M. and Löwen, H. (2008). Influence of hydrodynamic interactions on lane formation in oppositely charged driven colloids, *Eur. Phys. J. E* **26**, pp. 143–150.

Ribbe, A. E. (1997). Laser scanning microscopy, *Trends in polymer science* **5**, pp. 333–337.

Rice, R., Roth, R. and Royall, C. P. (2011). Polyhedral colloidal rocks: Low-dimensional networks, (submitted to Soft Matter).

Richert, R. (2002). Heterogeneous dynamics in liquids: fluctuations in space and time, *J. Phys.: Condens. Matter* **14**, pp. R703–R738.

Robbins, M. O., Kremer, K. and Grest, G. S. (1988). Phase diagram and dynamics of Yukawa systems, *J. Chem. Phys.* **88**, pp. 3286–3312.

Rojas, L. F., Urban, C., Schurtenberger, P., Gisler, T. and von Grünberg, H. H. (2002). Reappearance of structure in colloidal suspensions, *Europhys. Lett.* **60**, pp. 802–808.

Roorda, S., van Dillen, T., Polman, A., Graf, C., van Blaaderen, A. and Kooi, B. J. (2004). Aligned gold nanorods in silica made by ion irradiation of core–shell colloidal particles, *Adv. Mater.* **16**, pp. 235–237.

Rosakis, A. J., Samudrala, O. and Coker, D. (1999). Cracks faster than the shear wave speed, *Science* **284**, pp. 1337–1340.

Rosakis, P. (2001). Supersonic dislocation kinetics from an augmented Peierls model, *Phys. Rev. Lett.* **86**, pp. 95–98.

Rose-Petruck, C., Jimenez, R., Guo, T., Cavalleri, A., Siders, C. W., Raksi, F., Squier, J. A., Walker, B. C., Wilson, K. R. and Barty, C. P. J. (1999). Picosecond-milliangstrom lattice dynamics measured by ultrafast X-ray diffraction, *Nature* **398**, pp. 310–312.

Rosenberg, M., Mendis, D. A. and Sheehan, D. P. (1996). UV-induced Coulomb crystallization of dust grains in high-pressure gas, *IEEE Trans. Plasma Sci.* **24**, pp. 1422–1430.

Rosenberg, M., Mendis, D. A. and Sheehan, D. P. (1999). Positively charged dust crystals induced by radiative heating, *IEEE Trans. Plasma Sci.* **27**, pp. 239–242.

Rosenfeld, Y. (1989). Free-energy model for the inhomogeneous hard-sphere fluid mixture and density-functional theory of freezing, *Phys. Rev. Lett.* **63**, pp. 980–983.

Rossi, L., Sacanna, S., Irvine, W. T. M., Chaikin, P. M., Pine, D. J. and Philipse, A. P. (2011). Cubic crystals from cubic colloids, *Soft Matter* **7**, pp. 4139–4142.

Roth, R. and Dietrich, S. (2000). Binary hard-sphere fluids near a hard wall, *Phys. Rev. E* **62**, pp. 6926–6936.

306   *Complex Plasmas and Colloidal Dispersions*

Rothermel, H., Hagl, T., Morfill, G. E., Thoma, M. H. and Thomas, H. M. (2002). Gravity compensation in complex plasmas by application of a temperature gradient, *Phys. Rev. Lett.* **89**, p. 175001.

Rowlinson, J. S. and Swinton, F. (1982). *Liquids and liquid mixtures* (Butterworths, London).

Royall, C. P., Aarts, D. G. A. L. and Tanaka, H. (2007a). Bridging length scales in colloidal liquids and interfaces from near-critical divergence to single particles, *Nature Physics* **9**, pp. 636–640.

Royall, C. P., Dzubiella, J., Schmidt, M. and van Blaaderen, A. (2007b). Nonequilibrium sedimentation of colloids on the particle scale, *Phys. Rev. Lett.* **98**, p. 188304.

Royall, C. P., Leunissen, M. E., Hynninen, A.-P., Dijkstra, M. and van Blaaderen, A. (2006). Re-entrant melting and freezing in a model system of charged colloids, *J. Chem. Phys.* **124**, p. 244706.

Royall, C. P., Leunissen, M. E. and van Blaaderen, A. (2003). A new colloidal model system to study long-range interactions quantitatively in real space, *J. Phys.: Condens. Matter* **15**, pp. S3581–S3596.

Royall, C. P., Louis, A. A. and Tanaka, H. (2007c). Measuring colloidal interactions with confocal microscopy, *J. Chem. Phys.* **127**, 044507.

Royall, C. P., van Roij, R. and van Blaaderen, A. (2005). Extended sedimentation profiles in charged colloids: the gravitational length, entropy, and electrostatics, *J. Phys.: Condens. Matter* **17**, pp. 2315–2326.

Royall, C. P., Williams, S. R., Ohtsuka, T. and Tanaka, H. (2008). Direct observation of a local structural mechanism for dynamic arrest, *Nature Materials* **7**, pp. 556–561.

Royall, C. P., Williams, S. R. and Tanaka, H. (2011). Vitrification is continuous and gelation is quasi-discontinuous in sticky spheres, to be published.

Rubin-Zuzic, M., Morfill, G. E., Ivlev, A. V., Pompl, R., Klumov, B. A., Bunk, W., Thomas, H. M., Rothermel, H., Havnes, O. and Fouquet, A. (2006). Kinetic development of crystallization fronts in complex plasmas, *Nature Physics* **2**, pp. 181–185.

Russ, C., von Grünberg, H. H., Dijkstra, M. and van Roij, R. (2002). Three-body forces between charged colloidal particles, *Phys. Rev. E* **66**, p. 011402.

Russel, W. B., Saville, D. A. and Schowalter, W. R. (1989). *Colloidal dispersions* (Cambridge University Press, Cambridge).

Sacanna, S., Irvine, W. T. M., Chaikin, P. M. and Pine, D. J. (2010). Lock and key colloids, *Nature* **464**, pp. 575–578.

Sagui, C. and Desai, R. C. (1994). Kinetics of phase-separation in two-dimensional systems with competing interactions, *Phys. Rev. E* **49**, pp. 2225–2244.

Saigo, T. and Hamaguchi, S. (2002). Shear viscosity of strongly coupled Yukawa systems, *Phys. Plasmas* **9**, pp. 1210–1216.

Saija, F., Prestipino, S. and Giaquinta, P. (2006). Evaluation of phenomenological one-phase criteria for the melting and freezing of softly repulsive particles, *J. Chem. Phys.* **124**, p. 244504.

Salin, G. and Caillol, J. M. (2002). Transport coefficients of the Yukawa one-component plasma, *Phys. Rev. Lett.* **88**, p. 065002.

Salmon, J. B., Colin, A., Manneville, S. and Molino, F. (2003). Velocity profiles in shear-banding wormlike micelles, *Phys. Rev. Lett.* **90**, p. 228303.

Samarian, A. A., Chernyshev, A. V., Nefedov, A. P., Petrov, O. F., Mikhailov, Y. M., Mintsev, V. B. and Fortov, V. E. (2000). Structures of the particles of the condensed dispersed phase in solid fuel combustion products plasma, *JETP* **90**, pp. 817–822.

Samsonov, D., Elsaesser, A., Edwards, A., Thomas, H. M. and Morfill, G. E. (2008). High speed laser tomography system, *Rev. Sci. Instrum.* **79**, p. 035102.

Samsonov, D. and Goree, J. (1999). Instabilities in a dusty plasma with ion drag and ionization, *Phys. Rev. E* **59**, pp. 1047–1058.

Sanchez, P., Swift, M. R. and King, P. J. (2004). Stripe formation in granular mixtures due to the differential influence of drag, *Phys. Rev. Lett.* **93**, p. 184302.

Sanders, J. V. (1980). Close-packed structures of spheres of two different sizes. I. Observations on natural opal, *Phil. Mag. A* **42**, pp. 705–720.

Sandomirski, K., Allahyarov, E., Löwen, H. and Egelhaaf, S. U. (2011). Heterogeneous crystallization of hard sphere colloids near a wall, *Soft Matter* **7**, pp. 8050–8055.

Sanz, E., Valeriani, C., Frenkel, D. and Dijkstra, M. (2007). Evidence for out-of-equilibrium crystal nucleation in suspensions of oppositely charged colloids, *Phys. Rev. Lett.* **99**, p. 055501.

Saunders, B. R. and Vincent, B. (1999). Microgel particles as model colloids: theory, properties and applications, *Adv. Coll. Interf. Sci.* **80**, pp. 1–25.

Savage, J. R. and Dinsmore, A. D. (2009). Experimental evidence for two-step nucleation in colloids, *Phy. Rev. Lett.* **102**, p. 198302.

Schall, P., Cohen, I., Weitz, D. A. and Spaepen, F. (2004). Visualization of dislocation dynamics in colloidal crystals, *Science* **305**, pp. 1944–1948.

Schall, P., Weitz, D. A. and Spaepen, F. (2007). Structural rearrangements that govern flow in colloidal glasses, *Science* **318**, pp. 1895–1899.

Schirmacher, W., Diezemann, G. and Ganter, C. (1998). Harmonic vibrational excitations in disordered solids and the "Boson peak", *Phys. Rev. Lett.* **81**, pp. 136–139.

Schmidt, M. and Löwen, H. (1996). Freezing between two and three dimensions, *Phys. Rev. Lett.* **76**, pp. 4552–4555.

Schmidt, M. and Löwen, H. (1997). Phase diagram of hard spheres confined between two parallel plates, *Phys. Rev. E* **55**, pp. 7228–7241.

Schmiedeberg, M. and Stark, H. (2008). Colloidal ordering on a 2D quasicrystalline substrate, *Phys. Rev. Lett.* **101**, p. 218302.

Schmittmann, B. and Zia, R. K. P. (1998). Driven diffusive systems. An introduction and recent developments, *Phys. Rep.* **301**, pp. 45–64.

Schofield, A. B., Pusey, P. N. and Radcliffe, P. (2005). Colloidal dispersions, suspensions, and aggregates-stability of the binary colloidal crystals AB_2 and AB_{13}, *Phys. Rev. E* **72**, pp. 31407–31407.

Schweigert, V. A. (2001). Dielectric permittivity of a plasma in an external electric field, *Plasma Phys. Rep.* **27**, pp. 997–999.

Schweigert, V. A. and Peeters, F. M. (1999a). Enhanced stability of the square lattice of a classical bilayer Wigner crystal, *Phys. Rev. B* **60**, pp. 14665–14674.

Schweigert, V. A. and Peeters, F. M. (1999b). Melting of the classical bilayer Wigner crystal: influence of lattice symmetry, *Phys. Rev. Lett.* **82**, pp. 5293–5296.

Sciortino, F. (2008). Gel-forming patchy colloids and network glass formers: thermodynamic and dynamic analogies, *Eur. Phys. J. B* **64**, pp. 505–509.

Segre, P. N., Herbolzheimer, E. and Chaikin, P. M. (1997). Long-range correlations in sedimentation, *Phys. Rev. Lett.* **79**, pp. 2574–2577.

Selwyn, G. S., Singh, J. and Bennett, R. S. (1989). In situ laser diagnostic studies of plasma-generated particulate contamination, *J. Vac. Sci. Technol. A* **7**, pp. 2758–2765.

Senff, H. and Richtering, W. (1999). Temperature sensitive microgel suspensions: colloidal phase behavior and rheology of soft spheres, *J. Chem. Phys.* **111**, pp. 1705–1711.

Seul, M. and Andelman, D. (1995). Domain shapes and patterns - the phenomenology of modulated phases, *Science* **267**, pp. 476–483.

Sevonkaev, I., Goia, D. V. and Matijevic, E. (2008). Formation and structure of cubic particles of sodium magnesium fluoride (neighborite), *J. Coll. Interf. Sci.* **317**, pp. 130–136.

Sheppard, C. J. R. and Shotton, D. M. (1997). *Confocal laser scannig microscopy* (BIOS Scientific Publishers, Oxford, UK).

Shereda, L. T., Larson, R. G. and Solomon, M. J. (2010). Boundary-driven colloidal crystallization in simple shear flow, *Phys. Rev. Lett.* **105**, p. 228302.

Shimada, T., Murakami, T., Yukawa, S., Saito, K. and Ito, N. (2000). Simulational study on dimensionality dependence of heat conduction, *J. Phys. Society of Japan* **69**, pp. 3150–3153.

Shukla, P. K. and Eliasson, B. (2009). Colloquium: fundamentals of dust-plasma interactions, *Rev. Mod. Phys.* **81**, pp. 25–44.

Shukla, P. K. and Mamun, A. A. (2001). *Introduction to dusty plasma physics* (IOP Publishing, Bristol).

Sickafoose, A. A., Colwell, J. E., Horanyi, M. and Robertson, S. (2000). Photoelectric charging of dust particles in vacuum, *Phys. Rev. Lett.* **84**, pp. 6034–6037.

Siggia, E. D. (1979). Late stages of spinodal decomposition in binary-mixtures, *Phys. Rev. A* **20**, pp. 595–605.

Sillescu, H. (1999). Heterogeneity at the glass transition: a review, *J. Non-Cryst. Solids* **243**, pp. 81–108.

Simeonova, N. B., Dullens, R. P. A., Aarts, D. G. A. L., de Villeneuve, V. W. A., Lekkerkerker, H. N. W. and Kegel, W. K. (2006). Devitrification of colloidal glasses in real space, *Phys. Rev. E* **73**, p. 041401.

Smirnov, R. D., Pigarov, A. Y., Rosenberg, M., Krasheninnikov, S. I. and Mendis, D. A. (2007). Modelling of dynamics and transport of carbon dust particles in tokamaks, *Plasma. Phys. Controlled Fusion* **49**, pp. 347–371.

Smith, P. A., Petekidis, G., Egelhaaf, S. U. and Poon, W. C. K. (2007). Yielding and crystallization of colloidal gels under oscillatory shear, *Phys. Rev. E* **76**, p. 041402.

Sodha, M. S. and Guha, S. (1971). Physics of colloidal plasmas, *Adv. Plasma Phys.* **4**, pp. 219–309.

Solomon, M. J. and Varadan, P. (2001). Dynamic structure of thermoreversible colloidal gels of adhesive spheres, *Phys. Rev. E* **63**, p. 051402.

Solomon, M. J., Zeitoun, R., Ortiz, D., Sung, K. E., Deng, D., Shah, A., Burns, M. A., Glotzer, S. C. and Millunchick, J. M. (2010). Toward assembly of non-close-packed colloidal structures from anisotropic pentamer particles, *Macromolecular Rapid Communications* **31**, pp. 196–201.

Stangroom, J. E. (1983). Electrorheological fluids, *Phys. Technol.* **14**, pp. 290–296.

Steinhardt, P. J., Nelson, D. R. and Ronchetti, M. (1983). Bond-orientational order in liquids and glasses, *Phys. Rev. B* **28**, pp. 784–805.

Stevens, M. J. and Robbins, M. O. (1993). Melting of Yukawa systems – a test of phenomenological melting criteria, *J. Chem. Phys.* **98**, pp. 2319–2324.

Stillinger, F. H. (1995). A topographic view of supercooled liquids and glass-formation, *Science* **267**, pp. 1935–1939.

Stipp, A. and Palberg, T. (2007). Crystal growth kinetics in binary mixtures of model charged sphere colloids, *Phil. Mag. Lett.* **87**, pp. 899–908.

Stradner, A., Sedgwick, H., Cardinaux, F., Poon, W. C. K., Egelhaaf, S. U. and Schurtenberger, P. (2004). Equilibrium cluster formation in concentrated protein solutions and colloids, *Nature* **432**, pp. 492–495.

Strandburg, K. J. (1988). Two-dimensional melting, *Rev. Mod. Phys.* **60**, pp. 161–207.

Strandburg, K. J. (1992). *Bond-orientational order in condensed matter systems* (Springer, New York).

Strepp, W., Sengupta, S. and Nielaba, P. (2002). Phase transitions of soft disks in external periodic potentials: a Monte Carlo study, *Phys. Rev. E* **66**, p. 056109.

Su, C. H. and Lam, S. H. (1963). Continuum theory of spherical electrostatic probes, *Phys. Fluids* **6**, pp. 1479–1491.

Sütterlin, R. K., Wysocki, A., Ivlev, A. V., Räth, C., Thomas, H. M., Rubin-Zuzic, M., Goedheer, W. J., Fortov, V. E., Lipaev, A. M., Molotkov, V. I., Petrov, O. F., Morfill, G. E. and Löwen, H. (2009). Dynamics of lane formation in driven binary complex plasmas, *Phys. Rev. Lett.* **102**, p. 085003.

Swinkels, G. H. P. M., Kersten, H., Deutsch, H. and Kroesen, G. M. W. (2000). Microcalorimetry of dust particles in a radio-frequency plasma, *J. Appl. Phys.* **88**, pp. 1747–1755.

Takagiwa, H., Nishibori, E., Okada, N., Takata, M., Sakata, M. and Akimitsu, J. (2006). Relationship between superconductivity and crystal structure in NbB2+x, *Sci. Technol. Advanced Materials* **7**, pp. 22–25.

Tanaka, H. (1999). Two-order-parameter description of liquids. 1. A general model of glass transition covering its strong to fragile limit, *J. Chem. Phys.* **111**, pp. 3163–3174.

Tanaka, H. (2000). Viscoelastic phase seperation, *J. Phys.: Condens. Matter* **12**, pp. R207–R264.

Tanaka, H., Kawasaki, T., Shintani, H. and Watanabe, K. (2010). Critical-like behaviour of glass-forming liquids, *Nature Materials* **9**, pp. 324–331.

Tao, R. (1993). Electric-field-induced phase-transition in electrorheological fluids, *Phys. Rev. E* **47**, pp. 423–426.

Tarjus, G., Kivelson, S. A., Nussinov, Z. and Viot, P. (2005). The frustration-based approach of supercooled liquids and the glass transition: a review and critical assessment, *J. Phys.: Condens. Matter* **17**, pp. R1143–R1182.

Tarzia, M., Fierro, A., Nicodemi, M., Ciamarra, M. P. and Coniglio, A. (2005). Size segregation in granular media induced by phase transition, *Phys. Rev. Lett.* **95**, p. 078001.

Tehranian, S., Giovane, F., Blum, J., Xu, Y. L. and Gustafson, B. A. S. (2001). Photophoresis of micrometer-sized particles in the free-molecular regime, *Int. J. Heat Mass Transfer* **44**, pp. 1649–1657.

Ten Wolde, P. R. and Frenkel, D. (1997). Enhancement of protein crystal nucleation by critical density fluctuations, *Science* **277**, pp. 1975–1978.

Teng, L. W., Tu, P. S. and I, L. (2003). Microscopic observation of confinement-induced layering and slow dynamics of dusty-plasma liquids in narrow channels, *Phys. Rev. Lett.* **90**, p. 245004.

Tennekes, H. and Lumley, J. L. (1972). *A first course in turbulence* (MIT, Cambridge).

Thoma, M. H., Fink, M. A., Höfner, H., Kretschmer, M., Khrapak, S. A., Ratynskaia, S. V., Yaroshenko, V. V., Morfill, G. E., Petrov, O. F., Usachev, A. D., Zobnin, A. V. and Fortov, V. E. (2007). PK-4: complex plasmas in space - the next generation, *IEEE Trans. Plasma Sci.* **35**, pp. 255–259.

Thomas, E. (1999). Direct measurements of two-dimensional velocity profiles in direct current glow discharge dusty plasmas, *Phys. Plasmas* **6**, pp. 2672–2675.

Thomas, E., Williams, J. D. and Silver, J. (2004). Application of stereoscopic particle image velocimetry to studies of transport in a dusty (complex) plasma, *Phys. Plasmas* **11**, pp. L37–L40.

Thomas, H. M. and Morfill, G. E. (1996). Melting dynamics of a plasma crystal, *Nature* **379**, pp. 806–809.

Thomas, H. M., Morfill, G. E., Demmel, V., Goree, J., Feuerbacher, B. and Mohlmann, D. (1994). Plasma crystal – Coulomb crystallization in a dusty plasma, *Phys. Rev. Lett.* **73**, pp. 652–655.

Thomas, H. M., Morfill, G. E., Fortov, V. E., Ivlev, A. V., Molotkov, V. I., Lipaev, A. M., Hagl, T., Rothermel, H., Khrapak, S. A., Sütterlin, R. K. *et al.* (2008). Complex plasma laboratory PK-3 plus on the international space station, *New J. Phys.* **10**, p. 033036.

Thompson, C., Barkan, A., D'Angelo, N. and Merlino, R. L. (1997). Dust acoustic waves in a direct current glow discharge, *Phys. Plasmas* **4**, pp. 2331–2335.

Tobochnik, J. and Chester, G. V. (1982). Monte-Carlo study of melting in 2 dimensions, *Phys. Rev. B* **25**, pp. 6778–6798.

Totsuji, H. and Barrat, J.-L. (1988). Structure of a nonneutral classical plasma in a magnetic-field, *Phys. Rev. Lett.* **60**, pp. 2484–2487.

Totsuji, H., Kishimoto, T. and Totsuji, C. (1997). Structure of confined Yukawa system (dusty plasma), *Phys. Rev. Lett.* **78**, pp. 3113–3116.

Toyotama, A. and Yamanaka, J. (2011). Heating-induced freezing and melting transitions in charged colloids, *Langmuir* **27**, pp. 1569–1572.

Trizac, E., Bocquet, L. and Aubouy, M. (2002). Simple approach for charge renormalization in highly charged macroions, *Phys. Rev. Lett.* **89**, p. 248301.

Trizac, E., Eldridge, M. D. and Madden, P. A. (1997). Stability of the AB crystal for asymmetric binary hard sphere mixtures, *Molecular Physics* **90**, pp. 675–678.

Tsytovich, V. N. (1997). Dust plasma crystals, drops, and clouds, *Phys. Usp.* **40**, pp. 53–94.

Tsytovich, V. N., Khodataev, Y. K., Morfill, G. E., Bingham, R. and Winter, J. (1998). Radiative dust cooling and dust agglomeration in plasmas, *Comments Plasma Phys. Controlled Fusion* **18**, pp. 281–291.

Tuckerman, M. E., Liu, Y., Ciccotti, G. and Martyna, G. J. (2001). Non-Hamiltonian molecular dynamics: generalizing Hamiltonian phase space principles to non-Hamiltonian systems, *J. Chem. Phys.* **115**, pp. 1678–1702.

Turnbull, D. (1950). Kinetics of heterogeneous nucleation, *J. Chem. Phys.* **18**, pp. 198–203.

Uhlenbeck, G. E. and Ornstein, L. S. (1930). On the theory of Brownian motion, *Phys. Rev.* **36**, pp. 832–841.

van Blaaderen, A., Imhof, A., Hage, W. and Vrij, A. (1992). Three-dimensional imaging of submicrometer colloidal particles in concentrated suspensions using confocal scanning laser microscopy, *Langmuir* **8**, pp. 1514–1517.

van Blaaderen, A. and Wiltzius, P. (1995). Real-space structure of colloidal hard-sphere glasses, *Science* **270**, pp. 1177–1179.

van Bruggen, M. P. B., Lekkerkerker, H. N. W., Maret, G. and Dhont, J. K. G. (1998). Long-time translational self-diffusion in isotropic and nematic dispersions of colloidal rods, *Phys. Rev. E* **58**, pp. 7668–7677.

van den Pol, E., Petukhov, A. V., Thies-Weesie, D. M. E., Byelov, D. V. and Vroege, G. J. (2009). Experimental realization of biaxial liquid crystal phases in colloidal dispersions of boardlike particles, *Phys. Rev. Lett.* **103**, p. 258301.

van der Beek, D., Davidson, P., Wensink, H. H., Vroege, G. J. and Lekkerkerker, H. N. W. (2008). Influence of a magnetic field on the nematic phase of hard colloidal platelets, *Phys. Rev. E* **77**, p. 031708.

van der Beek, D., Reich, H., van der Schoot, P., Dijkstra, M., Schilling, T., Vink, R., Schmidt, M., van Roij, R. and Lekkerkerker, H. N. W. (2006). Isotropic-nematic interface and wetting in suspensions of colloidal platelets, *Phys. Rev. Lett.* **97**, p. 087801.

van Duc, N., Hu, Z. and Schall, P. (2011). Single crystal growth and anisotropic crystal-fluid interfacial free energy in soft colloidal systems, *Phys. Rev. E* **84**, p. 011607.

van Kampen, N. G. (1981). *Stochastic processes in physics and chemistry* (Elsevier, Amsterdam).

van Megen, W., Mortensen, T. C. and Williams, S. R. (1998). Measurement of the self-intermediate scattering function of suspensions of hard spherical particles near the glass transition, *Phys. Rev. E* **58**, pp. 6073–6085.

van Roij, R. (2003). Defying gravity with entropy and electrostatics: sedimentation of charged colloids, *J. Phys.: Condens. Matter* **15**, pp. S3569–S3580.

van Roij, R. and Hansen, J.-P. (1997). Van der Waals-like instability in suspensions of mutually repelling charged colloids, *Phys. Rev. Lett.* **77**, pp. 3082–3085.

van Teeffelen, S., Likos, C. N. and Löwen, H. (2008). Colloidal crystal growth at externally imposed nucleation clusters, *Phys. Rev. Lett.* **100**, p. 108302.

van Winkle, D. H. and Murray, C. A. (1986). Layering transitions in colloidal crystals as observed by diffraction and direct-lattice imaging, *Phys. Rev. A* **34**, pp. 562–573.

Vaulina, O. S. and Dranzhevskii, I. E. (2007). Transport properties of quasi-two-dimensional dissipative systems with a screened Coulomb potential, *Plasma Phys. Rep.* **33**, pp. 494–502.

Vaulina, O. S., Khrapak, S. A. and Morfill, G. E. (2002a). Universal scaling in complex (dusty) plasmas, *Phys. Rev. E* **66**, p. 016404.

Vaulina, O. S., Khrapak, S. A., Nefedov, A. P. and Petrov, O. F. (1999). Charge-fluctuation-induced heating of dust particles in a plasma, *Phys. Rev. E* **60**, pp. 5959–5964.

Vaulina, O. S., Nefedov, A. P., Petrov, O. F. and Fortov, V. E. (2000). Instability of plasma-dust systems with a macroparticle charge gradient, *JETP* **91**, pp. 1147–1162.

Vaulina, O. S., Nefedov, A. P., Petrov, O. F. and Fortov, V. E. (2001). Transport properties of macroparticles in dust plasma induced by solar radiation, *JETP* **92**, pp. 979–985.

Vaulina, O. S., Nefedov, A. P., Petrov, O. F. and Fortov, V. E. (2002b). Diffusion in microgravity of macroparticles in a dusty plasma under solar radiation, *Phys. Rev. Lett.* **88**, p. 035001.

Vega, C., Paras, E. P. A. and Monson, P. A. (1992). Solid–fluid equilibria for hard dumbbells via Monte Carlo simulation, *J. Chem. Phys.* **96**, pp. 9060–9072.

Vermant, J. and Solomon, M. J. (2005). Flow-induced structure in colloidal suspensions, *J. Phys.: Condens. Matter* **17**, pp. R187–R216.

Verwey, E. J. W. and Overbeek, J. T. G. (1948). *Theory of the stability of lyophobic colloids* (Elsevier, Amsterdam).

Vissers, T. (2010). *Oppositely charged colloids out of equilibrium*, Ph.D. thesis, University of Utrecht.

Vissers, T., van Blaaderen, A. and Imhof, A. (2011a). Band formation in mixtures of oppositely charged colloids driven by an AC electric field, *Phys. Rev. Lett.* **106**, p. 228303.

Vissers, T., Wysocki, A., Rex, M., Löwen, H., Royall, C. P., Imhof, A. and van Blaaderen, A. (2011b). Lane formation in driven mixtures of oppositely charged colloids, *Soft Matter* **7**, pp. 2352–2356.

Vladimirov, S. V., Maiorov, S. A. and Ishihara, O. (2003). Molecular dynamics simulation of plasma flow around two stationary dust grains, *Phys. Plasmas* **10**, pp. 3867–3873.

Vladimirov, S. V., Ostrikov, K. and Samarian, A. A. (2005). *Physics and applications of complex plasmas* (Imperial College, London).

Vlasov, Y. A., Bo, X., Sturm, J. C. and Norris, D. J. (2001). On-chip natural assembly of silicon photonic bandgap crystals, *Nature* **414**, pp. 289–293.

Voigtmann, T., Puertas, A. M. and Fuchs, M. (2004). Tagged-particle dynamics in a hard-sphere system: mode-coupling theory analysis, *Phys. Rev. E* **70**, p. 061506.

Vrij, A. (1976). Polymers at interfaces and interactions in colloidal dispersions, *Pure Appl. Chem.* **48**, pp. 471–483.

Vrij, A., Penders, M. H. G. M., Rouw, P. W., Dekruif, C. G., Dhont, J. K. G., Smits, C. and Lekkerkerker, H. N. W. (1990). Phase-transition phenomena in colloidal systems with attractive and repulsive particle interactions, *Faraday Discuss.* **90**, pp. 31–40.

Walch, B., Horanyi, M. and Robertson, S. (1995). Charging of dust grains in plasma with energetic electrons, *Phys. Rev. Lett.* **75**, pp. 838–841.

Wang, C. L., Joyce, G. and Nicholson, D. R. (1981). Debye shielding of a moving test charge in plasma, *J. Plasma Phys.* **25**, pp. 225–231.

Watanabe, K. and Tanaka, H. (2008). Direct observation of medium-range crystalline order in granular liquids near the glass transition, *Phys. Rev. Lett.* **100**, p. 158002.

Wearie, D. and Hutzler, S. (2000). *The physics of foams* (Oxford University Press, Oxford, UK).

Weeks, E. R., Crocker, J. C., Levitt, A. C., Schofield, A. B. and Weitz, D. A. (2001). Three-dimensional direct imaging of structural relaxation near the colloidal glass transition, *Science* **287**, pp. 627–631.

Weeks, E. R. and Weitz, D. A. (2002). Properties of cage rearrangements observed near the colloidal glass transition, *Phys. Rev. Lett.* **89**, p. 095704.

Weeks, J. D., Chandler, D. and Andersen, H. C. (1971). Role of repulsive forces in determining equilibrium structure of simple liquids, *J. Chem. Phys.* **54**, pp. 5237–5248.

Weiss, J. A., Oxtoby, D. W., Grier, D. G. and Murray, C. A. (1995). Martensitic transition in a confined colloidal suspension, *J. Chem. Phys.* **103**, pp. 1180–1190.

Wen, W., Huang, X. and Sheng, P. (2008). Electrorheological fluids: structures and mechanisms, *Soft Matter* **4**, pp. 200–210.

Wensink, H. H. and Jackson, G. (2009). Generalized van der Waals theory for the twist elastic modulus and helical pitch of cholesterics, *J. Chem. Phys.* **130**, p. 234911.

Wette, P., Schöpe, H. J. and Palberg, T. (2005). Crystallization in charged two-component suspensions, *J. Chem. Phys.* **122**, p. 144901.

Wette, P., Schöpe, H. J. and Palberg, T. (2009). Enhanced crystal stability in a binary mixture of charged colloidal spheres, *Phys. Rev. E* **80**, p. 021407.

Whipple, E. C. (1981). Potentials of surfaces in space, *Rep. Prog. Phys.* **44**, pp. 1197–1250.

Whitby, M. and Quirke, N. (2007). Fluid flow in carbon nanotubes and nanopipes, *Nature Nanotechnology* **2**, pp. 87–94.

Widmer-Cooper, A. and Harrowell, P. (2006). Predicting the long-time dynamic heterogeneity in a supercooled liquid on the basis of short-time heterogeneities, *Phys. Rev. Lett.* **96**, p. 185701.

Wierenga, A. M., Lenstra, T. A. J. and Philipse, A. P. (1998). Aqueous dispersions of colloidal Gibbsite platelets: synthesis, characterisation and intrinsic viscosity measurements, *Colloids and Surfaces A: Physicochemical and Engineering Aspects* **134**, pp. 359–371.

Williams, J. D. (2011). Application of tomographic particle image velocimetry to studies of transport in complex (dusty) plasma, *Phys. Plasmas* **18**, p. 050702.

Williams, S. R., Royall, C. P. and Bryant, G. (2008). Crystallization of dense binary hard-sphere mixtures with marginal size ratio, *Phys. Rev. Lett.* **100**, p. 225502.

Winske, D. (2001). Nonlinear wake potential in a dusty plasma, *IEEE Trans. Plasma Sci.* **29**, pp. 191–197.

Wittkowski, R. and Löwen, H. (2012). Dynamical density functional theory for colloidal particles with arbitrary shape, *Molecular Physics* (in press).

Wojciechowski, K. W., Frenkel, D. and Brańka, A. C. (1991). Nonperiodic solid phase in a two-dimensional hard-dimer system, *Phys. Rev. Lett.* **66**, pp. 3168–3171.

Wood, W. W. and Jacobson, J. D. (1957). Preliminary results from a recalculation of the Monte Carlo equation of state of hard spheres, *J. Chem. Phys.* **27**, pp. 1207–1208.

Woodruff, D. P. (1980). *The solid-liquid interface* (Cambridge University Press, Cambridge).

Woon, W. Y. and I, L. (2004). Defect turbulence in quasi-2D creeping dusty-plasma liquids, *Phys. Rev. Lett.* **92**, p. 065003.

Wu, Y. L., Derks, D., van Blaaderen, A. and Imhof, A. (2009). Melting and crystallization of colloidal hard-sphere suspensions under shear, *Proc. Nat. Acad. Sci.* **106**, pp. 10564–10569.

Wysocki, A. and Löwen, H. (2004). Instability of a fluid-fluid interface in driven colloidal mixtures, *J. Phys.: Condens. Matter* **16**, pp. 7209–7224.

Wysocki, A. and Löwen, H. (2009). Oscillatory driven colloidal binary mixtures: axial segregation versus laning, *Phys. Rev. E* **79**, p. 041408.

Wysocki, A. and Löwen, H. (2011). Effects of hydrodynamic interactions in binary colloidal mixtures driven oppositely by oscillatory external fields, *J. Phys.: Condens. Matter* **23**, p. 284117.

Wysocki, A., Raeth, C., Ivlev, A. V., Sütterlin, R. K., Thomas, H. M., Khrapak, S. A., Zhdanov, S. K., Fortov, V. E., Lipaev, A. M., Molotkov, V. I., Petrov, O. F., Löwen, H. and Morfill, G. E. (2010). Kinetics of fluid demixing in complex plasmas: role of two-scale interactions, *Phys. Rev. Lett.* **105**, p. 045001.

Wysocki, A., Royall, C. P., Winkler, R., Gompper, G., Tanaka, H., van Blaaderen, A. and Löwen, H. (2009). Direct observation of hydrodynamic instabilities in driven non-uniform colloidal dispersions, *Soft Matter* **5**, pp. 1340–1344.

Xie, B. S., He, K. F. and Huang, Z. Q. (1999). Attractive potential in weak ion flow coupling with dust-acoustic waves, *Phys. Lett. A* **253**, pp. 83–87.

Yakubov, I. T. and Khrapak, A. G. (1989). Thermophysical and electrophysical properties of low temperature plasma with condensed disperse phase, *Sov. Tech. Rev. B. Therm. Phys.* **2**, pp. 269–337.

Yamamoto, R., Nakayama, Y. and Kim, K. (2009). Smoothed profile method to simulate colloidal particles in complex fluids, *Int. Journal Mod. Phys. C* **20**, pp. 1457–1465.

Yamanaka, J., Hayashi, Y., Ise, N. and Yamaguchi, T. (1997). Control of the surface charge density of colloidal silica by sodium hydroxide in salt-free and low-salt dispersions, *Phys. Rev. E.* **55**, pp. 3028–3036.

Yazaki, A., Kishimura, H., Kawano, H., Hironaka, Y., Nakamura, K. G. and Kondo, K. (2002). Picosecond time-resolved X-ray diffraction of a photoexcited silicon crystal, *Jpn. J. Appl. Phys.* **41**, pp. 1614–1615.

Yethiraj, A., Thijssen, J. H. J., Wouterse, A. and van Blaaderen, A. (2004). Large-area electric-field-induced colloidal single crystals for photonic applications, *Adv. Mater.* **16**, pp. 596–600.

Yethiraj, A. and van Blaaderen, A. (2003). A colloidal model system with an inteaction tunable from hard sphere to soft and dipolar, *Nature* **421**, pp. 513–517.

Yokoyama, S., Honda, Y., Morioka, H., Okamoto, S., Funakubo, H., Iijima, T., Matsuda, H., Saito, K., Yamamoto, T., Okino, H., Sakata, O. and Kimura, S. (2005). Dependence of electrical properties of epitaxial Pb(Zr,Ti)O-3 thick films on crystal orientation and Zr/(Zr+Ti) ratio, *J. Appl. Phys.* **98**, p. 094106.

Yoshida, H., Ito, K. and Ise, N. (1991). Localized ordered structure in polymer latex suspensions as studied by a confocal laser scanning microscope, *Phys. Rev. B* **44**, pp. 435–438.

Young, A. P. (1979). Melting and the vector Coulomb gas in 2 dimensions, *Phys. Rev. B* **19**, pp. 1855–1866.

Yunker, P., Zhang, X., Aptowicz, K. B. and Yodh, A. G. (2009). Irreversible rearrangements, correlated domains, and local structure in aging glasses, *Phys, Rev. Lett.* **103**, p. 115701.

Zahn, K., Lenke, R. and Maret, G. (1994). Friction coefficient of rod-like chains of spheres at very-low Reynolds-numbers. 1. Experiment, *Journal de Physique II* **4**, pp. 555–560.

Zahn, K., Lenke, R. and Maret, G. (1999). Two-stage melting of paramagnetic colloidal crystals in two dimensions, *Phys. Rev. Lett.* **82**, pp. 2721–2724.

Zahn, K. and Maret, G. (2000). Dynamic criteria for melting in two dimensions, *Phys. Rev. Lett.* **85**, pp. 3656–3659.

Zahn, K., Mendez-Alcaraz, J. M. and Maret, G. (1997). Hydrodynamic interactions may enhance the self-diffusion of colloidal particles, *Phys. Rev. Lett.* **79**, pp. 175–178.

Zausch, J., Horbach, J., Laurati, M., Egelhaaf, S. U., Brader, J. M., Voigtmann, T. and Fuchs, M. (2008). From equilibrium to steady state: the transient dynamics of colloidal liquids under shear, *J. Phys.: Condens. Matter* **20**, p. 404210.

Zerrouki, D., Baudry, J., Pine, D. J., Chaikin, P. M. and Bibette, J. (2008). Chiral colloidal clusters, *Nature* **455**, pp. 380–382.

Zhakhovskii, V. V., Molotkov, V. I., Nefedov, A. P., Torchinskii, V. M., Khrapak, A. G. and Fortov, V. E. (1997). Anomalous heating of a system of dust particles in a gas-discharge plasma, *JETP Lett.* **66**, pp. 419–425.

Zhang, Z. and Glotzer, S. C. (2004). Self-assembly of patchy particles, *Nano Lett.* **4**, pp. 1407–1413.

Zhdanov, S. K., Nunomura, S., Samsonov, D. and Morfill, G. E. (2003). Polarization of wave modes in a two-dimensional hexagonal lattice using a complex (dusty) plasma, *Phys. Rev. E* **68**, p. 035401.

Zhdanov, S. K., Ivlev, A. V. and Morfill, G. E. (2005). Non-Hamiltonian dynamics of grains with spatially varying charges, *Phys. Plasmas* **12**, p. 072312.

Zheng, X. H. and Earnshaw, J. C. (1998). On the Lindemann criterion in 2D, *Europhys. Lett.* **41**, pp. 635–640.

Zhu, X., Birringer, R., Herr, U. and Gleiter, H. (1987). X-ray-diffraction studies of the structure of nanometer-sized crystalline materials, *Phys. Rev. B* **35**, pp. 9085–9090.

Zobnin, A. V., Nefedov, A. P., Sinel'shchikov, V. A. and Fortov, V. E. (2000). On the charge of dust particles in a low-pressure gas discharge plasma, *JETP* **91**, pp. 483–487.

Zorn, R. (2003). Microscopic dynamics of glass-forming polymers, *J. Phys.: Condens. Matter* **15**, pp. R1025–R1046.

Zuzic, M., Ivlev, A. V., Goree, J., Morfill, G. E., Thomas, H. M., Rothermel, H., Konopka, U., Sütterlin, R. K. and Goldbeck, D. D. (2000). Three-dimensional strongly coupled plasma crystal under gravity conditions, *Phys. Rev. Lett.* **85**, pp. 4064–4067.

Zykova-Timan, T., Rozas, R. E., Horbach, J. and Binder, K. (2009). Computer simulation studies of finite-size broadening of solid-liquid interfaces: from hard spheres to nickel, *J. Phys.: Condens. Matter* **21**, p. 464102.

Index

320   *Complex Plasmas and Colloidal Dispersions*